Medicinal Chemistry

The INSTANT NOTES series

Series editor
B.D. Hames
School of Biochemistry and Molecular Biology, University of Leeds, Leeds, UK

Animal Biology
Ecology
Genetics
Microbiology
Chemistry for Biologists
Immunology
Biochemistry 2nd edition
Molecular Biology 2nd edition
Neuroscience
Psychology
Developmental Biology
Plant Biology

Forthcoming title
Bioinformatics

The INSTANT NOTES Chemistry series
Consulting editor: Howard Stanbury

Organic Chemistry
Inorganic Chemistry
Physical Chemistry
Medicinal Chemistry

Forthcoming title
Analytical Chemistry

Medicinal Chemistry

Graham Patrick

Department of Chemistry and
Chemical Engineering, Paisley
University, Paisley, Scotland

First published 2001

A CIP catalogue record for this book is available from the British Library.

ISBN 1 85996 207 6

BIOS Scientific Publishers Ltd
9 Newtec Place, Magdalen Road, Oxford OX4 1RE, UK
Tel. +44 (0)1865 726286. Fax +44 (0)1865 246823
World Wide Web home page: http://www.bios.co.uk/

Distributed exclusively in the United States, its dependent territories, Canada, Mexico, Central and South America, and the Caribbean by Springer-Verlag New York Inc, 175 Fifth Avenue, New York, USA, by arrangement with BIOS Scientific Publishers, Ltd, 9 Newtec Place, Magdalen Road, Oxford OX4 1RE, UK

Production Editor: Andrea Bosher
Typeset by Phoenix Photosetting, Chatham, Kent, UK
Printed by Biddles Ltd, Guildford, UK, www.biddles.co.uk

Contents

Abbreviations vii
Preface ix
About the author x

Section A – The science of medicinal chemistry **1**
A1 Introduction 1
A2 From concept to market 4

Section B – Drug targets **7**
B1 Enzymes 7
B2 Receptors 13
B3 Carrier proteins 24
B4 Structural proteins 27
B5 Nucleic acids 29
B6 Lipids 35
B7 Carbohydrates 38

Section C – Pharmacokinetics **41**
C1 Pharmacokinetics 41
C2 Drug absorption 42
C3 Drug distribution 45
C4 Drug metabolism 48
C5 Drug excretion 51
C6 Drug administration 54
C7 Drug dosing 58

Section D – Biological testing and bioassays **61**
D1 Testing drugs 61
D2 Testing drugs *in vitro* 62
D3 Testing drugs *in vivo* 70

Section E – Drug discovery **75**
E1 The lead compound 75
E2 Natural sources of lead compounds 77
E3 Synthetic sources of lead compounds 82

Section F – Synthesis **85**
F1 Synthetic considerations 85
F2 Stereochemistry 90
F3 Combinatorial synthesis 100

Section G – Structure–activity relationships **105**
G1 Definition of structure-activity relationships 105
G2 Binding interactions 107
G3 Functional groups as binding groups 110
G4 The pharmacophore 120
G5 Quantitative structure–activity relationships 125

Section H – Target orientated drug design 141
H1 Aims of drug design 141
H2 Computer aided drug design 143
H3 Simplification of complex molecules 148
H4 Conformational restraint 151
H5 Extra binding interactions 155
H6 Enhancing existing binding interactions 160

Section I – Pharmacokinetic orientated drug design 167
I1 Drug solubility 167
I2 Drug stability 171

Section J – Patenting and manufacture 177
J1 Patenting and chemical development 177
J2 Optimization of reactions 181
J3 Scale-up issues 188
J4 Process development 193
J5 Specifications 197

Section K – Preclinical testing and clinical trials 203
K1 Toxicology 203
K2 Pharmacology and pharmaceutical chemistry 205
K3 Drug metabolism studies 207
K4 Clinical trials 212
K5 Regulatory affairs 216

Section L – Case study: inhibitors of EGF-receptor kinase 221
L1 Epidermal growth factor receptor 221
L2 Testing procedures 224
L3 From lead compound to dianilino-phthalimides 226
L4 Modeling studies 233
L5 4-(Phenylamino)pyrrolopyrimidines 239
L6 Pyrazolopyrimidines 241

Section M – History of medicinal chemistry 245
M1 The age of herbs, potions and magic 245
M2 The nineteenth century 247
M3 A fledgling science (1900–1930) 251
M4 The dawn of the antibacterial age (1930–1945) 254
M5 The antibiotic age (1945–1970s) 258
M6 The age of reason (1970s to present) 266

Further reading 271

Index 273

Abbreviations

AMP	adenosine monophosphate
6-APA	6-aminopenicillanic acid
ATP	adenosine triphosphate
CNS	central nervous system
DAG	diacylglycerol
DNA	deoxyribonucleic acid
DMSO	dimethylsulfoxide
EGF	epidermal growth factor
EGF-R	epidermal growth factor receptor
EP	enzyme-bound product
ES	enzyme–substrate (complex)
FDA	Food and Drugs Agency
GABA	γ-aminobutyric acid
GCP	good clinical practice
GDP	guanosine diphosphate
GLP	good laboratory practice
GMP	good manufacturing practice
GTP	guanosine triphosphate
HBA	hydrogen bond acceptor
HBD	hydrogen bond donor
HPLC	high performance liquid chromatography
IND	Investigational Exemption to a New Drug Application
IP_3	inositol trisphosphate
i.v.	intravenous
MAOI	monoamine oxidase inhibitor
mRNA	messenger RNA
NDA	New Drug Application
NMR	nuclear magnetic resonance
PIP_2	phosphatidylinositol diphosphate
PLC	phospholipase C
QSAR	quantitative structure–activity relationship
RNA	ribonucleic acid
rRNA	ribosomal RNA
SAR	structure–activity relationship
tRNA	transport RNA

PREFACE

This textbook provides a comprehensive set of basic notes in medicinal chemistry, which will be suitable for undergraduate students studying a module in medicinal chemistry as part of a science, pharmacy or medical course. The book concentrates on the fundamental principles of medicinal chemistry and assumes no more than an elementary background of chemistry or biology. It also serves as a useful 'first dip' into the subject for those students wishing to study medicinal chemistry itself.

Medicinal chemistry is an exciting new science, which has only come of age in the last 10–20 years. It is a truly multidisciplinary subject involving such subject specialties as organic chemistry, pharmacology, biochemistry, physiology, microbiology, toxicology, genetics and computer modeling. Indeed, most pharmaceutical companies organize research teams in such a way that scientists of different disciplines interact with each other on a daily basis in order to fight the battle against disease. The very breadth of knowledge required by a medicinal chemist is both a challenge and a reward. Mastering an understanding of such a breadth of subject areas is no straightforward task, but by the same token there is ample intellectual stimulation in understanding the battle against disease at the molecular level and in designing molecular 'soldiers' to win that battle.

This book attempts to condense the essentials of medicinal chemistry into a manageable text, which is student friendly and does not cost an arm and a leg. It does this by concentrating purely on the basics of the subject without going into exhaustive detail or repetitive examples. Furthermore, keynotes at the start of each topic summarize the essential facts covered and help focus the mind on the essentials.

Medicinal chemistry is a peculiar subject in that it feeds off so many other subjects. Understanding disease at the physiological, cellular and molecular levels is crucial if one is to design a suitable drug, and therefore knowledge of the relevant physiology, biochemistry, and pharmacology is of immense aid. However, the rapid advances made in two particular scientific areas are worth emphasizing. Molecular biology and genetic engineering have produced a deluge of potential new targets for drug design and have unraveled the structures and mechanisms of traditional targets, while advances in computers and computer aided design have allowed medicinal chemists to take full advantage of this newly earned knowledge.

The first four sections of this book serve as an introduction to the science of medicinal chemistry. Subsequent sections then follow the identifiable stages that have to be negotiated by any drug candidate on its journey from initial 'brainstorm' to the market place – a journey which takes many years and will see you graduated before it is! There are many difficult hurdles for the novice drug to overcome. The drug has to be 'tuned' so that it recognizes a specific target in the body and does not go flying off 'attacking' all and sundry. It has to gain access to the target by being absorbed into the blood supply, and in doing so it must be sturdy enough to ward off the many attacks that will be made on it by the body's defenses. It must also be controlled in its 'aggression', being a mild and harmless visitor during most of its body tour, but ruthless and efficient when it reaches its target. Designing such characteristics is not straightforward and a drug's behavior involves many tests and trials, both in the lab and in the clinic before it 'comes of age'.

It is hoped that students will find this textbook useful in their studies and that once they have grasped what medicinal chemistry is all about, they will read more widely and enter this truly exciting world of molecular medicine.

About the Author

Graham Patrick studied chemistry at Glasgow University where he gained a BSc Honours (1st) and won the Mackay Smith prize. He was awarded a Carnegie scholarship and successfully completed a PhD degree on the biosynthesis of gliotoxin. Since then he has held postdoctoral research posts at Strathclyde University and the Australian National University, and has had industrial experience working with pharmaceutical firms such as Glaxo, Beechams, and Organon Pharmaceuticals. He lectured at the Department of Chemistry at Leeds University and is currently lecturing in chemistry and medicinal chemistry at Paisley University, where he is also the course leader for medicinal chemistry.

Dr Patrick has written several undergraduate textbooks including *An Introduction to Medicinal Chemistry* (2nd edition, 2001, Oxford University Press), two self-learning texts on basic organic chemistry, and *Instant Notes in Organic Chemistry* (BIOS Scientific Publishers Ltd., 2000). He has several research publications in the area of organic synthesis, medicinal chemistry and bio-organic chemistry, and has also written several reviews. His current research interests are the design and synthesis of novel antifungal and antimalarial agents.

Recently, he has collaborated with the Borders Educational Council of Scotland in the production of video and CD lectures covering aspects of medicinal chemistry for school courses.

A1 INTRODUCTION

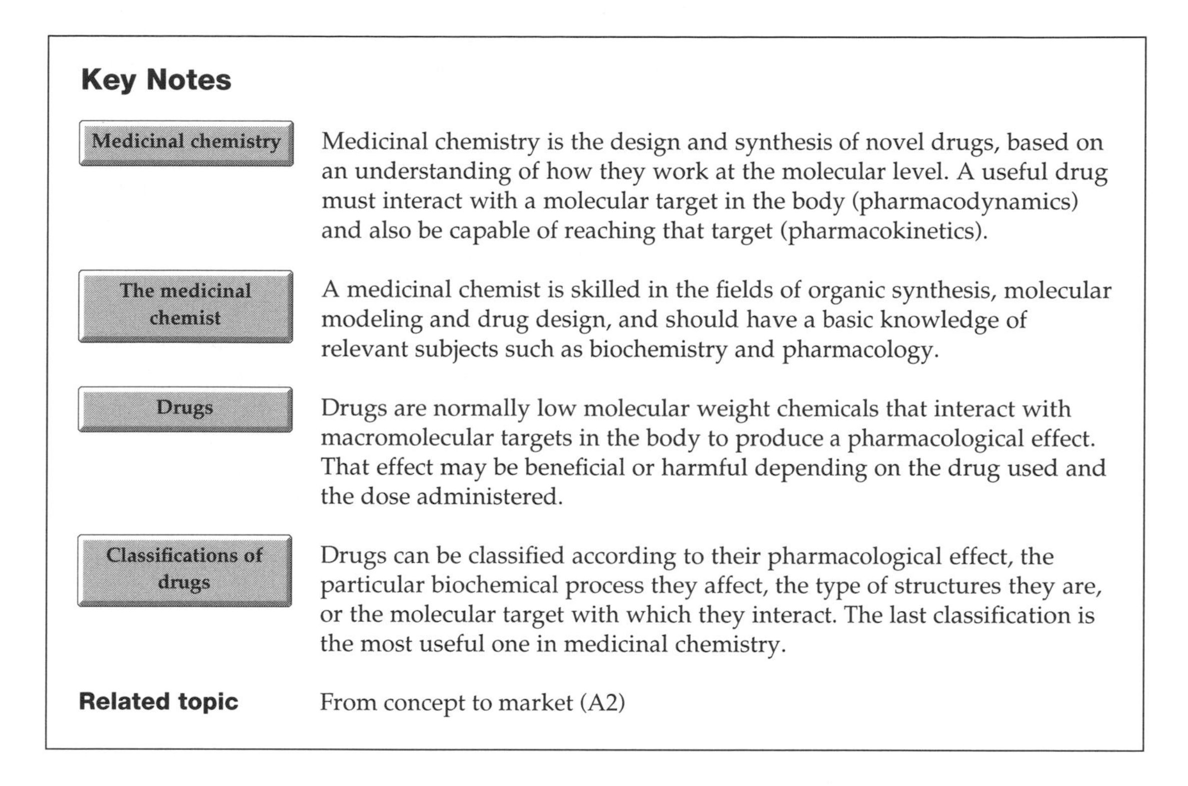

Key Notes

Medicinal chemistry	Medicinal chemistry is the design and synthesis of novel drugs, based on an understanding of how they work at the molecular level. A useful drug must interact with a molecular target in the body (pharmacodynamics) and also be capable of reaching that target (pharmacokinetics).
The medicinal chemist	A medicinal chemist is skilled in the fields of organic synthesis, molecular modeling and drug design, and should have a basic knowledge of relevant subjects such as biochemistry and pharmacology.
Drugs	Drugs are normally low molecular weight chemicals that interact with macromolecular targets in the body to produce a pharmacological effect. That effect may be beneficial or harmful depending on the drug used and the dose administered.
Classifications of drugs	Drugs can be classified according to their pharmacological effect, the particular biochemical process they affect, the type of structures they are, or the molecular target with which they interact. The last classification is the most useful one in medicinal chemistry.
Related topic	From concept to market (A2)

Medicinal chemistry

The science of medicinal chemistry involves the design and synthesis of novel drugs based on an understanding of how drugs work in the body at the molecular level. There are two major considerations that have to be considered in any drug design project. First of all, drugs interact with molecular targets in the body, and so it is important to choose the correct target for the desired pharmaceutical effect. It is then a case of designing a drug that will interact as powerfully and selectively as possible for that target – an area of medicinal chemistry known as **pharmacodynamics**. (Drug targets will be discussed in more detail in Section B.) Secondly, a drug has to travel through the body in order to reach its target, so it is important to design the drug so that it is able to carry out that journey. This is an area known as **pharmacokinetics** and is discussed in Section C.

Medicinal chemistry has come of age in the last 20 years. Before that, advances were often made as a result of trial and error, intuition or pure luck. Large numbers of analogs were synthesized based on the structure of a known active compound (defined as the **lead compound**), but little was known about the detailed mechanism of drug action or the structures of the targets with which they interacted. Advances in the biological sciences have now resulted in a much better understanding of drug targets and the mechanisms of drug action. As a result, drug design is as much 'target oriented' as 'lead compound oriented'.

The medicinal chemist

Medicinal chemistry is an interdisciplinary science that, by its very nature, encompasses the sciences of chemistry, biochemistry, physiology, pharmacology, and molecular modeling, to name but a few. A good understanding of these subject areas is useful, but it is unlikely that any one person could be master of all. Thus, the pharmaceutical industry relies on multidisciplinary teams of scientists who are specialists in their own fields and can work together on a particular project.

The chief role of the medicinal chemist is to design and synthesize the target structures required. Therefore, the medicinal chemist is an essential member of any drug design team since he or she has to identify whether proposed target structures are likely to be stable and whether they can be synthesized or not. Traditionally, the pharmaceutical industry has recruited graduates with a chemistry degree since this is the best method of acquiring the synthetic organic chemistry skills required for medicinal chemistry. However, it is often the case that graduates with a conventional chemistry degree have little background in the biological sciences and have had to acquire that background 'on the job'. In recent years, many universities have started to offer medicinal chemistry degrees that are specifically designed to prepare chemistry graduates for the pharmaceutical industry. Such degrees contain the important core topics required for a conventional chemistry degree (i.e. physical, inorganic and organic chemistry), but also include topics such as drug design, pharmacology, molecular modeling, combinatorial synthesis, bio-organic and bio-inorganic chemistry.

Drugs

Drugs are chemicals that are normally of low molecular weight (~100–500) and which interact with macromolecular targets to produce a biological response. That biological response may be therapeutically useful in the case of medicines, or harmful in the case of poisons. Most drugs used in medicine are potential poisons if taken in doses higher than those recommended.

Classifications of drugs

There are several ways in which drugs can be classified. First, drugs can be classified according to their pharmacological effect – for example, analgesics are drugs which have a pain-killing effect. This classification is useful for doctors wishing to know the arsenal of drugs available to tackle a particular problem, but it is not satisfactory for a medicinal chemist as there are many different targets and mechanisms by which drugs can have an analgesic effect. Therefore, it is not possible to identify a common feature which is shared by all analgesics. For example, **aspirin** and **morphine** act on different targets and have no structural relationship (*Fig. 1*). Other examples of drugs that are classified in this way

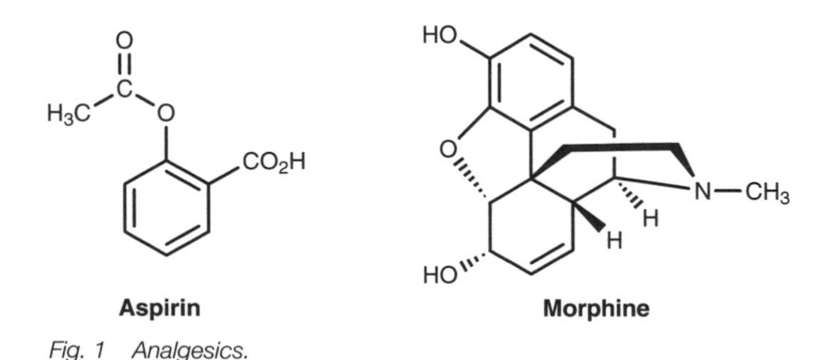

Fig. 1 Analgesics.

are antidepressants, cardiovascular drugs, anti-asthmatics, and anti-ulcer agents.

Second, drugs can be classified depending on whether they act on a particular biochemical process. For example, antihistamines act by inhibiting the action of the inflammatory agent **histamine** in the body. Although this classification is more specific than the above, it is still not possible to identify a common feature relating all antihistamines. This is because there are various ways in which the action of histamine can be inhibited. Other examples of this kind of classification are cholinergic or adrenergic drugs.

A third method of classifying drugs is by their chemical structure (*Fig. 2*). Drugs classified in this way share a common structural feature and often share a similar pharmacological activity. For example, **penicillins** all contain a β-lactam ring and kill bacteria by the same mechanism. As a result, this classification can sometimes be useful in medicinal chemistry. However, it is not foolproof. **Sulfonamides** have a similar structure and are mostly antibacterial. However, some sulfonamides are used for the treatment of diabetes. Similarly, **steroids** all have a tetracyclic structure, but the pharmacological effect of different steroids can be quite different.

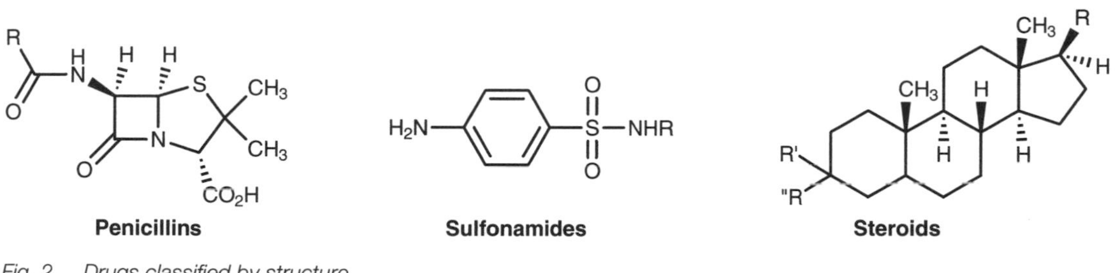

Fig. 2. Drugs classified by structure.

Finally, classifying drugs according to their molecular target is the most useful classification as far as the medicinal chemist is concerned, since it allows a rational comparison of the structures involved. For example, anticholinesterases are compounds that inhibit an enzyme called **acetylcholinesterase**. They have the same mechanism of action and so it is valid to compare the various structures and identify common features.

A2 From concept to market

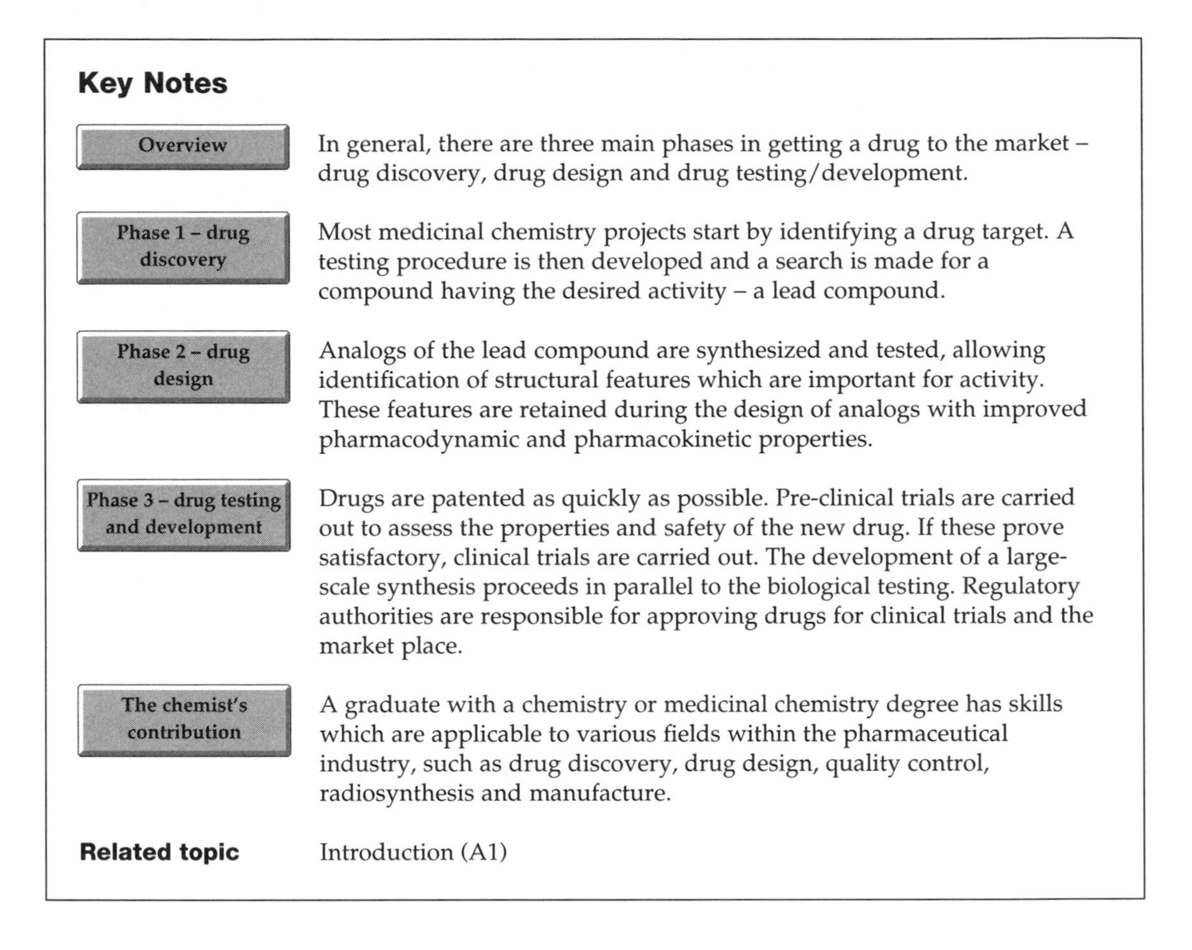

Key Notes

Overview — In general, there are three main phases in getting a drug to the market – drug discovery, drug design and drug testing/development.

Phase 1 – drug discovery — Most medicinal chemistry projects start by identifying a drug target. A testing procedure is then developed and a search is made for a compound having the desired activity – a lead compound.

Phase 2 – drug design — Analogs of the lead compound are synthesized and tested, allowing identification of structural features which are important for activity. These features are retained during the design of analogs with improved pharmacodynamic and pharmacokinetic properties.

Phase 3 – drug testing and development — Drugs are patented as quickly as possible. Pre-clinical trials are carried out to assess the properties and safety of the new drug. If these prove satisfactory, clinical trials are carried out. The development of a large-scale synthesis proceeds in parallel to the biological testing. Regulatory authorities are responsible for approving drugs for clinical trials and the market place.

The chemist's contribution — A graduate with a chemistry or medicinal chemistry degree has skills which are applicable to various fields within the pharmaceutical industry, such as drug discovery, drug design, quality control, radiosynthesis and manufacture.

Related topic — Introduction (A1)

Overview

In general, there are three phases involved in discovering a new drug and getting it to market. Phase 1 is drug discovery, which involves finding an active compound for a particular target. Phase 2 is drug design, where the properties of that active compound are improved such that it is potent and selective for its target and can also reach that target. Phase 3 involves all the testing procedures and development work that have to be carried out on the drug in order to get it to the market.

Phase 1 – drug discovery

Nowadays, most medicinal chemistry projects start by identifying a suitable drug target (Section B). Knowledge of the physiological role played by that target allows the researcher to propose what effect a drug would have if it interacted with the target. Drug targets are usually biological macromolecules such as carbohydrates, lipids, proteins and nucleic acids. The most common targets are proteins followed by nucleic acids. Once a target has been chosen, suitable testing methods have to be developed (Section D) which will demonstrate whether potential drugs have the desired activity. It is then a case of finding a structure that will interact with that target. Such a structure is known as a lead compound (Section E) and is the starting point for drug design.

Phase 2 – drug design

Once a lead compound has been identified, the medicinal chemist will devise synthetic routes that will allow the synthesis of various analogs (Section F). Having produced a series of analogs, the activities of these compounds are compared and certain structural features that are more important to activity than others are identified (structure activity relationships – Section G). These features are retained in the design of further analogs, which will interact more effectively or more selectively with their target (Section H). As well as this, it is necessary to design analogs that have the correct pharmacokinetic properties to reach their target (Section I).

Phase 3 – drug testing and development

As soon as a potentially useful drug is discovered, it is patented (Section J). The potential drug must then be thoroughly tested for any side effects or toxicity (Section K). Drug metabolism studies are also carried out to identify what metabolites are formed. These metabolites are then tested for activity and side effects. If the drug passes these tests, it is put forward for clinical trials (Section K). Finally, the drug can be marketed. At the same time as the pre-clinical and clinical trials are taking place, work is carried out to develop a large-scale synthesis of the compound (section J). The various tests and development work carried out on a new drug must be properly controlled and documented such that they adhere to the requirements laid down by the various regulatory authorities (Section K). Otherwise, the drug may not be allowed to enter clinical trials or the market.

The chemist's contribution

Chemists and medicinal chemists make important contributions at various stages of the drug research program. The medicinal chemist is most involved with drug discovery and drug design. Development chemists are involved in designing a large-scale synthesis for a drug. Organic synthetic chemists are involved in synthesizing radiolabeled drugs for drug metabolism studies, while analytical chemists are involved in quality control, ensuring that the drug satisfies purity specifications. Many of the practical skills acquired in a chemistry or medicinal chemistry degree are invaluable to the pharmaceutical industry. Practical techniques such as extraction, chromatography, distillation and crystallization are vital if a lead compound is to be isolated and purified from a natural extract or a synthetic mixture. Synthetic skills are required in order to synthesize analogs of a lead compound in order to carry out structure activity relationships. Analytical skills, such as the ability to interpret nuclear magnetic resonance (NMR) spectra, are important in determining the structure of a lead compound or a synthetic analog. These skills are just as relevant to the development chemist devising a large-scale synthesis.

B1 ENZYMES

Key Notes

Enzymes

Enzymes are proteins that catalyze the body's chemical reactions. The starting material for an enzyme-catalyzed reaction is known as a substrate.

Active site

The active site is a hollow or cleft on the enzyme surface where the substrate binds and the reaction takes place. The substrate is bound to the active site by intermolecular interactions. The active site contains amino acid residues, which act as nucleophiles or acid/base catalysts in the reaction mechanism.

Mechanisms of catalysis

Serine and cysteine can act as nucleophiles in a reaction mechanism, while histidine can act as an acid/base catalyst. Substrate binding weakens important bonds and constrains the substrate in a specific conformation such that it will undergo reaction.

Enzyme inhibitors

Competitive inhibitors compete with the natural substrate for the active site. Noncompetitive inhibitors bind to allosteric binding sites and distort the active site so that it can no longer bind the natural substrate. Reversible inhibitors bind by noncovalent interactions, whereas irreversible inhibitors are linked to the enzyme through covalent bonds.

Enzyme selectivity

Drugs should be as selective as possible for the target enzyme or isozyme.

Related topics

Binding interactions (G2)
Functional groups as binding groups (G3)
Epidermal growth factor receptor (L1)
Modeling studies (L4)
The dawn of the antibacterial age (1930–1945) (M4)

Enzymes

Enzymes are protein structures that act as the body's catalysts. A **catalyst** aids a chemical reaction by lowering the activation energy of the reaction. This speeds up the rate at which the reaction reaches equilibrium but does not affect the equilibrium itself. Therefore, an enzyme can catalyze a reaction in either direction depending on the relative ratio of compounds present. The chemical that undergoes an enzyme-catalyzed reaction is known as a **substrate**. The substrate is bound to the enzyme to form an enzyme–substrate complex (ES), which then undergoes reaction to form the enzyme-bound product (EP) (*Fig. 1*). The product is then released and the enzyme is free to bind another substrate molecule.

Since enzymes are proteins, they are made up of amino acid subunits linked together by peptide bonds. There are 20 essential amino acids in human biochemistry, five of which are shown in *Fig. 2*.

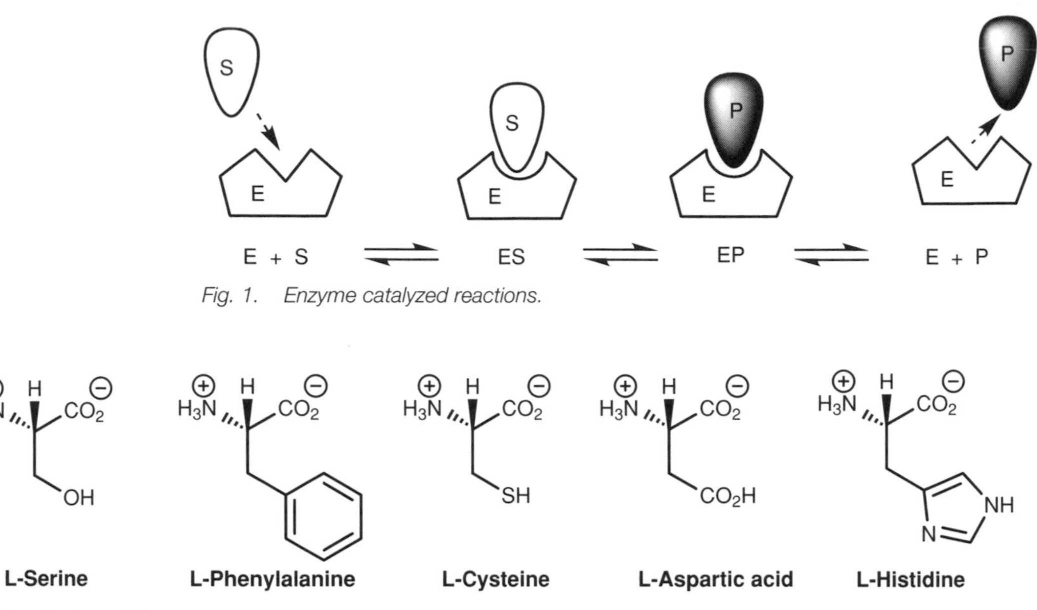

Fig. 1. Enzyme catalyzed reactions.

Fig. 2. Amino acids.

Active site

The **active site** of an enzyme is usually a hollow or cleft on the protein surface into which the substrate can fit and bind (*Fig. 3*). The substrate is usually bound to amino acids present in the binding site by a variety of interactions, such as hydrogen bonding, ionic bonding, van der Waals interactions or dipole–dipole interactions. For example, a substrate might bind to a serine residue by H-bonding, to an aspartate residue by ionic binding and to a phenylalanine residue by van der Waals interactions. These binding interactions must be strong enough to hold the substrate long enough for the enzyme-catalyzed reaction to take place, but weak enough to allow the product to depart once it is formed. This is important when it comes to designing inhibitors since one could introduce extra binding interactions such that the inhibitor 'sticks' to the binding site and blocks it.

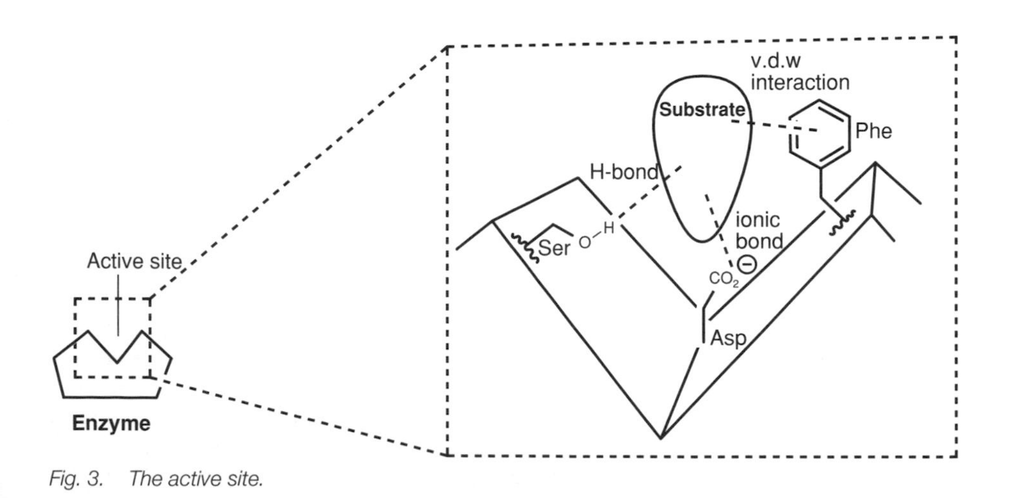

Fig. 3. The active site.

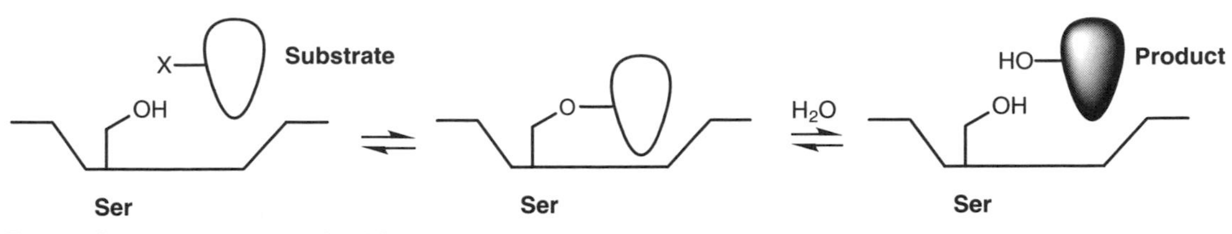

Fig. 4. Serine acting as a nucleophile.

The active site also contains amino acids, which assist in the reaction mechanism. Nucleophilic amino acids, such as serine or cysteine are commonly involved in enzyme-catalyzed mechanisms and will form a temporary covalent bond with the substrate as part of the reaction mechanism (*Fig. 4*).

The amino acid histidine is commonly involved as an acid/base catalyst. This is because the imidazole ring of the histidine residue can easily equilibrate between the ionized and nonionized forms, allowing the amino acid to act both as a source and as a 'sink' for protons (*Fig. 5*).

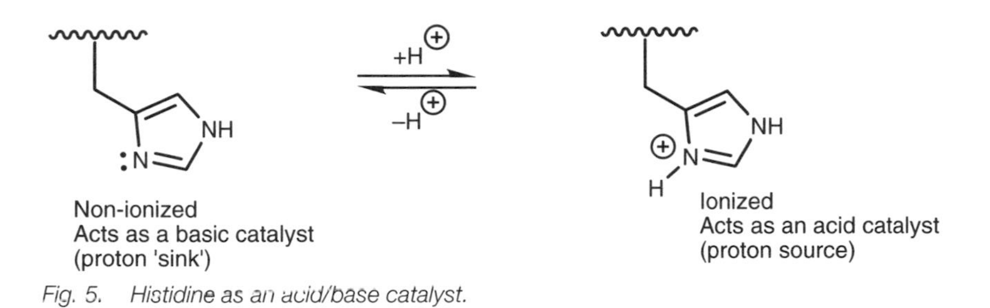

Fig. 5. Histidine as an acid/base catalyst.

Mechanisms of catalysis

There are several reasons why enzymes catalyze reactions. We have already mentioned that amino acids in the active site can aid the enzyme mechanism by acting as nucleophiles or acid/base catalysts. Another reason why enzymes act as catalysts is the binding process itself. The active site is not the ideal shape for the substrate, and when binding takes place it changes shape in order to accommodate the substrate and to maximize the bonding forces between the substrate and the active site. This is known as an **induced fit** (*Fig. 6*).

However, these binding processes also mean that the substrate is forced to adopt a specific conformation (not necessarily the most stable conformation) in

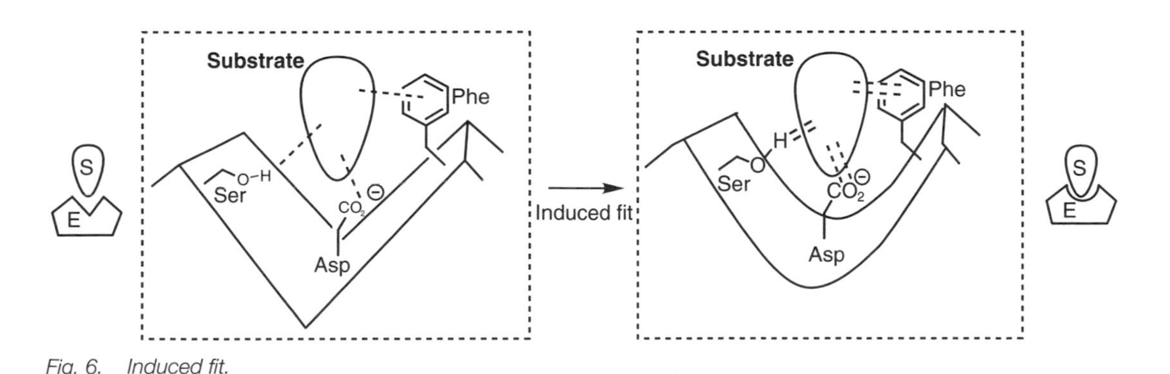

Fig. 6. Induced fit.

order to bind effectively. Constraining the substrate into a specific conformation usually holds the molecule in the ideal position for further reaction with the aforesaid nucleophilic and catalytic amino acids. Another feature of the binding between substrate and enzyme is that important bonds in the substrate may be strained and weakened, allowing the reaction mechanism to proceed more easily. An example of an enzyme-catalyzed reaction is the hydrolysis of the neurotransmitter **acetylcholine** by the enzyme **acetylcholinesterase** (*Fig. 7*). In this reaction, acetylcholine is bound in the active site such that it is held in position for nucleophilic attack by a serine residue. A histidine residue is also positioned ideally to act as an acid/base catalyst. Hydrogen bonding between the ester group of the acetylcholine and a tyrosine residue in the active site also serves to weaken the ester linkage, allowing it to be cleaved more easily. The mechanism shown in *Fig. 8* illustrates the roles played by serine and histidine.

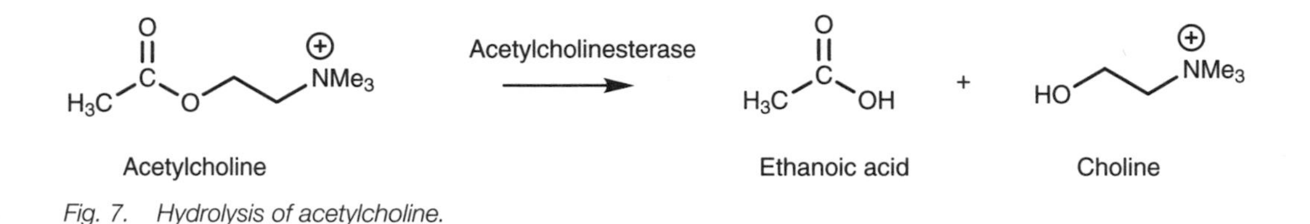

Fig. 7. Hydrolysis of acetylcholine.

Enzyme inhibitors

Enzyme inhibitors are drugs that inhibit the catalytic activity of an enzyme. Inhibitors can be defined as being competitive or noncompetitive. **Competitive inhibitors** compete with the natural substrate for the active site (*Fig. 9*). The

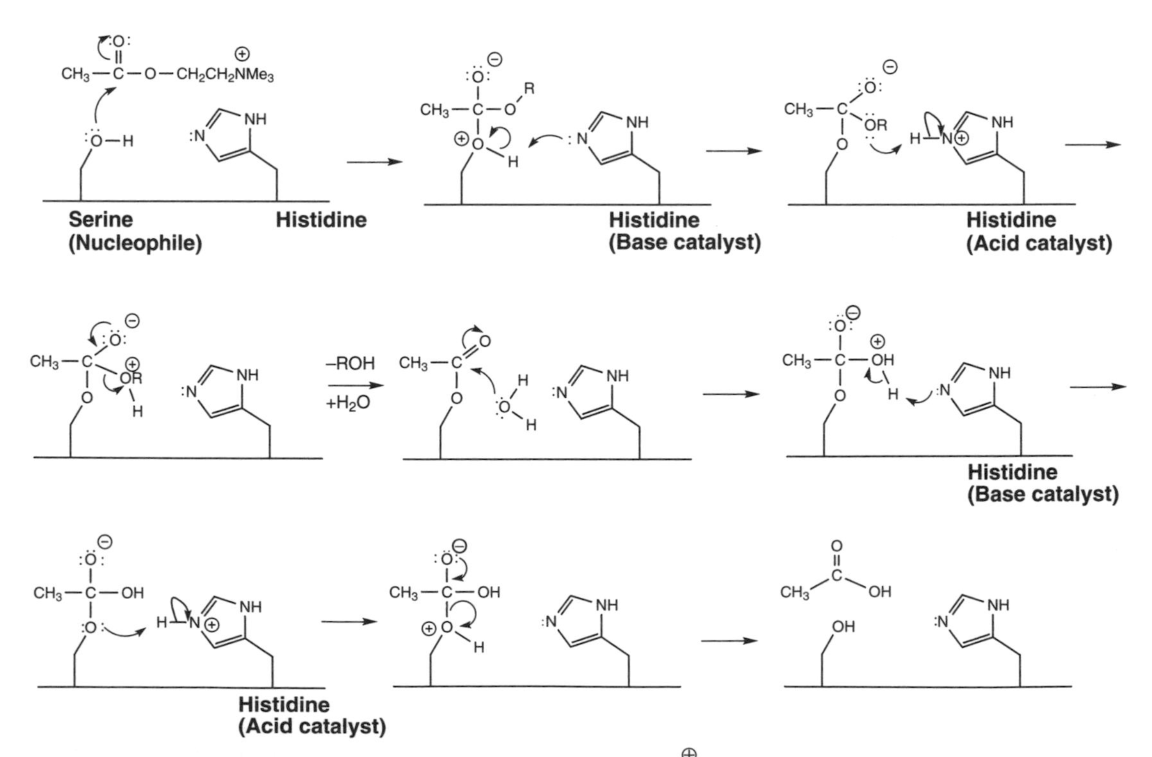

Fig. 8. Mechanism of hydrolysis by acetylcholinesterase ($R = CH_2CH_2\overset{\oplus}{N}Me_3$).

greater the concentration of the inhibitor, the less chance there is for the natural substrate to enter the active site. Conversely, the greater the concentration of natural substrate, the less efficient is the inhibition.

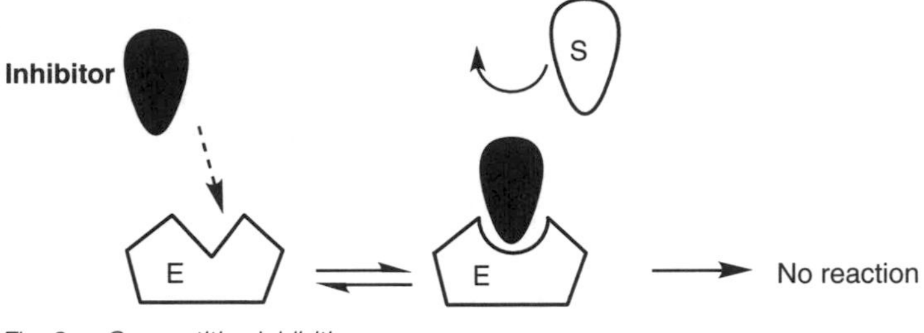

Fig. 9. Competitive inhibition.

Noncompetitive inhibitors do not compete with the natural substrate for the active site. Such inhibitors usually bind to a different region of the enzyme and in doing so produce an induced fit, which changes the enzyme's shape such that the active site is no longer recognizable to the substrate (*Fig. 10*). Increasing the amount of substrate will have no effect on the level of inhibition.

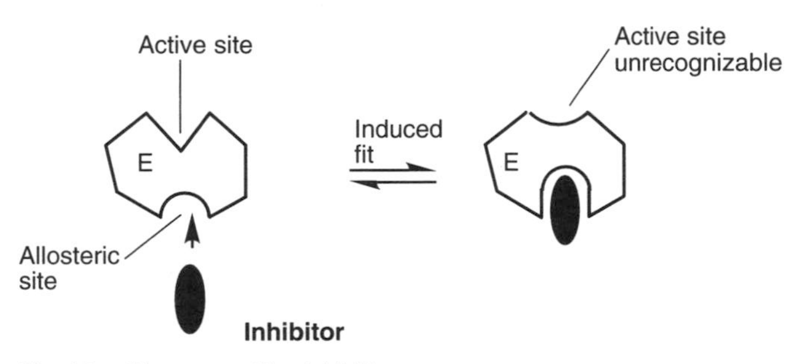

Fig. 10. Noncompetitive inhibition.

The binding site used by a noncompetitive inhibitor is called an **allosteric site** and is often present in an enzyme at the start of a biosynthetic pathway (*Fig. 11*). The allosteric site binds the final product of the pathway and provides a feedback control for the pathway. When levels of the final product are low, the allosteric site is unoccupied and the enzyme is active. When levels are high, the allosteric site is occupied, thus switching off the enzyme and the biosynthesis. Thus, the final biosynthetic product can be used as a lead compound for the design of inhibitors that will bind to an allosteric site.

Both competitive and noncompetitive inhibitors can be reversible or irreversible. **Reversible inhibitors** interact with an enzyme through noncovalent

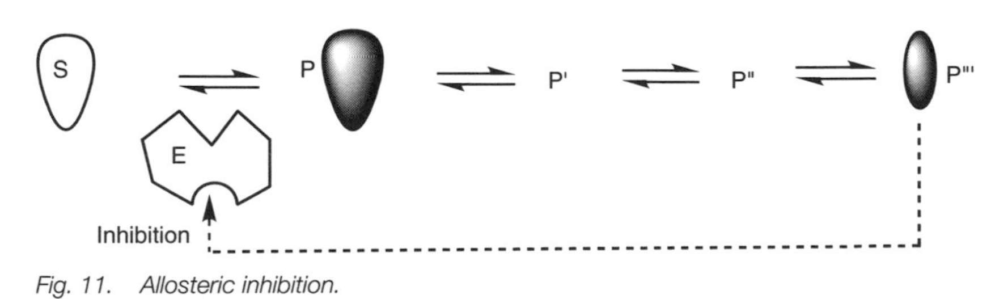

Fig. 11. Allosteric inhibition.

intermolecular interactions (i.e. hydrogen bonding, ionic bonding and van der Waals interactions). There is an equilibrium between the enzyme–inhibitor complex and the free enzyme and unbound inhibitor. The position of the equilibrium will depend on the strength of the intermolecular interactions. If the intermolecular interactions are extremely strong, the equilibrium could be so far over to the enzyme–inhibitor complex that inhibition is effectively irreversible. **Irreversible inhibitors** generally contain reactive electrophilic groups which are susceptible to attack from nucleophilic amino acids present in the enzyme, resulting in the formation of covalent bonds (*Fig. 12*). Such bonds are strong and not easily broken and the enzyme remains permanently blocked. The only way that the body can sort out this problem is to degrade the enzyme–inhibitor complex and to synthesize new enzyme.

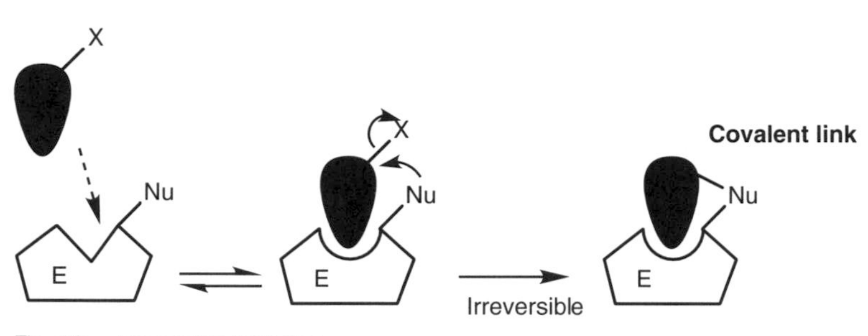

Fig. 12. Irreversible inhibition.

Enzyme selectivity

Inhibitors should be as selective as possible for the target enzyme in order to avoid side effects. There are also several instances where a particular enzyme can exist in different forms known as isozymes. The catalytic reaction is identical, but the amino acid composition between **isozymes** is slightly different. If this variation is in the binding site, it is possible to design drugs that will be isozyme-selective. Specific isozymes are more prevalent in one type of cell or tissue than another, so an isozyme-selective inhibitor is more likely to have a specific effect and have fewer side effects.

B2 RECEPTORS

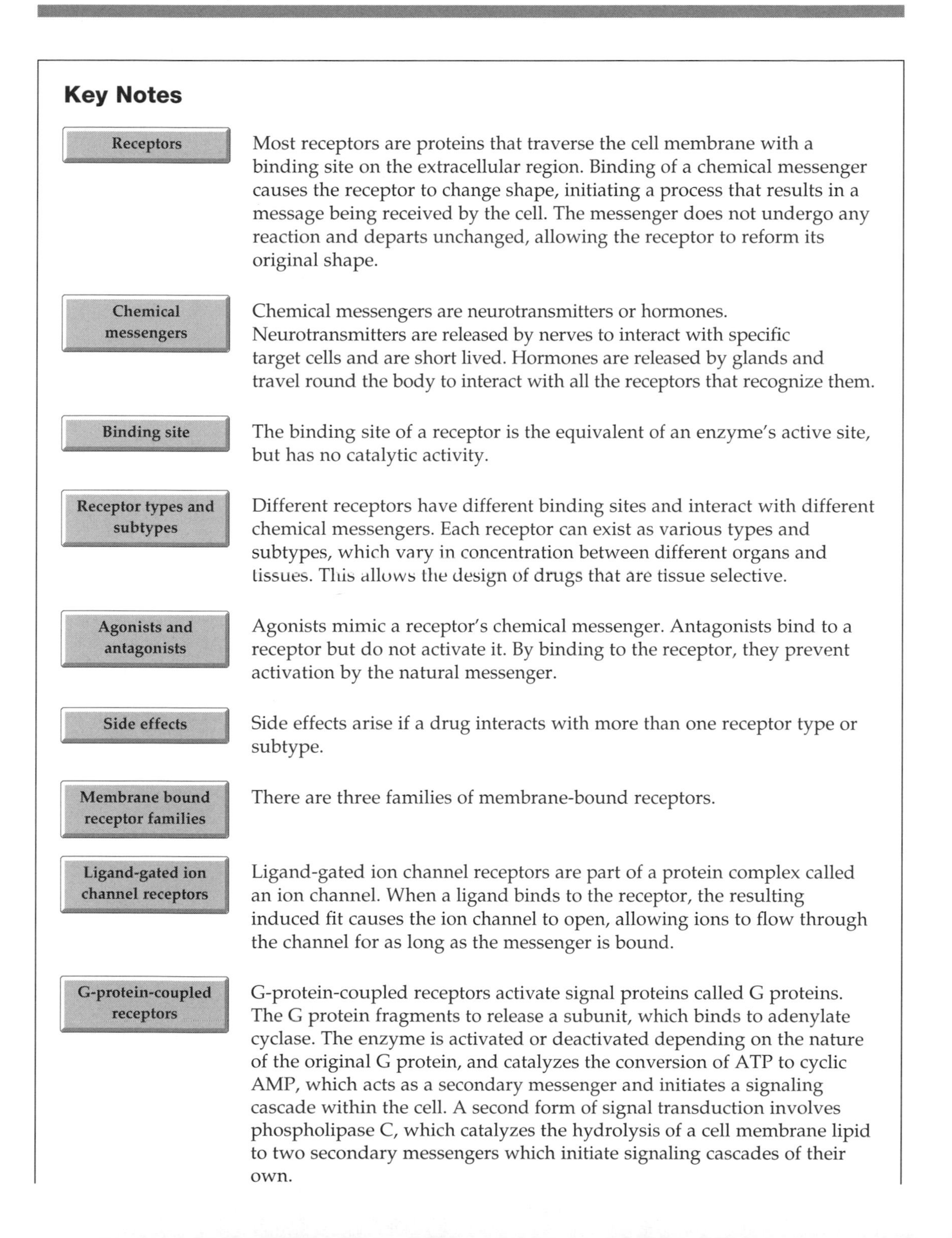

Key Notes

Receptors — Most receptors are proteins that traverse the cell membrane with a binding site on the extracellular region. Binding of a chemical messenger causes the receptor to change shape, initiating a process that results in a message being received by the cell. The messenger does not undergo any reaction and departs unchanged, allowing the receptor to reform its original shape.

Chemical messengers — Chemical messengers are neurotransmitters or hormones. Neurotransmitters are released by nerves to interact with specific target cells and are short lived. Hormones are released by glands and travel round the body to interact with all the receptors that recognize them.

Binding site — The binding site of a receptor is the equivalent of an enzyme's active site, but has no catalytic activity.

Receptor types and subtypes — Different receptors have different binding sites and interact with different chemical messengers. Each receptor can exist as various types and subtypes, which vary in concentration between different organs and tissues. This allows the design of drugs that are tissue selective.

Agonists and antagonists — Agonists mimic a receptor's chemical messenger. Antagonists bind to a receptor but do not activate it. By binding to the receptor, they prevent activation by the natural messenger.

Side effects — Side effects arise if a drug interacts with more than one receptor type or subtype.

Membrane bound receptor families — There are three families of membrane-bound receptors.

Ligand-gated ion channel receptors — Ligand-gated ion channel receptors are part of a protein complex called an ion channel. When a ligand binds to the receptor, the resulting induced fit causes the ion channel to open, allowing ions to flow through the channel for as long as the messenger is bound.

G-protein-coupled receptors — G-protein-coupled receptors activate signal proteins called G proteins. The G protein fragments to release a subunit, which binds to adenylate cyclase. The enzyme is activated or deactivated depending on the nature of the original G protein, and catalyzes the conversion of ATP to cyclic AMP, which acts as a secondary messenger and initiates a signaling cascade within the cell. A second form of signal transduction involves phospholipase C, which catalyzes the hydrolysis of a cell membrane lipid to two secondary messengers which initiate signaling cascades of their own.

Tyrosine kinase-linked receptors

Tyrosine kinase-linked receptors are proteins that act both as receptor and enzyme. Binding of a chemical messenger activates a kinase enzyme on the intracellular region of the protein, resulting in the phosphorylation of tyrosine residues. These regions act as binding sites for signal proteins and enzymes, initiating a signaling cascade which ultimately results in gene expression and protein synthesis.

Intracellular receptors

Some receptors are present within the cell and so the chemical messenger must be hydrophobic to cross the cell membrane. Activation of the estrogen receptor leads to a receptor-ligand complex, which enters the nucleus and switches on transcription, leading to the synthesis of proteins.

Related topics

Enzymes (B1)
Binding interactions (G2)
Functional groups as binding groups (G3)
Epidermal growth factor receptor (L1)

Receptors

Receptors are proteins that are crucial to the body's communication process. They act as the cell's 'letter boxes' and receive messages from chemical messengers called neurotransmitters or hormones. The majority of receptors are embedded in the cell membrane, traversing it such that there are extracellular and intracellular regions (*Fig. 1*). On the extracellular region, there is a hollow, or cleft, called the **binding site**. Chemical messengers fit into these binding sites and are bound by intermolecular forces in the same way that substrates are bound to the active sites of enzymes. Similarly, the binding process involves an induced fit whereby the receptor alters shape to accommodate the messenger. However, unlike enzymes, no reaction takes place. The messenger binds, but since this is an equilibrium process, it can depart again unchanged, allowing the receptor to return to its original shape. This might appear to be a pointless exercise, but the change of shape induced by binding has a 'knock on' effect, which results in a message being transmitted into the cell. Thus, the chemical messenger gives a message to a cell without entering the cell. This will be covered in more detail below. The study of how drugs interact with receptors is known as **pharmacodynamics**.

Chemical messengers

Chemical messengers can be classed as neurotransmitters or hormones. **Neurotransmitters** are released from nerve endings and are crucial to the

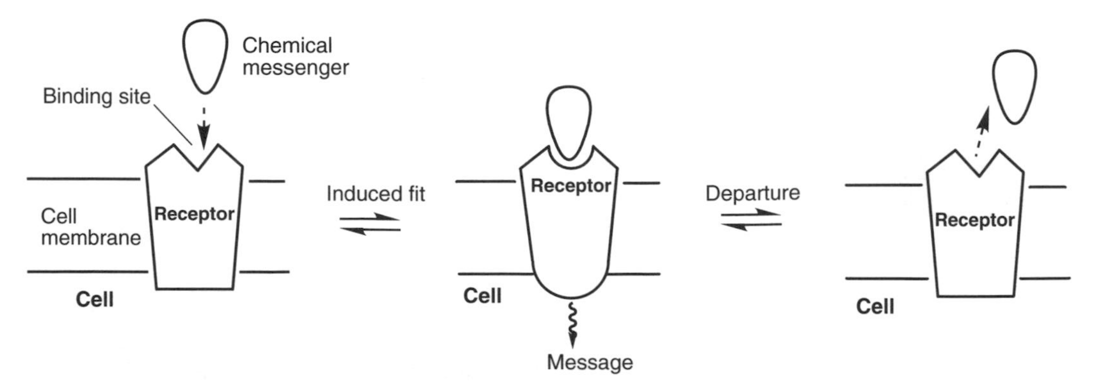

Fig. 1. Membrane bound receptors.

mechanism by which nerves transmit messages to cells (*Fig. 2*). Nerves do not make direct contact with their target cells and the separation between them is called the **synaptic gap**. When a nerve is active, it releases a neurotransmitter, which diffuses across the synaptic gap and binds with a receptor in the cell membrane of the target cell.

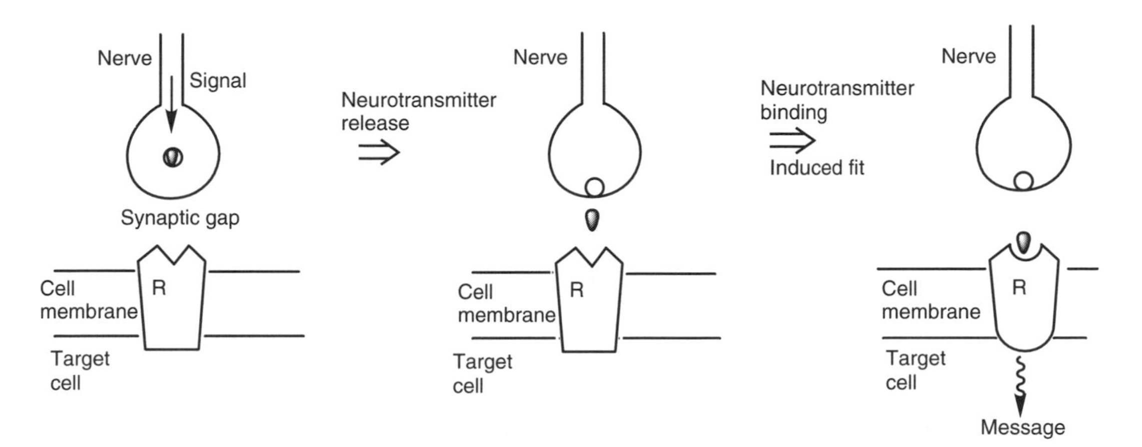

Fig. 2. Neurotransmission.

Neurotransmitters are usually small molecules, such as **acetylcholine**, **norepinephrine** (noradrenaline), **dopamine** and **serotonin** (*Fig. 3*). They bind briefly to their target receptor, pass on their message and depart unchanged to be quickly deactivated such that the message is only received once. This deactivation also means that neurotransmitters do not activate receptors on more distant cells.

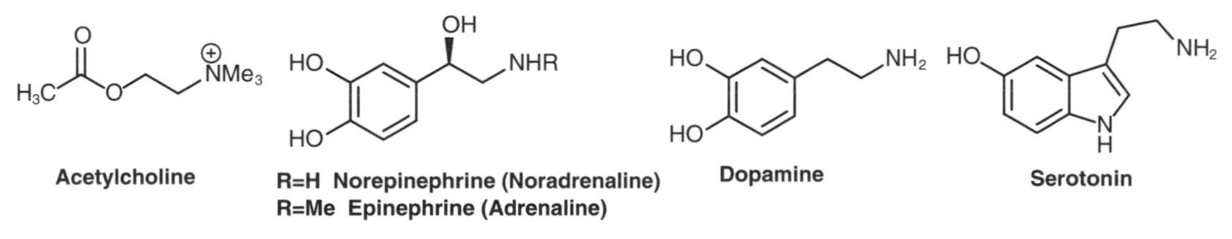

Fig. 3. Neurotransmitters and hormones.

Hormones are chemicals released from glands or cells that enter the bloodstream in order to travel round the body, activating all the receptors that recognize them. Since they have long journeys to undertake, they are not swiftly deactivated. **Epinephrine** (or adrenaline) is a hormone released from the adrenal medulla in situations of stress or danger, and which prepares the body for physical exercise. Other examples include the **steroids**, which have diverse actions in the body. The mechanism by which a receptor is activated is the same regardless of whether the messenger is a neurotransmitter or a hormone. In both cases, the chemical messenger binds to the receptor and causes it to change shape, resulting in a message being received by the cell.

Binding site

The **binding site** of a receptor is the equivalent of an enzyme's active site. The binding process is also identical, involving the same kind of intermolecular binding forces and induced fit. However, the binding site of a receptor does not catalyze any reaction and the messenger can depart unchanged.

Receptor types and subtypes

There are a large number of different receptors in the body that interact with different chemical messengers. Three examples are receptors that are activated by acetylcholine (**cholinergic receptors**), receptors that are activated by epinephrine (**adrenergic receptors**), and those that are activated by dopamine (**dopaminergic receptors**). The various receptors show selectivity for one chemical messenger over another because they have binding sites of different shape, structure and amino acid composition. However, receptors that interact with one specific chemical messenger are not identical. For example, there are two **types** of adrenergic receptor – **α- and β-adrenergic receptors**. These receptors both bind epinephrine, but there are slight differences in their binding sites. This makes no difference as far as epinephrine is concerned, but it is very important when it comes to drug design because it is possible to design drugs which will bind better to one type of adrenergic receptor than the other. Moreover, the different receptor types are not evenly distributed around the body. For example, the heart has more β- than α-adrenergic receptors. This means that drugs that are selective for β-adrenergic receptors will act on the heart rather than on tissues which are rich in α-adrenergic receptors. Such selectivity is crucial in designing drugs with fewer side effects.

There are even subtle differences in the binding sites within a particular receptor type. For example, there are three **subtypes** of the β-adrenergic receptor (β_1-, β_2- and β_3-subtypes). This provides the opportunity to design drugs with even greater selectivity. For example, lung tissue has a predominance of β_2-adrenergic receptors, while heart tissue has a predominance of β_1-adrenergic receptors. Anti-asthmatic drugs that mimic epinephrine have been designed to interact with β_2-adrenergic receptors rather than β_1-adrenergic receptors, so they interact with the adrenergic receptors in the lungs rather than those in the heart, thus lowering cardiovascular side effects.

Agonists and antagonists

Agonists are drugs that bind to the receptor binding site and mimic the natural messenger by 'switching on' the receptor. Binding produces the induced fit required to activate the receptor. Such drugs are useful if there is a lack of the natural chemical messenger. **Antagonists** are drugs that bind to the receptor-binding site, but do not activate the receptor, either because they fail to cause the required induced fit or because they distort the receptor in a different way. A bound antagonist prevents the natural messenger from binding, so the receptor can no longer receive messages. Antagonists are useful in blocking messages if there is a surplus of the normal messenger.

Not all agonists and antagonists bind to a receptor's binding site. Some antagonists bind to a different region of the receptor and distort the protein such that the normal binding site is no longer recognized. This is equivalent to the allosteric inhibition of an enzyme. In some receptors, drugs can bind to an **allosteric binding site** and enhance the activity of the natural messenger. For example, the **benzodiazepine tranquilizers** act by potentiating the action of an inhibitory neurotransmitter called **γ-aminobutyric acid** (GABA) at the GABA receptor.

Side effects

Side effects are usually caused when a drug binds to more than one type of receptor. Therefore, a major goal in drug design is to design drugs which will be as selective for one type (or subtype) of receptor as possible. For example, the **serotonin receptor** is the target for some antidepressant drugs. However, side effects can arise if these drugs interact with receptors for histamine or acetylcholine.

Membrane bound receptor families

All membrane bound receptors belong to one of the three following families:

- ligand-gated ion channels;
- G-protein-coupled receptors;
- Kinase-linked receptors.

The mechanism by which a receptor conveys a signal to the cell depends on which family the receptor belongs to, and is called **signal transduction**.

Ligand-gated ion channel receptors

In this family of receptors, the receptor is an integral part of an ion channel (*Fig. 4*). **Ion channels** are protein complexes that traverse the cell membrane and form a tunnel through it. The function of an ion channel is to allow the flow of ions across the cell membrane and there are specific ion channels for sodium, potassium, calcium and chloride ions. Without ion tunnels, ions could not cross the fatty cell membranes and this would have devastating effects on the chemistry of the cell, since many enzyme-catalyzed reactions are dependent on ionic concentration. However, the ion channels cannot be permanently open, since the uncontrolled flow of ions across the cell membrane would be as devastating as if they did not cross at all. Thus, ion channels are normally closed and are only opened when signaled to do so. This is where the receptor comes in, controlling whether the ion channel is open or closed.

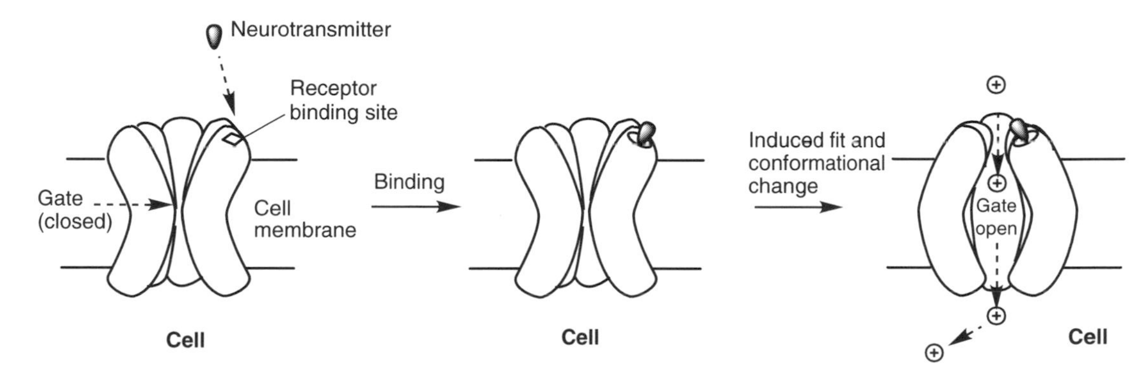

Fig. 4. Ligand-gated ion channel receptors.

In the resting state, the ion channel is closed and the receptor's binding site is unoccupied. The signal to open the ion channel comes from the chemical messenger, which binds to the receptor and causes it to change shape as previously described. This change in shape is the start of a domino effect, which travels through the entire ion channel. The proteins making up the ion channel are normally positioned such that they seal the ion channel, but once the messenger binds to the receptor, the proteins alter their positions relative to each other, resulting in the channel opening up. Ions can then flow through the channel. Once the chemical messenger leaves the receptor, the receptor and the ion channel reform their original shapes and block off the flow of ions. Such receptors are called **ligand-gated ion channel receptors** because the trigger for the process is a chemical messenger (the ligand), and the effect is to open up the 'gate' sealing the ion channel. Since the receptor is an integral part of the ion channel, the effects of a receptor binding to its messenger are felt almost immediately by the cell as ions flow through the channel. Therefore, this mode of signal transmission is used when speedy communication is vital (e.g. when one nerve signals to another).

G-protein-coupled receptors

The second family of receptors is the **G-protein-coupled receptors** (*Fig. 5*), so named because the receptor conveys a signal to the cell via a signaling protein called a G protein. The G-protein-coupled receptor traverses the cell membrane with the binding site for the chemical messenger on the extracellular portion of the receptor. However, there is a second binding site on the intracellular portion of the receptor, which is specific for the G protein. This binding site is closed when the receptor is in the resting state.

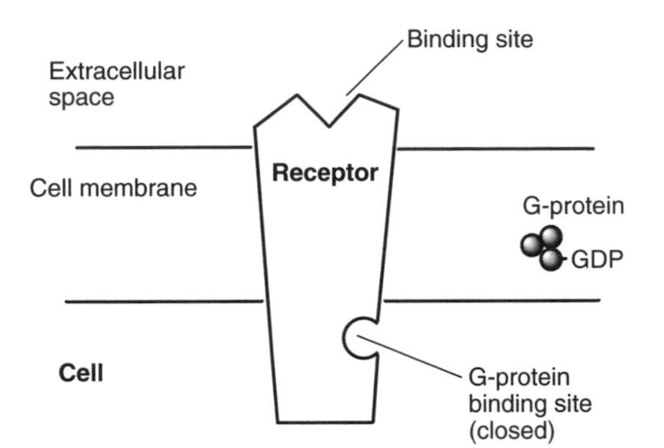

Fig. 5. G-protein-coupled receptors.

The **G protein** is made up of three subunits (α, β, and γ) and is free to move through the cell membrane. It also has a binding site which binds a nucleotide called **guanosine diphosphate** (GDP) (*Fig. 6*). In this state, the G protein is 'inert'.

The method by which a receptor of this family conveys a message to the cell involves several stages (*Fig. 7*).

1. A chemical messenger binds to the receptor and causes it to change shape. This change in shape opens up the G-protein binding site.
2. The G protein binds to this binding site and the interaction between the receptor and the G protein results in the G protein itself changing shape.
3. The change of shape in the G protein alters the binding site for GDP such that it favors **guanosine triphosphate** (GTP) over GDP. As a result, GDP departs the G protein and GTP binds in its place.
4. The binding interaction between the G protein and GTP causes the G protein to change shape yet again. This destabilizes the structure such that the α-subunit (along with GTP) splits off from the β- and γ-subunits. The α-subunit and the β,γ-dimer now leave the receptor.

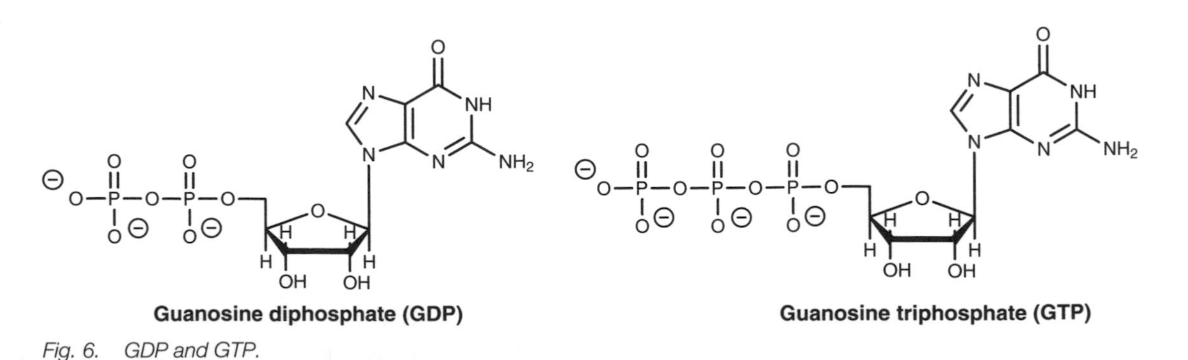

Fig. 6. GDP and GTP.

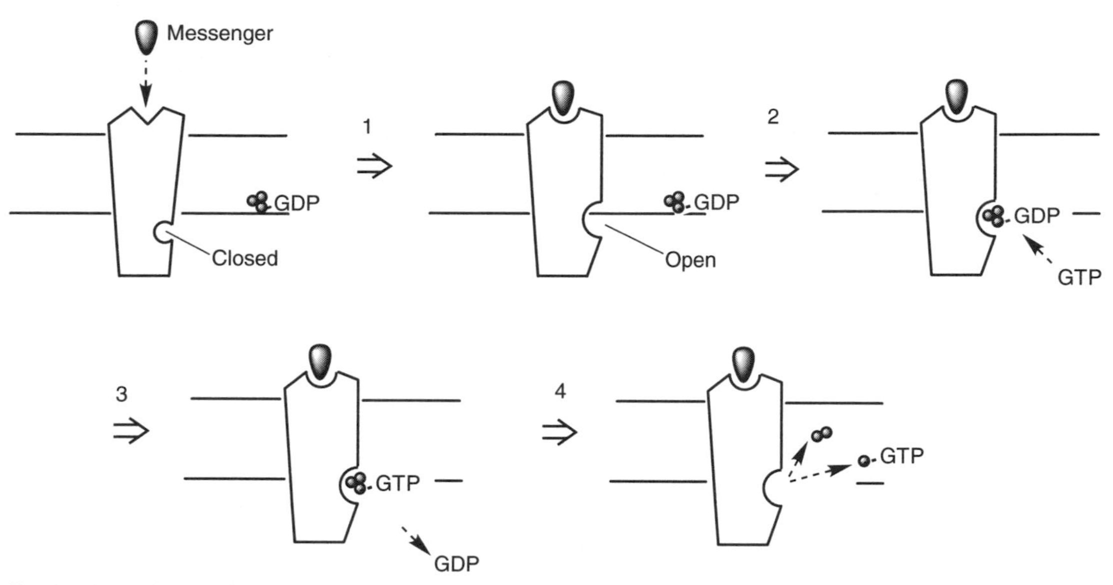

Fig. 7. Activation of a G protein.

The receptor can now bind another G protein and so the process repeats itself for as long as the chemical messenger is bound to the receptor. This means that one chemical messenger can trigger the fragmentation of several G proteins.

The α-subunit of the G protein now acts as a signaling unit. It floats through the membrane until it reaches a membrane-bound enzyme called **adenylate cyclase** (*Fig. 8*). This enzyme has a binding site, which recognizes the α-subunit and binds it. The resulting induced fit causes the enzyme to change shape and its active site is opened. The reaction catalyzed by this enzyme is the conversion of adenosine triphosphate (ATP) to **cyclic AMP** (cyclic adenosine monophosphate) (*Fig. 9*). The reaction continues as long as the α-subunit is bound to the enzyme, which means that one α-subunit can lead to the synthesis of several cyclic AMP molecules – another amplification process. Cyclic AMP is produced in the cytoplasm of the cell and is an example of a **secondary messenger**. It triggers the start of a signaling cascade within the cell in which the signal is further amplified, leading to several different enzymes in the cell being activated or deactivated.

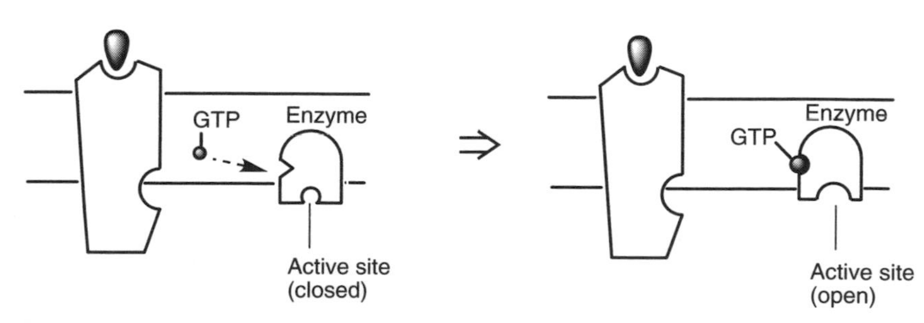

Fig. 8. Activation of adenylate cyclase.

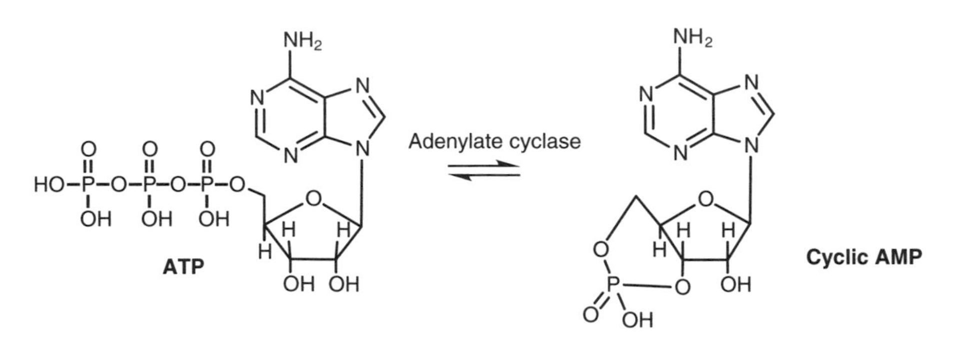

Fig. 9. Conversion of ATP to cyclic AMP.

The example given above shows the effect of a stimulatory G protein (G_s) where the α-subunit activates adenylate cyclase; but there is a different G protein called a G_i protein whose α-subunit inhibits adenylate cyclase. The G_s and G_i proteins are activated by different types of receptors, which are in turn activated by different chemical messengers. Thus, adenylate cyclase is under dual control. Whether the enzyme increases or decreases in activity will depend on whether the stimulatory or inhibitory G protein is in the ascendancy, which in turn depends on which kind of receptor is activated more strongly.

There is another type of G protein which works by activating a different membrane-bound enzyme called **phospholipase C** (PLC). The reaction catalyzed by this enzyme hydrolyses a component of the cell membrane itself, **phosphatidylinositol diphosphate** (PIP_2), to generate two secondary messengers called **inositol trisphosphate** (IP_3) and **diacylglycerol** (DG) which proceed to initiate signaling cascades of their own (*Fig. 10*).

Tyrosine kinase-linked receptors

Tyrosine kinase-linked receptors can be viewed as a receptor and enzyme rolled into one (*Fig. 11*). The receptor protein traverses the cell membrane and has an extracellular region and an intracellular region. The extracellular region acts as

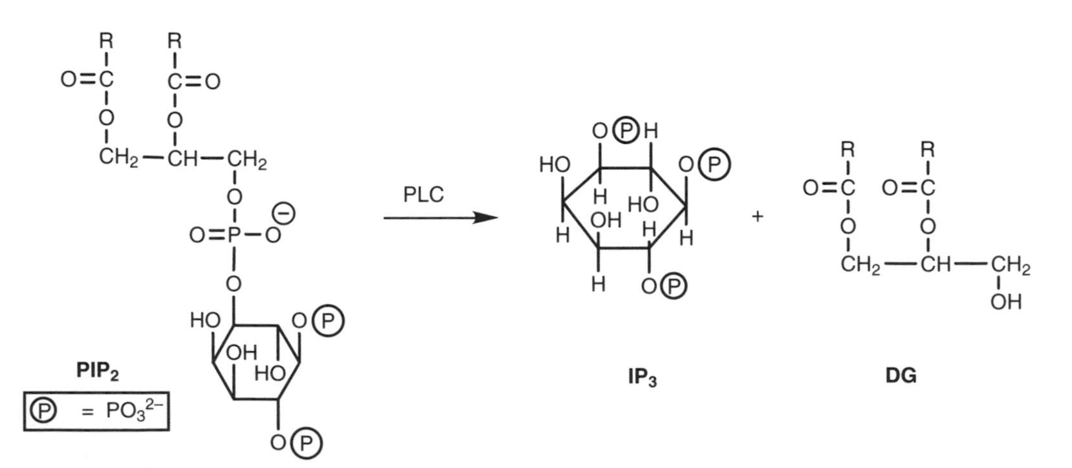

Fig. 10. Hydrolysis of phosphatidylinositol diphosphate.

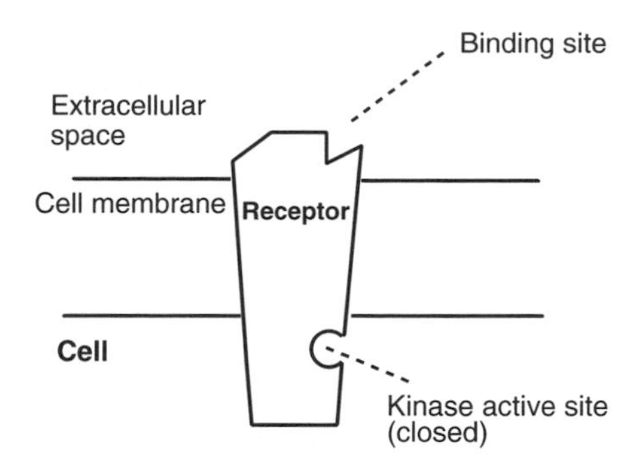

Fig. 11. Tyrosine kinase-linked receptors.

the receptor and has a binding site waiting to receive its chemical messenger. The intracellular region acts as a tyrosine kinase enzyme and has an active site, which is closed when the receptor is in the resting state.

The reaction catalyzed by this enzyme is the **phosphorylation** of tyrosine residues in protein substrates (*Fig. 12*). In order to carry out the reaction, **adenosine triphosphate** (ATP) is required to supply the phosphate unit. ATP is itself dephosphorylated to **adenosine diphosphate** (ADP) (*Fig. 13*).

The chemical messengers that activate kinase-linked receptors include hormones, such as **insulin**, **growth factors** and **cytokines**. To illustrate what happens when a tyrosine kinase-linked receptor is activated, we shall look at a

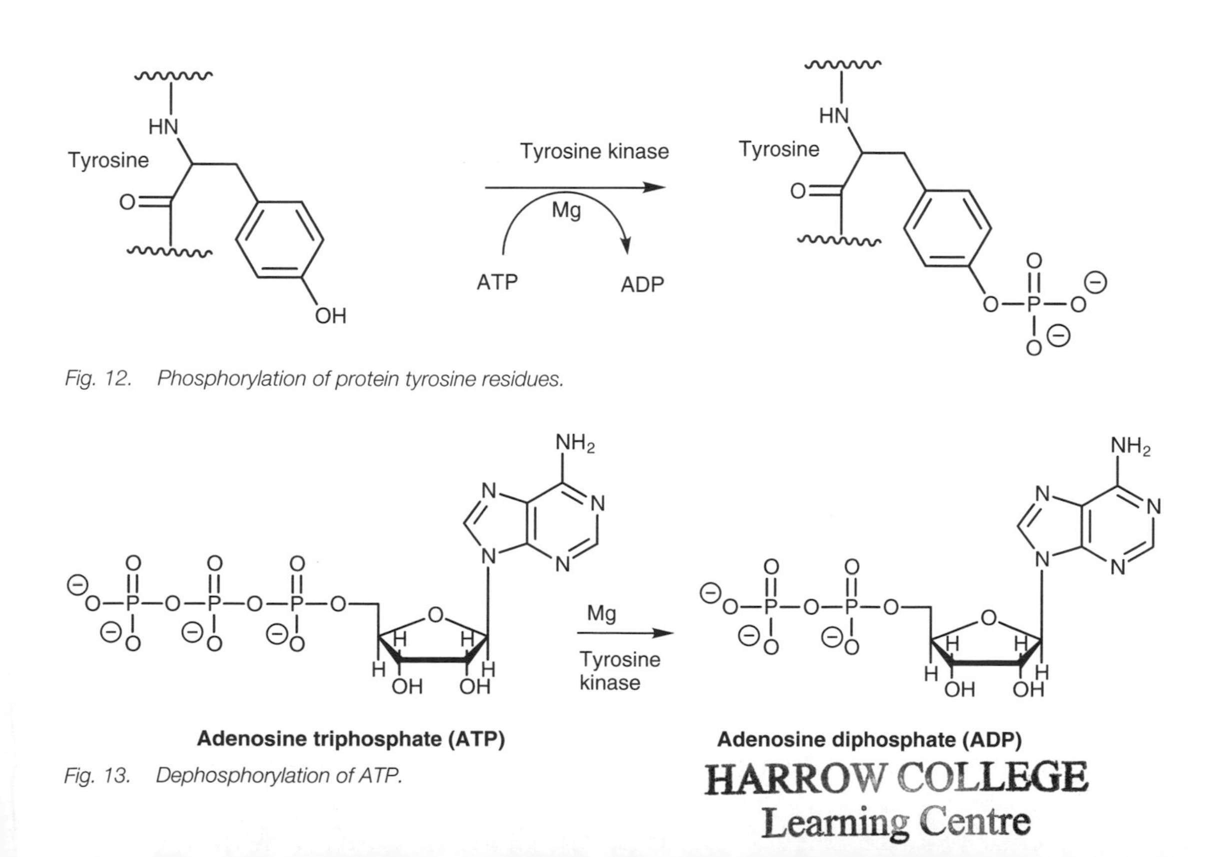

Fig. 12. Phosphorylation of protein tyrosine residues.

Fig. 13. Dephosphorylation of ATP.

specific example – the **epidermal growth factor receptor** (EGF-R), so called because the chemical messenger is the hormone **epidermal growth factor** (EGF) (*Fig. 14*). Epidermal growth factor is a protein that can act as a **bivalent ligand**. This means that one molecule can bind to two separate receptors resulting in **receptor dimerization**. Both receptors in the dimer change shape and this opens up the kinase active site in each receptor. The kinase enzyme of each half of the dimer can then catalyze the phosphorylation of tyrosine residues in protein substrates. Since the most accessible protein is the other half of the dimer, each enzyme catalyzes phosphorylation of the tyrosine residues of its partner.

Phosphorylation of the EGF-R dimer initiates a rather complex cascade effect which begins with the binding of various proteins and enzymes to the phosphorylated regions. Some of these proteins act as signal proteins, which are activated and travel to other parts of the cell in order to initiate further processes. Others are enzymes, which are activated to catalyze further reactions within the cell. The full story of this process is too complex to be included in a text of this sort. Suffice it to say that the end effect is the activation of proteins called **transcription proteins**, which interact with DNA and initiate gene expression. This switches on the synthesis of cell proteins, resulting in cell growth and cell division. It has been observed that several cancers are associated with the unregulated activity of kinase receptors such as the EGF receptor. Therefore, the design of compounds that switch off the tyrosine kinase activity of such receptors may be of interest in fighting cancer. A case study on this topic is given in Section L.

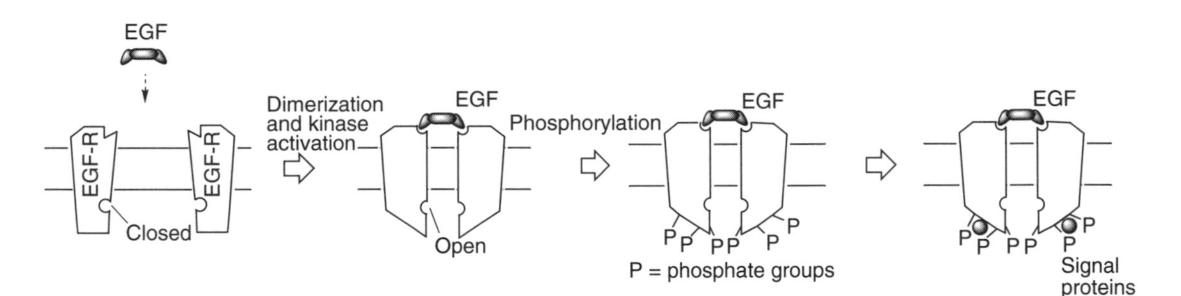

Fig. 14. Activation of the epidermal growth factor receptor.

Intracellular receptors

Not all receptors are membrane bound. There are a group of receptors that exist within the cell which interact with chemical messengers such as **steroids, thyroid hormones** and **retinoids**. The chemical messengers involved have to be sufficiently hydrophobic in order to cross the cell membrane to reach their target receptors. The **estrogen receptor** is an example of an **intracellular receptor**, which exists in the cytoplasm as a complex with another protein (*Fig. 15*). When the female hormone **estrogen** enters the cell, it binds to the receptor–protein complex resulting in dissociation of the protein–receptor complex. The receptor–ligand complex then enters the nucleus and binds to specific DNA sequences, thus activating transcription and the generation of mRNA, which is then translated to form various functional and structural proteins.

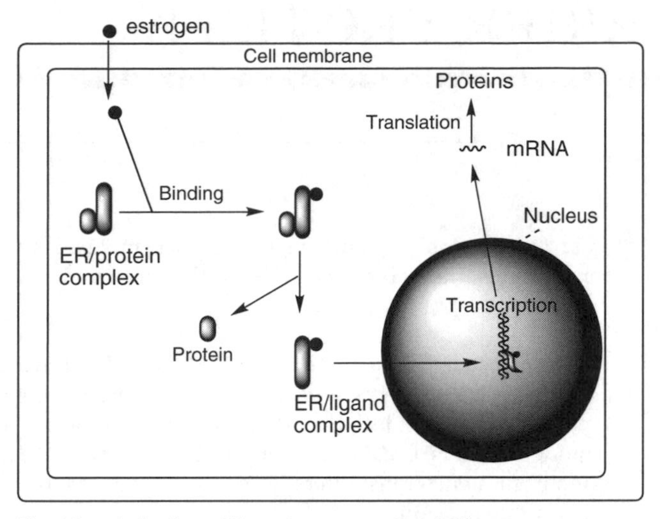

Fig. 15. Activation of the estrogen receptor (ER).

B3 Carrier proteins

Key Notes

Function

Carrier proteins transport important polar molecules across the cell membrane. They do so by enclosing the polar molecule in a hydrophilic cavity.

Carrier protein blockers

Carrier protein blockers are drugs that either bind to a carrier protein and prevent it from accepting its natural guest, or compete with the natural guest for transport into the cell. Drugs such as the tricyclic antidepressants, cocaine and amphetamine hinder the uptake of important neurotransmitters from nerve synapses, resulting in increased neurotransmission.

Drug 'smuggling'

Some polar drugs can be 'smuggled' across cell membranes by carrier proteins if the drug is attached to a natural guest molecule.

Related topics

Drug absorption (C2) | Drug distribution (C3)

Function

Carrier proteins float freely through cell membranes 'visiting' the outer and inner surfaces of the membrane (*Fig. 1*). They have an outer surface of hydrophobic amino acids, which can interact with the fatty cell membrane by van der Waals interactions. In the center, there is a hydrophilic cavity, which can accommodate polar molecules.

The function of a carrier protein is to 'smuggle' important polar molecules across the cell membrane. It does this by 'wrapping up' the polar molecule inside the hydrophilic cavity, transporting the concealed molecule across the membrane, then releasing it into the cell. Carrier proteins are crucial in transporting many of the building blocks required for the cell's survival. For example, there are carrier proteins that are specific for the transport of the amino acid building blocks required for protein synthesis. Other carrier proteins are used to transport the nucleic acid bases required for nucleic acid synthesis.

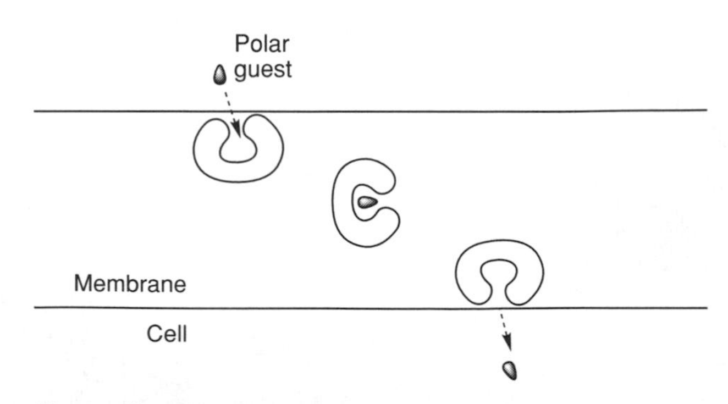

Fig. 1. Function of carrier proteins.

Without carrier proteins, these polar molecules would not be able to cross the hydrophobic cell membrane, the synthesis of proteins and nucleic acids would cease and the cell would die. Carrier proteins are also important in transporting important neurotransmitters such as **norepinephrine** and **dopamine** back into the nerves from which they were released.

Carrier protein blockers

Some drugs prevent carrier proteins transporting their natural guest (*Fig. 2*). For example, the **tricyclic antidepressants** work by inhibiting the uptake of norepinephrine by its carrier protein. These carrier proteins are responsible for transporting norepinephrine from the synaptic gap back into the nerve from which it was released. Inhibition of this process means that the released norepinephrine remains in the synapse for a longer period of time and reactivates the adrenergic receptors of the target cell. As a result, adrenergic activity increases. **Cocaine** acts by inhibiting the carrier proteins for norepinephrine in the peripheral nervous system as well as the carrier proteins for dopamine in the central nervous system. The former activity is responsible for the physical effects of cocaine (e.g. suppression of hunger), while the latter is responsible for the psychological effects (e.g. euphoria).

Some drugs compete with the natural guest for its carrier protein. For example, **amphetamine** is transported into nerves by the carrier protein for norepinephrine. This means that it takes the carrier protein longer to transport norepinephrine back into the nerve, resulting in greater receptor activation and greater adrenergic activity.

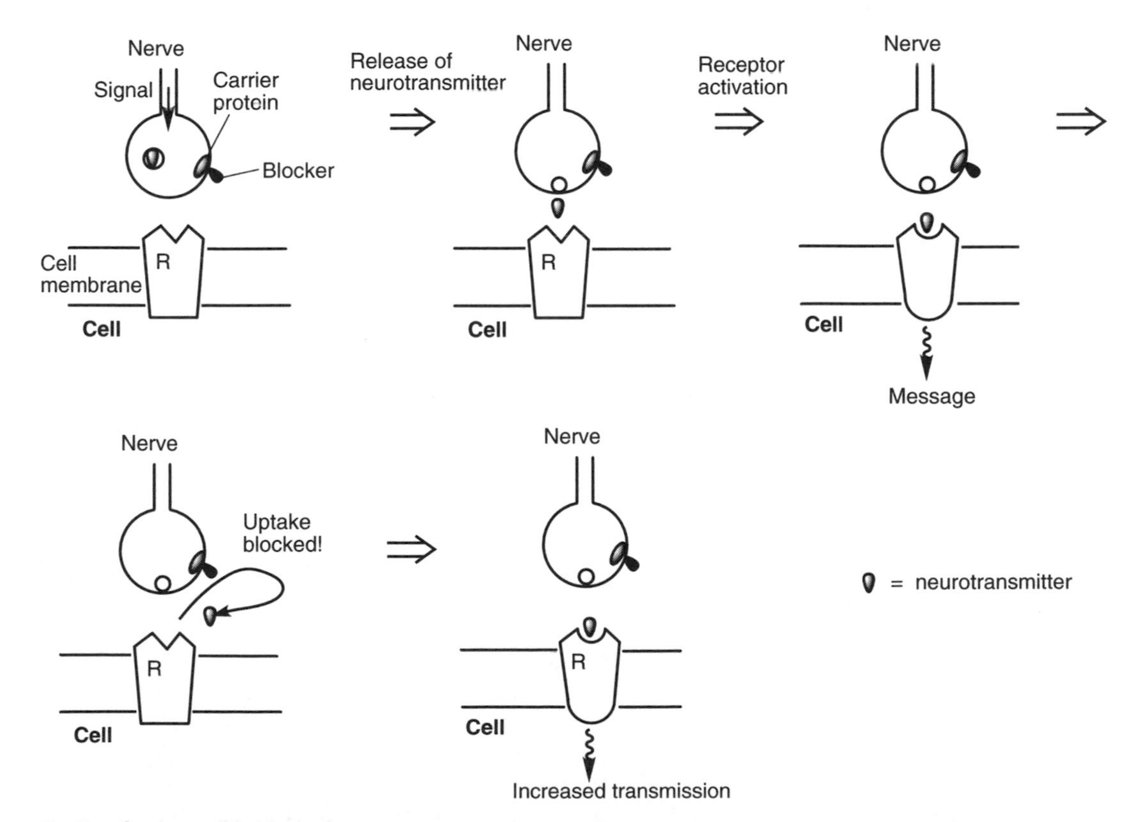

Fig. 2. Carrier protein blockade.

Drug 'smuggling'

Carrier proteins provide a method by which polar drugs can be transported across fatty cell membranes. The drug is attached to a natural guest such as a nucleic acid base. The latter is recognized by its carrier protein, which then 'smuggles' the compound into the cell. **Uracil mustard** (*Fig. 3*) is an anti-cancer drug that is smuggled into cells in this way. The uracil portion of the drug is recognized by the transport protein, while the mustard portion is an alkylating agent which reacts with DNA and prevents DNA from functioning. Some drugs are sufficiently similar in structure to the natural guest that they are accepted by carrier proteins. For example, **L-dopa** (*Fig. 3*) is used for the treatment of Parkinson's disease. Because it is an amino acid, it is smuggled across the blood–brain barrier by amino acid transport proteins.

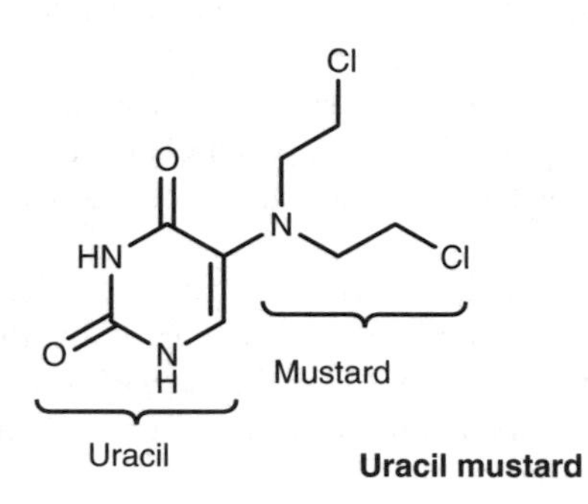

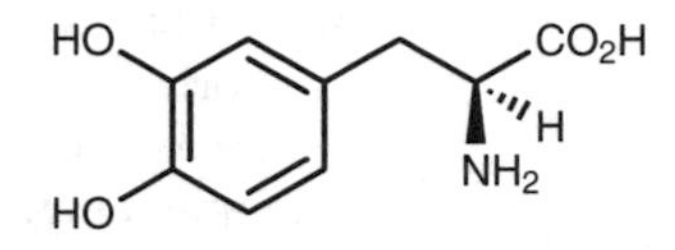

Fig. 3. Uracil mustard and L-dopa.

B4 STRUCTURAL PROTEINS

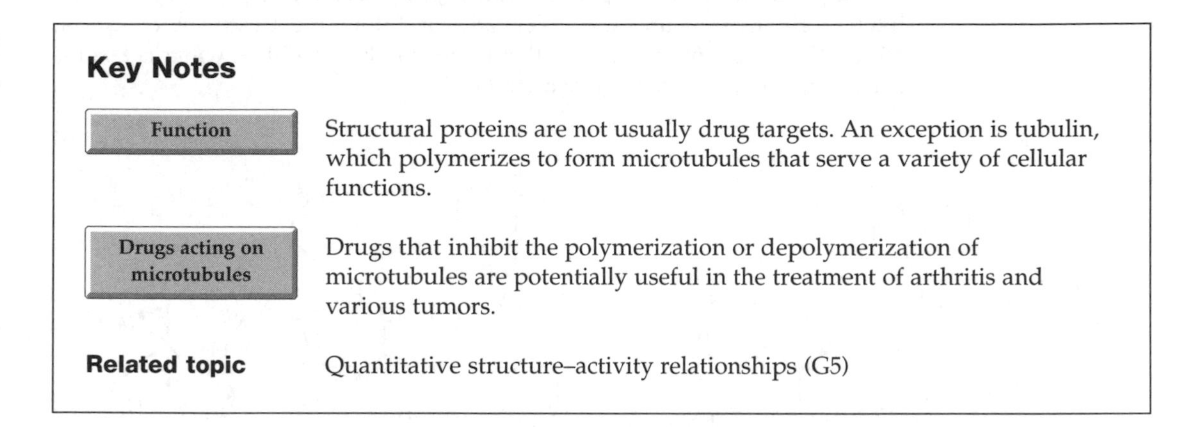

Key Notes

Function	Structural proteins are not usually drug targets. An exception is tubulin, which polymerizes to form microtubules that serve a variety of cellular functions.
Drugs acting on microtubules	Drugs that inhibit the polymerization or depolymerization of microtubules are potentially useful in the treatment of arthritis and various tumors.
Related topic	Quantitative structure–activity relationships (G5)

Function

Structural proteins include cellulose in plants and collagen in animals. In general, structural proteins are not important drug targets. An exception is a protein called **tubulin**, which can polymerize into small tubes called **microtubules** (*Fig. 1*). These microtubules have various cellular functions and are important to the structural integrity and mobility of cells.

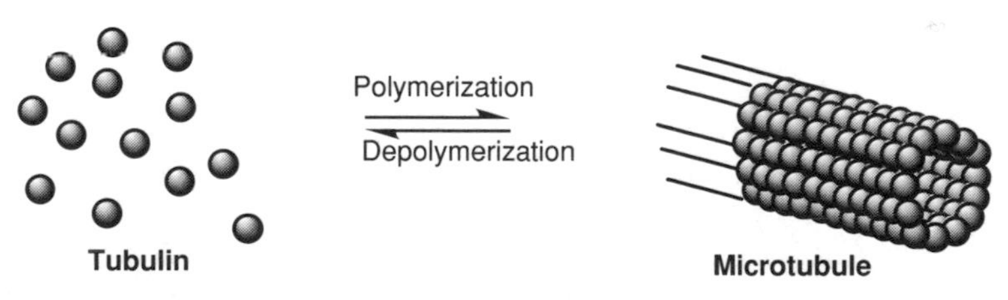

Fig. 1. Tubulin and microtubules. Reproduced from An Introduction to Medicinal Chemistry, *G.L. Patrick, 2001, by permission of Oxford University Press.*

Microtubules are also important to the process of **cell division** (*Fig. 2*). When a cell is about to divide, the microtubules are depolymerized to their tubulin monomer units. These are then repolymerized in order to form a spindle structure, which pushes the two daughter cells apart and is the framework by which the chromosomes are transferred to each of the daughter cells.

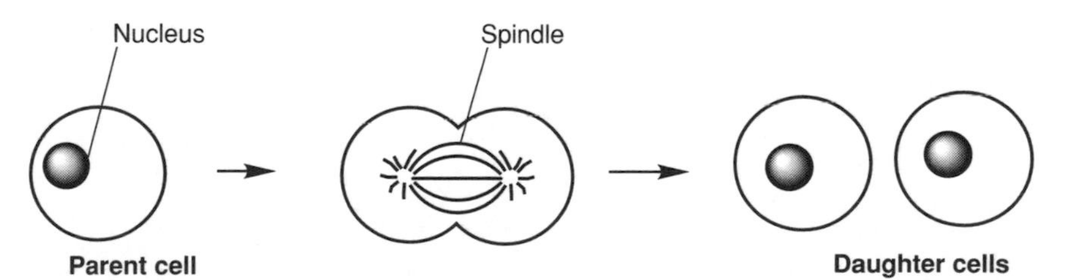

Fig. 2. Cell division. Reproduced from An Introduction to Medicinal Chemistry, *G.L. Patrick, 2001, by permission of Oxford University Press.*

Drugs acting on microtubules

Drugs that depolymerize microtubules could be useful in the treatment of arthritis by reducing the mobility of inflammatory cells called **neutrophils**, thus hindering them from entering joints.

Drugs that inhibit the depolymerization or repolymerization of microtubules inhibit cell division and are potentially useful in the treatment of cancer. **Vincristine** is an anticancer drug that inhibits polymerization, whereas **taxol** is a drug that inhibits depolymerization (*Fig. 3*).

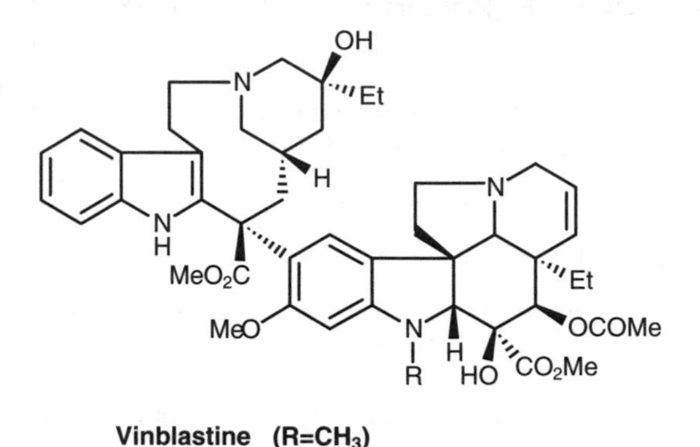

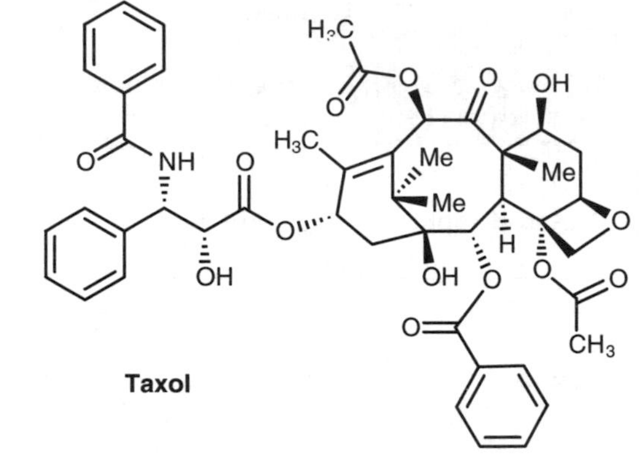

Fig. 3. Drugs acting on microtubules.

B5 Nucleic acids

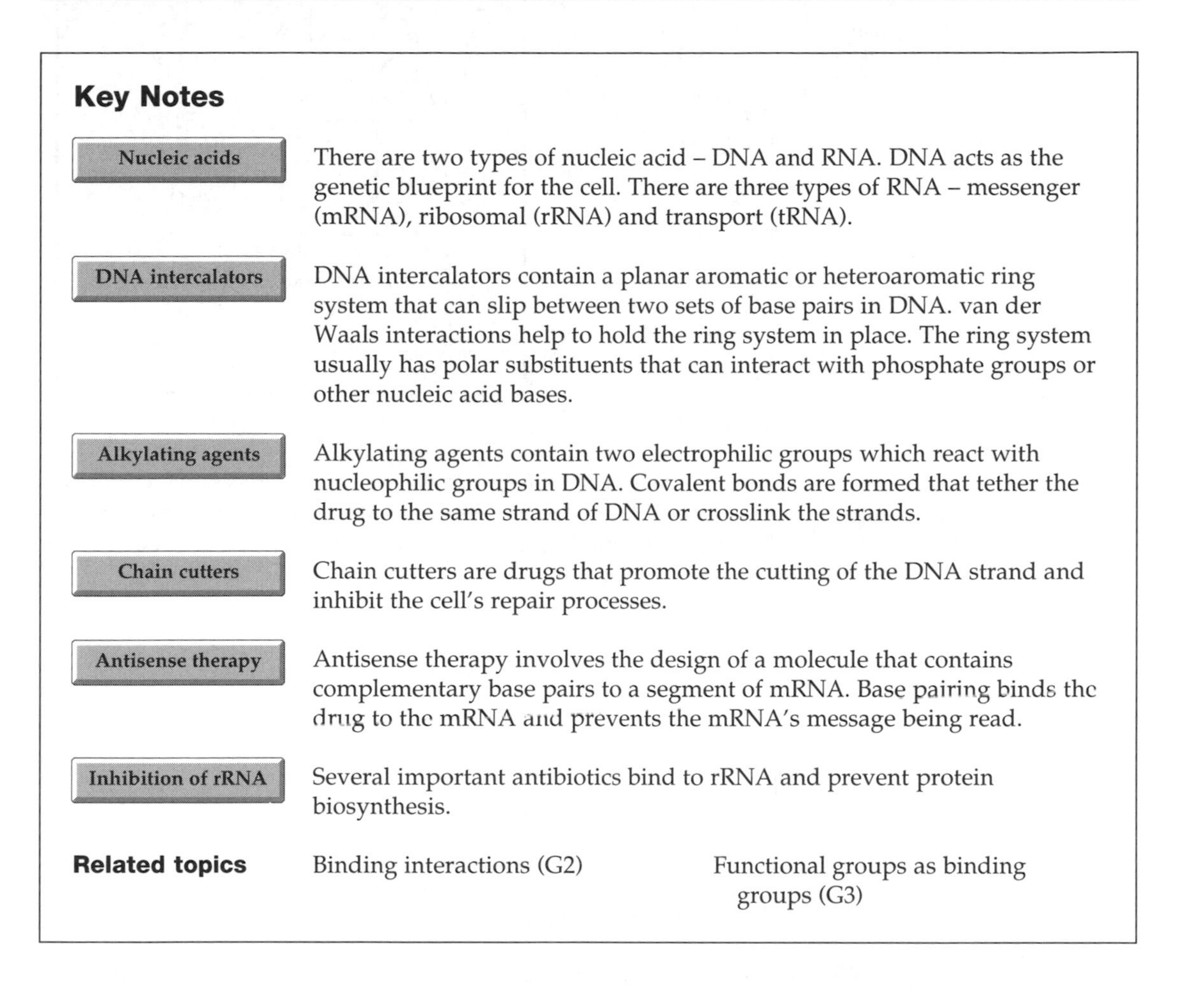

Key Notes

Nucleic acids There are two types of nucleic acid – DNA and RNA. DNA acts as the genetic blueprint for the cell. There are three types of RNA – messenger (mRNA), ribosomal (rRNA) and transport (tRNA).

DNA intercalators DNA intercalators contain a planar aromatic or heteroaromatic ring system that can slip between two sets of base pairs in DNA. van der Waals interactions help to hold the ring system in place. The ring system usually has polar substituents that can interact with phosphate groups or other nucleic acid bases.

Alkylating agents Alkylating agents contain two electrophilic groups which react with nucleophilic groups in DNA. Covalent bonds are formed that tether the drug to the same strand of DNA or crosslink the strands.

Chain cutters Chain cutters are drugs that promote the cutting of the DNA strand and inhibit the cell's repair processes.

Antisense therapy Antisense therapy involves the design of a molecule that contains complementary base pairs to a segment of mRNA. Base pairing binds the drug to the mRNA and prevents the mRNA's message being read.

Inhibition of rRNA Several important antibiotics bind to rRNA and prevent protein biosynthesis.

Related topics Binding interactions (G2) Functional groups as binding groups (G3)

Nucleic acids

Nucleic acids are targets for several important drugs, including various antimicrobial and anticancer agents. There are two types of nucleic acid – **deoxyribonucleic acid** (DNA) and **ribonucleic acid** (RNA). DNA is the genetic blueprint for the cell and contains all the information required for the biosynthesis of the cell's proteins. It consists of two polymeric oligonucleotide strands, which form a double helix (*Fig. 1*). Each strand of the helix is made up of a deoxyribose sugar–phosphate backbone, with a nucleic acid base linked to each sugar moiety. The double helix is held together by hydrogen bonds, which link the nucleic acid bases of one strand to the bases of the other strand. The nucleic acid pairings are such that **adenine** is paired with **thymine**, while **guanine** is linked with **cytosine**. This means that one DNA strand is complementary to the other and explains how genetic information can be passed on from cell to cell, and from generation to generation. Unraveling of the DNA helix means that each strand can act as a template for the synthesis of two identical DNA molecules, which will be transferred to daughter cells. The nucleic acid bases and the order in which they occur in the DNA molecule determine the genetic code.

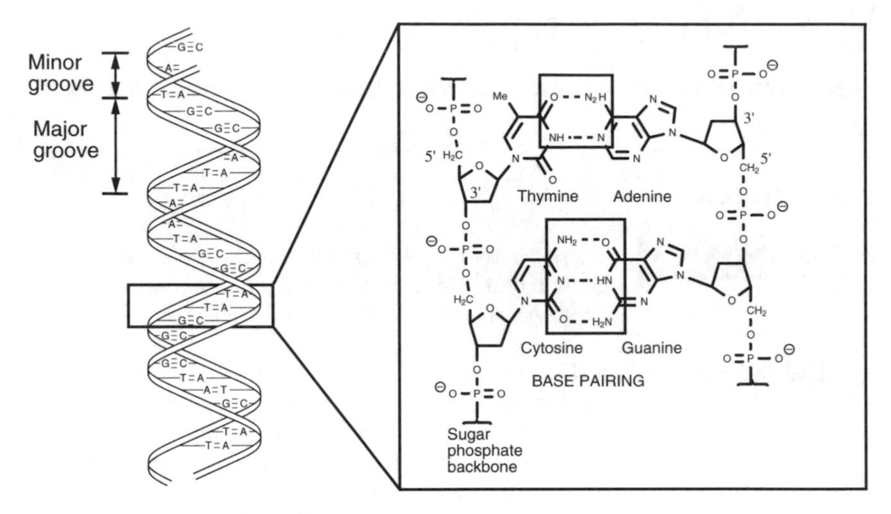

Fig. 1. Deoxyribonucleic acid.

RNA molecules have a similar primary structure to DNA. However, the sugar is **ribose** rather than **deoxyribose**, and the nucleic acid base **uracil** is present rather than thymine (*Fig. 2*). There are also other less common nucleic acid bases present in smaller quantities.

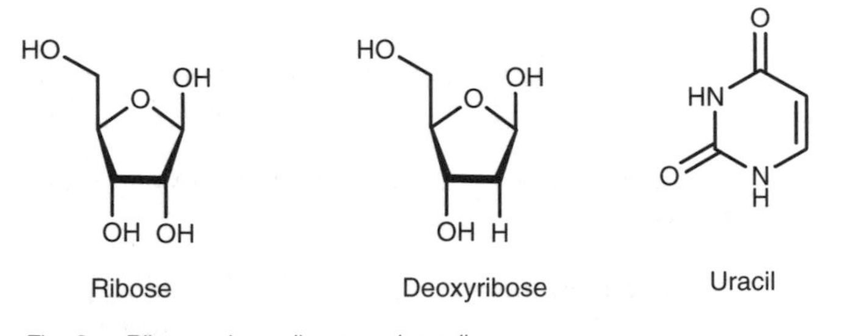

Fig. 2. Ribose, deoxyribose and uracil.

RNA is a single polymer and does not form a double helix. There are three types of RNA – messenger RNA (mRNA), ribosomal RNA (rRNA) and transfer RNA (tRNA), all of which are required for the biosynthesis of proteins. **Messenger RNA** acts as a 'photocopy' of specific regions of the DNA and carries the genetic code for a specific protein. **Ribosomal RNA** is present in ribosomes which function as 'factories' for protein synthesis. **Transfer RNA** is the 'adapter', which is used to interpret the genetic code and has two binding regions (*Fig. 3*). One binding region consists of three nucleic acid bases, which are complementary to a specific triplet of nucleic acid bases on mRNA, while the other binding region recognizes a specific amino acid. In this way, a specific triplet of nucleic acid bases corresponds to a particular amino acid. There are different tRNA molecules for each of the natural amino acids, each recognizing a different nucleic acid base triplet.

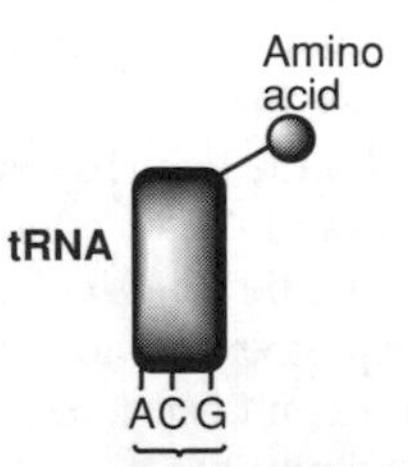

Fig. 3. Transport RNA (tRNA).

During protein biosynthesis, rRNA attaches to one end of an mRNA molecule and then travels along the length of the strand (*Fig. 4*). As it travels along the strand, the nucleic acid bases on the mRNA are read as triplets. The tRNA that recognizes that triplet is bound and brings the amino acid coded by that triplet.

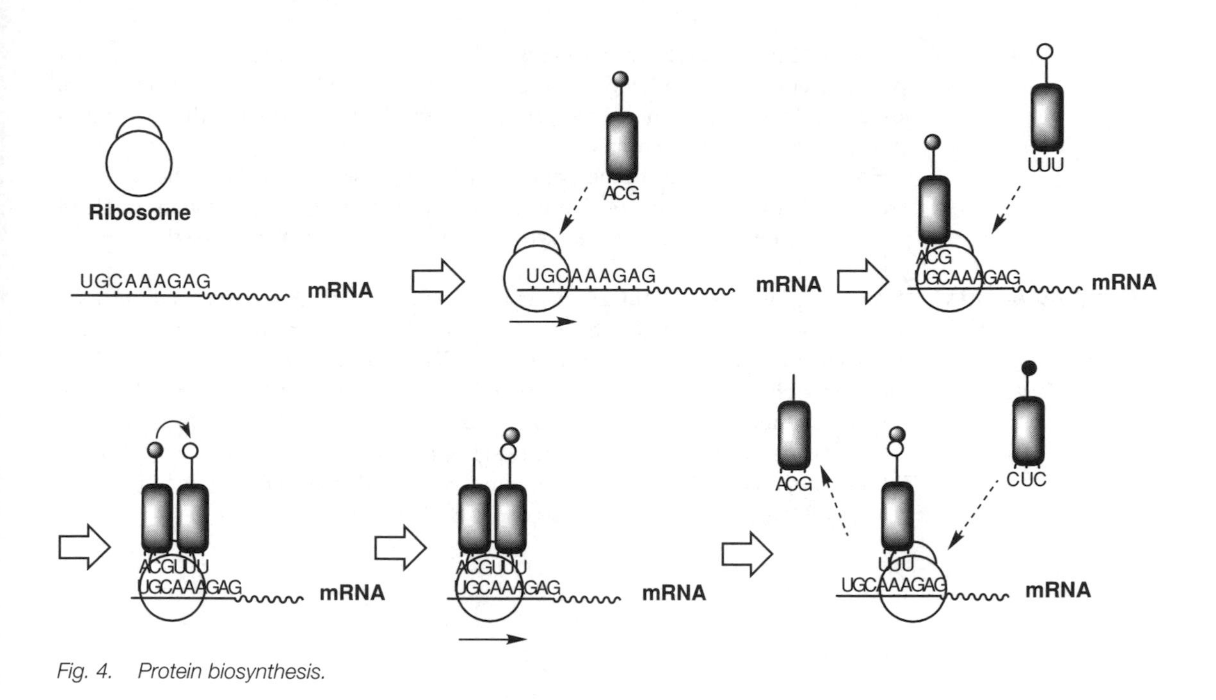

Fig. 4. *Protein biosynthesis.*

The protein is constructed on tRNA and transferred from one tRNA to the next until the full protein has been completed.

DNA intercalators

Intercalating drugs bind to DNA by inserting themselves between the stacked base pairs. Intercalation distorts the DNA double helix and prevents DNA from being copied, thus blocking protein synthesis. Drugs acting in this way have been found to be useful antibacterial and antitumor agents.

In order to slip between the stacked bases, the drug must be planar and have the correct dimensions. It must also be hydrophobic, so that there are favorable intermolecular interactions between the drug and the base pairs above and below it. These requirements are met by aromatic or heteroaromatic structures, such as the antibacterial agent **proflavine** (*Fig. 5*). The tricyclic system is the

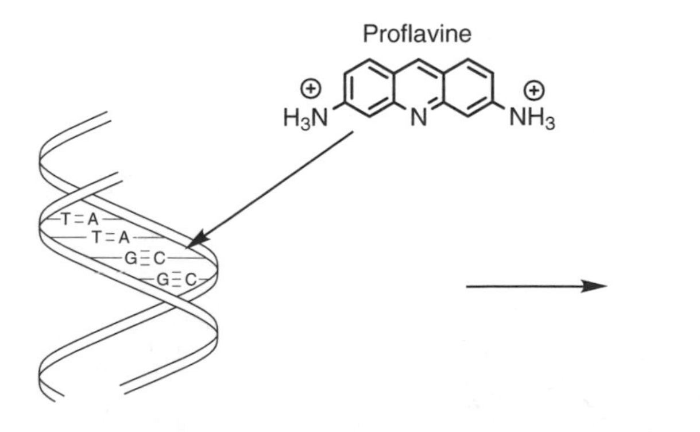

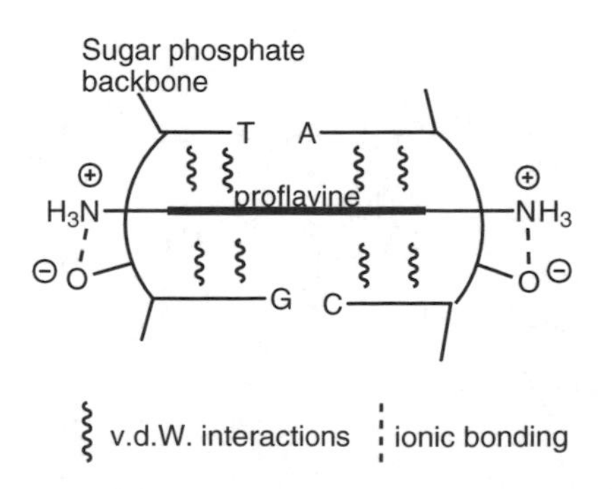

Fig. 5. *DNA intercalation.*

correct size to be inserted and can interact with the nucleic acid bases by van der Waals interactions. Proflavine also contains two ionized amino groups that form ionic bonds with the phosphate groups on the DNA backbone, thus strengthening the binding interactions.

Some intercalators sit in the grooves that are present in the DNA helix. There are two distinct grooves, one minor and one major. The dimensions of these grooves are important as several drugs show a preference for one or the other.

Alkylating agents

Alkylating agents contain an electrophilic functional group such as an alkyl halide. Reaction of an alkyl halide with a nucleophilic group on DNA (e.g. the nitrogens of a guanine unit) result in a nucleophilic substitution reaction where the nucleophile displaces the halide and forms a covalent bond with the drug (*Fig. 6*). If there are two electrophilic groups present in the drug, the reaction occurs twice resulting in **cross-linking** within a strand or between strands. Either way, cross-linking disrupts the normal functions of DNA. Uracil mustard (Section B3) is an example of a drug that acts in this manner.

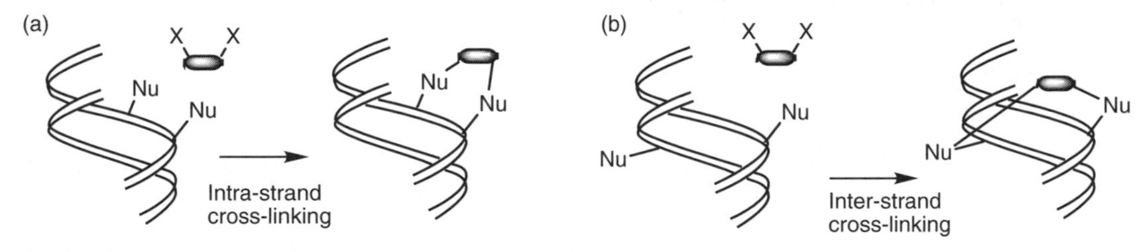

Fig. 6. Cross-linking of DNA by an alkylating agent.

Chain cutters

Some drugs react with DNA to cut the DNA chain. **Calicheamicin γ_1^I** is an antitumor agent that was isolated from a bacterium (*Fig. 7*). It binds to the minor groove of DNA and cuts the DNA chain by producing highly reactive radical species. The driving force behind this reaction is the formation of an aromatic ring from the unusual enediyne system.

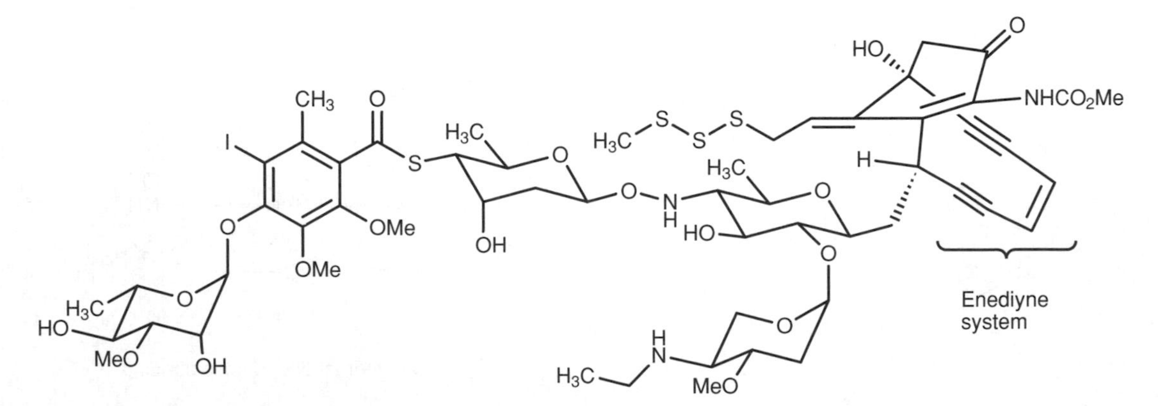

Fig. 7. Calicheamicin γ_1^I.

The reaction starts with a nucleophile attacking the trisulfide group (*Fig. 8*). The sulfur that is released then undergoes a Michael addition with a reactive α,β-unsaturated ketone. The resulting product cycloaromatizes to produce an aromatic diradical species which 'snatches' two hydrogens from DNA. As a result, DNA becomes a diradical. Reaction with oxygen then leads to chain cutting.

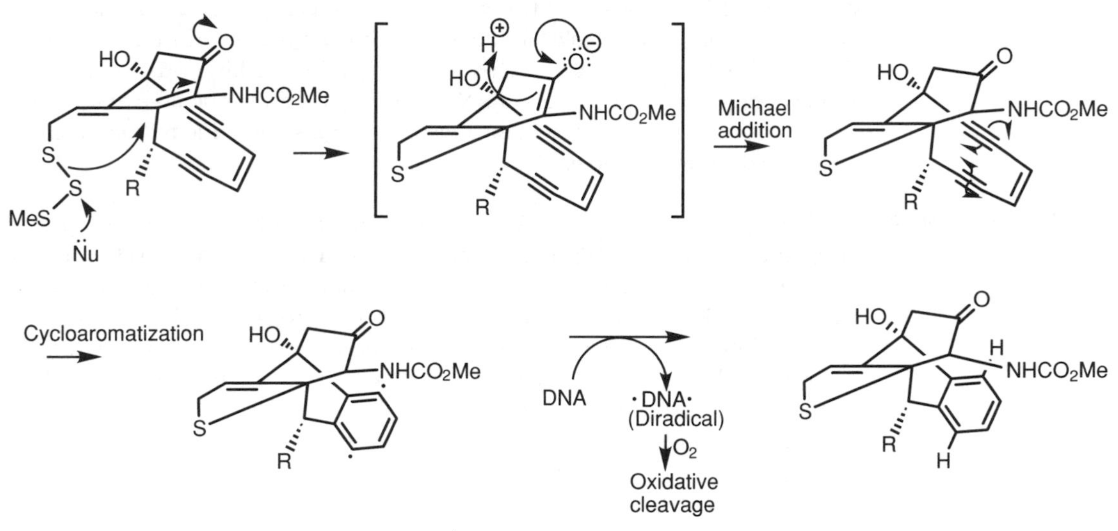

Fig. 8. Mechanism of action of calicheamicin γ_1^I.

Antisense therapy In recent years, drugs have been designed that will bind to mRNA. The strategy is called **antisense therapy** as it involves the synthesis of oligonucleotides that contain complimentary base pairs to a segment of the target mRNA (*Fig. 9*). The rationale is that the synthetic oligonucleotide should bind to the complimentary segment of mRNA by base pairing. Once bound, the antisense drug prevents mRNA from being decoded, and blocks the synthesis of a specific protein. If the protein is a receptor or enzyme, then less of it is synthesized and the activity

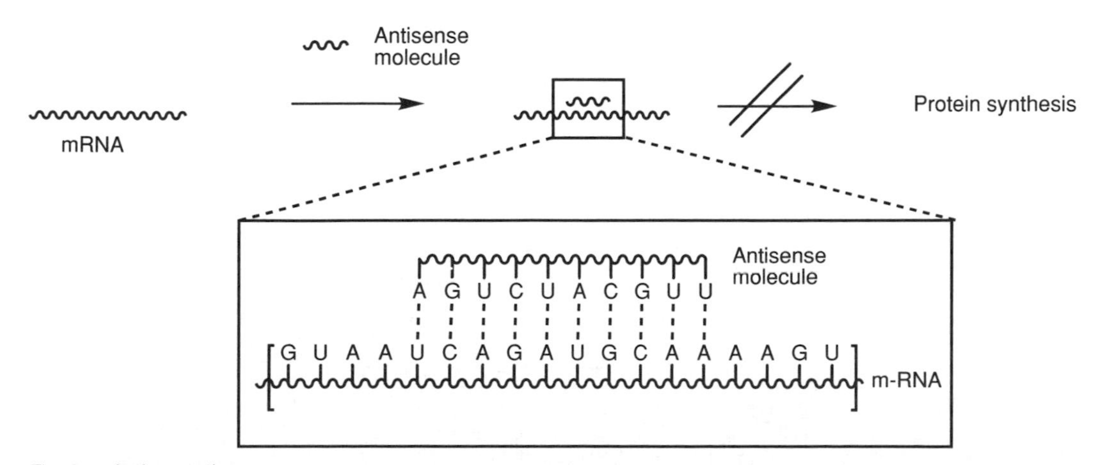

Fig. 9. Antisense therapy.

associated with the protein is reduced. Therefore, antisense therapy has the same overall effect as using an enzyme inhibitor or a receptor antagonist. However, the potential of antisense therapy for target selectivity is much greater as it is theoretically possible to design an antisense drug that would knock out one specific isozyme or receptor subtype.

Antisense therapy has great potential. However, there are various difficulties that have to be overcome. For example, the sugar–phosphate backbone of oligonucleotides is easily hydrolyzed, hence these compounds are not orally active; therefore, more stable backbones have to be designed. Careful consideration also has to be given to which segment of mRNA is targeted. The segment should be unique to the target mRNA, but it should also be accessible to the drug. If the mRNA is folded in such a way that a segment is hidden, then an antisense drug will fail to bind.

Inhibition of rRNA

Ribosomal RNA is the target for some important antibacterial agents such as **streptomycin**, the **tetracyclines**, **chloramphenicol** (*Fig. 10*) and **erythromycin**. All of these agents bind to ribosomal RNA, and in doing so, prevent protein biosynthesis. For example, chloramphenicol binds to the r-RNA of ribosomes then inhibits ribosomal movement along mRNA, probably by inhibiting the mechanism by which the peptide chain is transferred from one t-RNA to another. Chloramphenicol is the drug of choice against typhoid and is also used in severe bacterial infections which are insensitive to other antibacterial agents. It has also found widespread use against eye infections. However, the drug is quite toxic, especially to bone marrow.

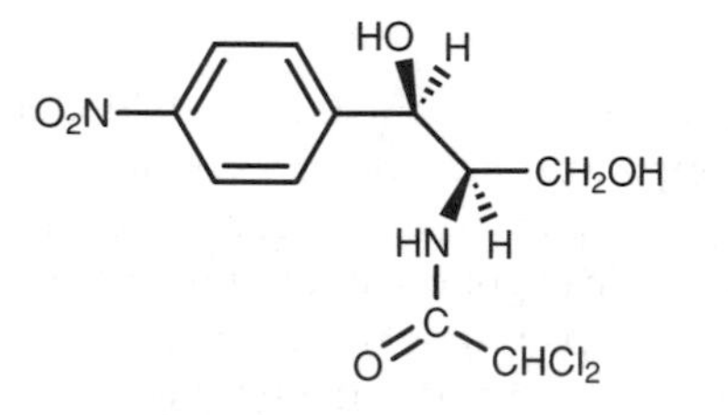

Fig. 10. Chloramphenicol.

B6 LIPIDS

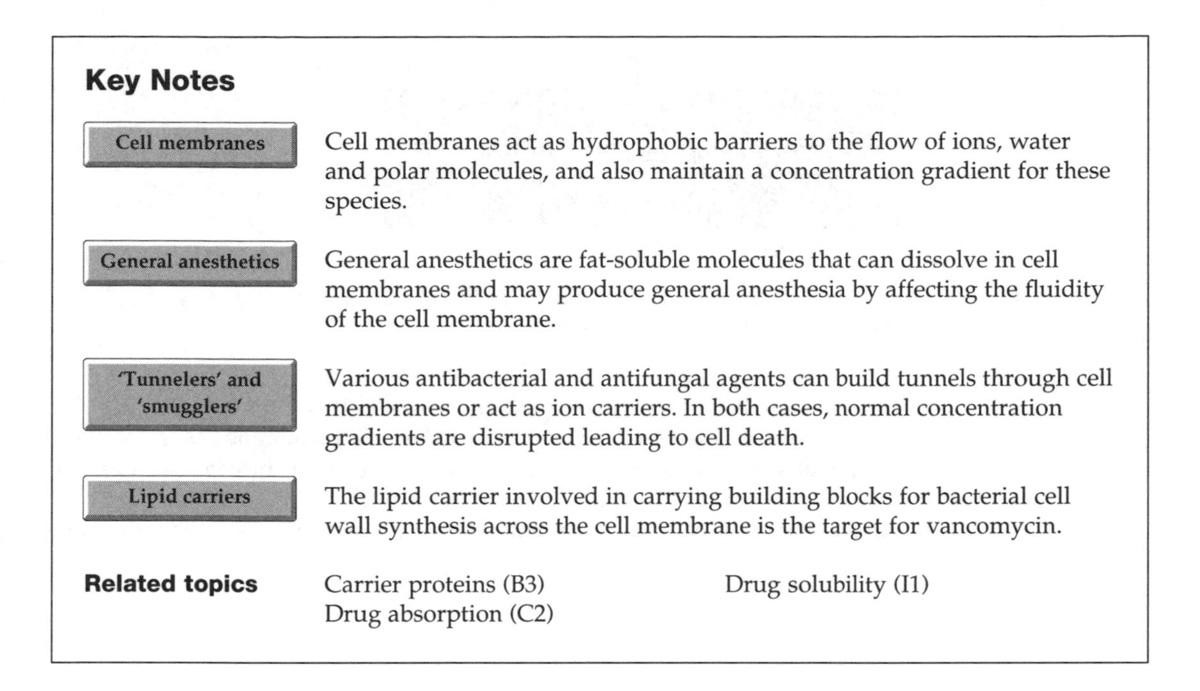

Key Notes

Cell membranes	Cell membranes act as hydrophobic barriers to the flow of ions, water and polar molecules, and also maintain a concentration gradient for these species.
General anesthetics	General anesthetics are fat-soluble molecules that can dissolve in cell membranes and may produce general anesthesia by affecting the fluidity of the cell membrane.
'Tunnelers' and 'smugglers'	Various antibacterial and antifungal agents can build tunnels through cell membranes or act as ion carriers. In both cases, normal concentration gradients are disrupted leading to cell death.
Lipid carriers	The lipid carrier involved in carrying building blocks for bacterial cell wall synthesis across the cell membrane is the target for vancomycin.

Related topics Carrier proteins (B3) Drug absorption (C2) Drug solubility (I1)

Cell membranes

Cell membranes consist of a **phospholipid bilayer**, which acts as a hydrophobic barrier. Water and ions can only cross this barrier through ion channels, which are controlled by receptors. Polar molecules can only cross using carrier proteins. As a result, there are concentration gradients across the cell membrane for various ions and polar molecules. For example, there is a greater concentration of potassium ions within cells than in the fluid surrounding the cell. Conversely, the concentration of sodium ions is greater outside the cell than inside. The maintenance of these concentration gradients is crucial to several important functions such as the transmission of nerve signals along nerves.

Cell membranes also act as barriers to any drugs that are intended to act on a target within the cell. In order to access the cell, the drug must be sufficiently hydrophobic to cross the membrane. Alternatively, it should be designed so that it can be accepted by a carrier protein.

General anesthetics

It is generally thought that **general anesthetics** disrupt cell membrane structure, making it more fluid. Support for this theory is provided by the fact that general anesthetics have little similarity in structure, but are hydrophobic in character and easily dissolve in the cell membrane. However, it has been argued that general anesthetic activity may not be solely related to this property, and that a receptor protein may be involved.

'Tunnelers' and 'smugglers'

Compounds that disrupt the cell membrane can have devastating effects. For example, there are various antibacterial and antifungal compounds that destroy cells by building helical structures through the cell membrane, thus forming a tunnel through which ions and small polar molecules can flow in an uncontrolled manner (*Fig. 1*).

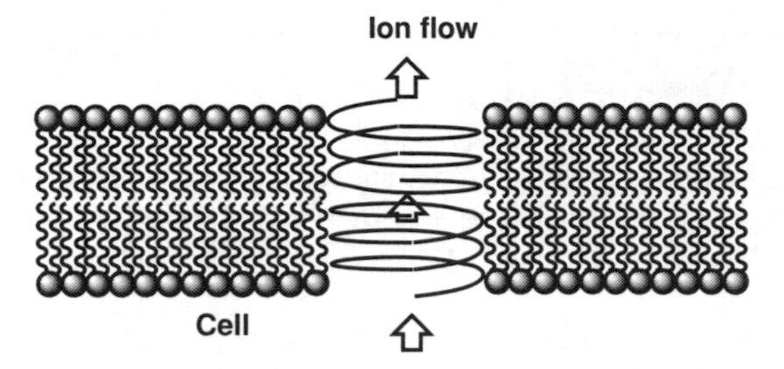

Fig. 1. Tunneling. Reproduced from An Introduction to Medicinal Chemistry, *G.L. Patrick, 2001, by permission of Oxford University Press.*

Other antifungal compounds are cyclic with a polar center, and they are able to complex an ion then carry it across the cell membrane. Most of these compounds are useless as medicines because they affect microbial and animal cells alike. However, the antibiotic **polymyxin B** (*Fig. 2*) shows some selectivity against bacterial cells over animal cells and has been used against some *Pseudomonas* infections.

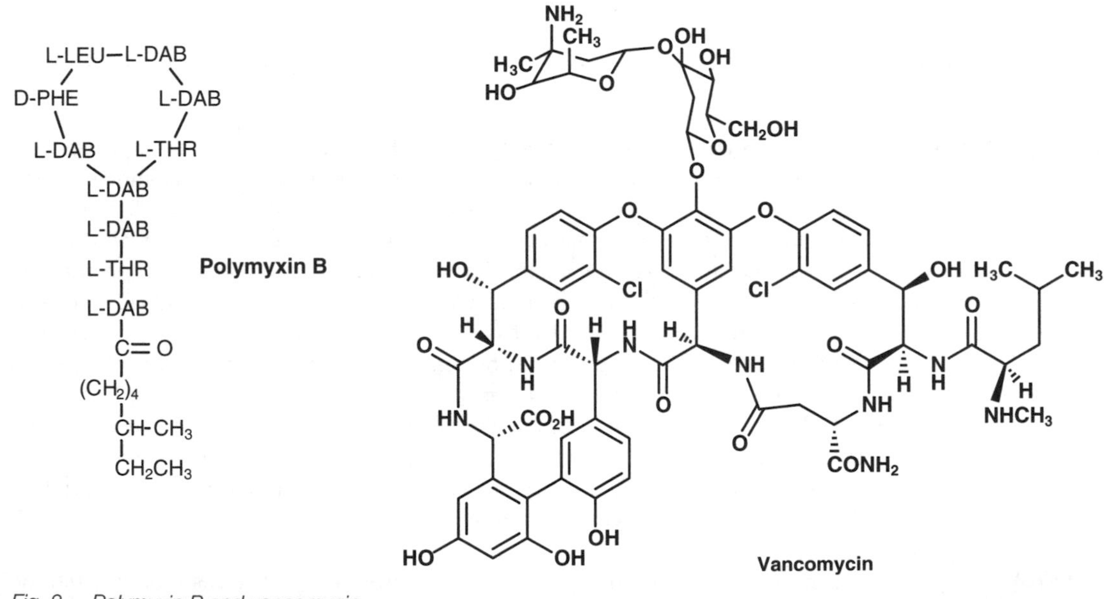

Fig. 2. Polymyxin B and vancomycin.

Lipid carriers

Lipid carriers are not major drug targets. However, **vancomycin** (*Fig. 2*) is an important antibacterial drug that *does* act on a lipid carrier. Vancomycin is a glycopeptide which inhibits the construction of bacterial cell walls and which is the main stand-by drug for penicillin resistant strains of bacteria. The specific target for vancomycin is a lipid carrier in bacteria which is responsible for carrying the building blocks required for cell wall synthesis across the cell membrane (*Fig. 3*). Each building block is a disaccharide/peptide structure, which has been constructed in the cytoplasm and vancomycin prevents the release of that building block from the lipid carrier.

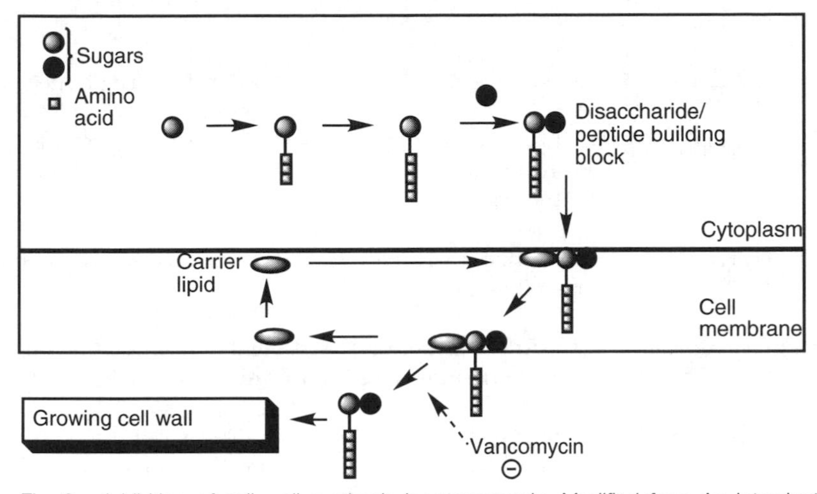

Fig. 3. Inhibition of cell wall synthesis by vancomycin. Modified from An Introduction to Medicinal Chemistry, *G.L. Patrick, 2001, published by Oxford University Press.*

B7 CARBOHYDRATES

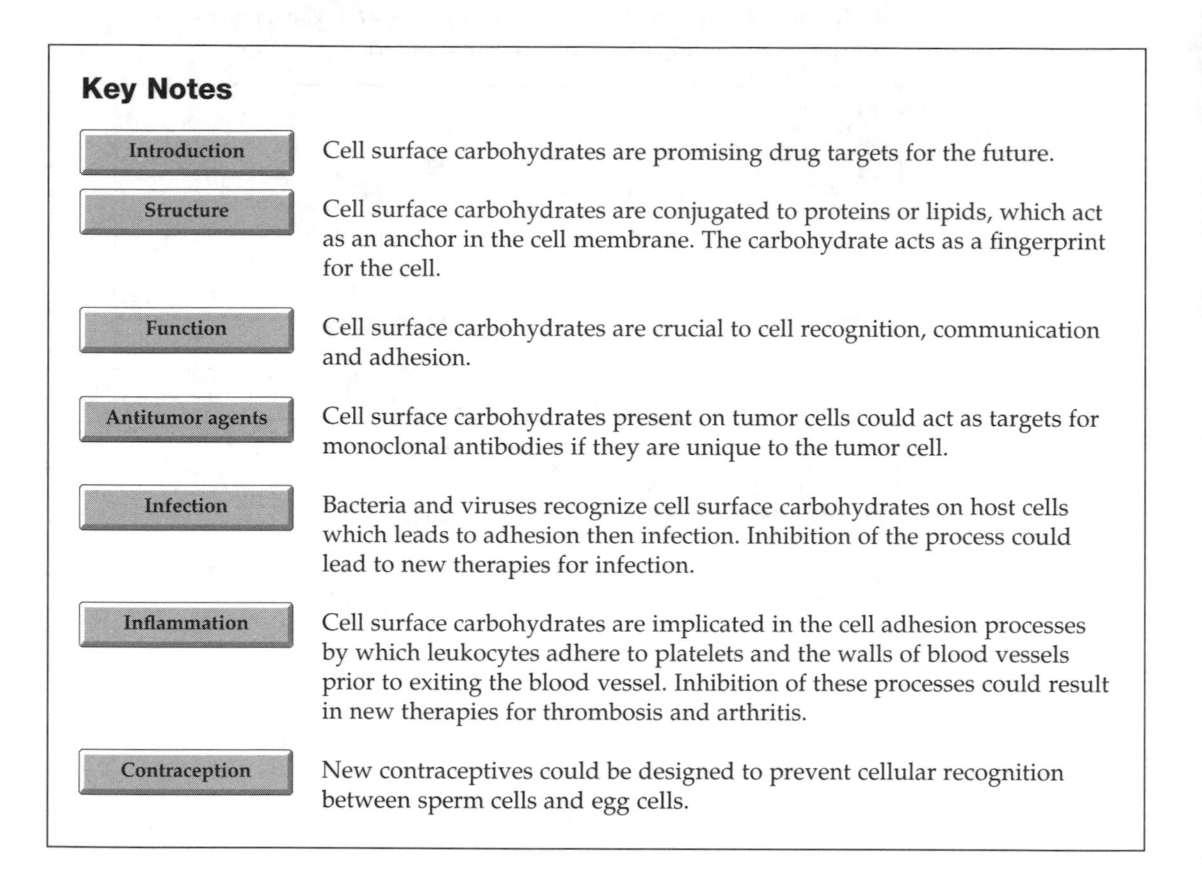

Key Notes

Introduction	Cell surface carbohydrates are promising drug targets for the future.
Structure	Cell surface carbohydrates are conjugated to proteins or lipids, which act as an anchor in the cell membrane. The carbohydrate acts as a fingerprint for the cell.
Function	Cell surface carbohydrates are crucial to cell recognition, communication and adhesion.
Antitumor agents	Cell surface carbohydrates present on tumor cells could act as targets for monoclonal antibodies if they are unique to the tumor cell.
Infection	Bacteria and viruses recognize cell surface carbohydrates on host cells which leads to adhesion then infection. Inhibition of the process could lead to new therapies for infection.
Inflammation	Cell surface carbohydrates are implicated in the cell adhesion processes by which leukocytes adhere to platelets and the walls of blood vessels prior to exiting the blood vessel. Inhibition of these processes could result in new therapies for thrombosis and arthritis.
Contraception	New contraceptives could be designed to prevent cellular recognition between sperm cells and egg cells.

Introduction

Carbohydrates have generally been neglected as drug targets. However, a better understanding of their structure and function suggests that these compounds will be important drug targets for the future. Carbohydrates have many different roles, but those present on the surface of cell membranes are of most interest.

Structure

Cell surface carbohydrates are known as **glycoconjugates** since the carbohydrate is linked to a protein or lipid that is embedded in the cell membrane (*Fig. 1*). The protein or lipid acts as an anchor for the carbohydrate, which sticks out from the surface of the cell to act as a 'fingerprint' for the cell. The carbohydrate itself is made up of various sugar molecules linked together in a chain. Carbohydrates are ideally suited for such a role since large variations in structure are possible by linking different sugar monomers together in different ways.

Function

Cell surface carbohydrates are crucial to cell recognition and communication. They are also important to various adhesion processes whereby cells 'stick' to other cells. Consequently, cell surface carbohydrates could be potential drug

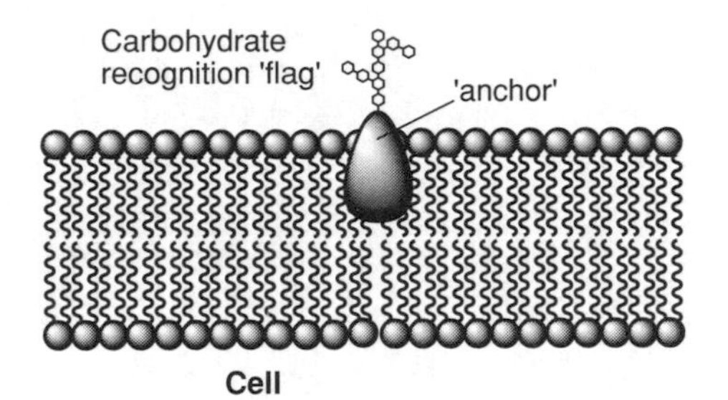

Fig. 1. Cell surface carbohydrates.

targets for the treatment of a number of diseases including cancer, stroke, autoimmune disease, inflammation, arthritis, thrombosis, and genetic disease. The following are some examples of where carbohydrates might prove to be useful targets.

Antitumor agents

It has been observed that the cell surface carbohydrates present on tumor cells are different in structure and composition from those which are normally present. Such carbohydrates act as **antigens**, so it should be possible to produce **monoclonal antibodies** which will recognize and bind to them, thus directing the immune system against the tumor. Antibodies could also be used to target anticancer drugs more specifically to tumor cells. It is important to identify tumor-specific antigens before attempting any of the antibody strategies mentioned. For example, the antigens expressed in tumor cells may be unusual for that type of cell, but may be normal for a different type of cell.

Infection

Cell surface glycoproteins are important to the mechanisms involved in infection. Proteins on the surface of bacteria and viruses recognize particular carbohydrates on host cells. The invading bacterium or virus then adheres to the host cell initiating the process of infection. If this recognition and adherence process could be inhibited, then new antibacterial and antiviral therapies could be devised. Bacteria also have their own carbohydrate antigens, which allows the prospect of carbohydrate-based antibacterial vaccines.

Inflammation

Glycoproteins called **selectins** are embedded in the endothelial cells which make up the lining of blood vessels. These molecules bind with cell surface carbohydrates expressed on circulating **leukocytes**, which then initiates a process whereby the leukocyte adheres to the wall of the blood vessel and is then extruded from the vessel to a site of inflammation. Inhibiting this process could lead to novel anti-inflammatory and anti-arthritic agents. Selectins are also implicated in the adhesion of leukocytes to blood platelets. Inhibition of this process could lead to antithrombotic agents.

Contraception

It has been proposed that the fertilization of an egg by a sperm cell is initiated by proteins on the surface of the sperm cell recognizing carbohydrates on the cell surface of the egg. Inhibitors of this process could provide novel contraceptives.

C1 PHARMACOKINETICS

Key Notes

Introduction

Pharmacokinetics is the study of what happens to a drug when it is administered to a patient. There are four main factors to be considered – absorption, distribution, metabolism and excretion.

The rule of five

In general, most orally active drugs have a molecular weight less than 500, a log *P* value less than 5, no more than five hydrogen bond donor groups and no more than 10 hydrogen bond acceptor groups.

Related topics

From concept to market (A2)
Drug absorption (C2)
Drug distribution (C3)
Drug metabolism (C4)
Drug excretion (C5)
Drug administration (C6)
Drug dosing (C7)
Drug solubility (I1)
Drug stability (I2)
Drug metabolism studies (K3)
Clinical trials (K4)

Introduction

It is possible to design molecules which interact extremely effectively with their targets, but this does not necessarily mean that these compounds will be clinically active. This is because a drug has to be administered to a patient, and then travel through the body where it is exposed to a variety of factors that can either remove, destroy or prevent it reaching the required target site. There are four main factors to be taken into account – absorption, distribution, metabolism, and excretion. **Pharmacokinetics** is the study of these various factors and can be viewed as what the body does to the drug, rather than what the drug does to the body.

The rule of five

It is important to consider pharmacokinetics as soon as possible during a drug design program. There is no point perfecting a compound with superb drug–target interactions if it has no chance of reaching its target. For that reason, many research programs will impose the following requirements on target structures:

- a molecular weight less than 500;
- no more than five hydrogen bond donor groups;
- no more than 10 hydrogen bond acceptor groups;
- a calculated log *P* value (see Section G5) less than +5.

This is called the '**rule of five**' because the number 5 (or multiples of it) is relevant in each case. The reasons for these limits will become clear in the sections to follow. However, in general, the larger and more polar the molecule, the less chance that it will be orally active.

C2 DRUG ABSORPTION

Key Notes

Absorption

Most drugs have to pass through the cells lining the gut wall to reach the blood supply. This means they have to cross a cell membrane on two occasions.

The cell membrane

The cell membrane is a phospholipid bilayer, which acts as a hydrophobic barrier to the passage of water, ions and polar molecules.

Drug solubility

Drugs must have a balance of hydrophilic and hydrophobic character if they are to be soluble in the gut and the blood supply, and also cross cell membranes. Drugs usually obey the rule of five to be orally active. Some highly polar drugs can be smuggled through cell membranes by carrier proteins or by pinocytosis, but, failing that, they have to be injected. Drugs targeted against gut infections are made highly polar so that they are not absorbed.

Related topics

Carrier proteins (B3)
Pharmacokinetics (C1)
Drug solubility (I1)

Absorption

In order to reach a specific target, an orally administered drug has to negotiate several barriers. One of these is the gut wall. Most orally administered drugs are absorbed in the upper intestine and they must pass through the gut wall in order to reach the blood supply. If the drug has a low molecular weight (less than 200), it can 'squeeze' through small gaps between the cells of the gut wall. However, the majority of drugs have molecular weights higher than 200 and have to pass from the gastrointestinal tract into the blood supply by traveling through the cells lining the gut wall. Since cells are bounded by cell membranes, it is clear that such drugs have to pass through a cell membrane on two occasions.

The cell membrane

The **cell membrane** consists mainly of molecules called **phospholipids**. The structures of these phospholipids vary, but in each case there is a polar head group containing an ionized phosphate group to which are attached two long hydrophobic, hydrocarbon chains. **Phosphatidylcholine** (*Fig. 1*) is one example of a phospholipid commonly found in cell membranes.

In the cell membrane, the phospholipid molecules are arranged in two layers measuring about 80 Å in thickness. The hydrophobic chains point inwards to the center of the membrane while the polar head groups are positioned on the inner and outer surfaces of the membrane, forming a **phospholipid bilayer** (*Fig. 2*). This means that the polar head groups can interact favorably with the aqueous environments inside and outside the cell, while the hydrophobic tails are kept away from water and interact with each other through van der Waals interactions. The phospholipid bilayer is a fluid structure and there are various

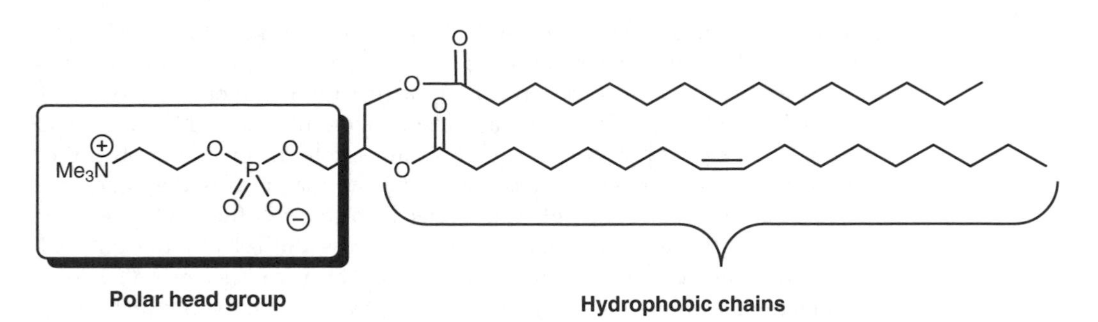

Fig. 1. Phosphatidylcholine.

other structures, such as glycoproteins, receptors and enzymes which are embedded in the membrane, but which can float about like icebergs in the ocean. However, the important thing to note here is that the center of the cell membrane is hydrophobic and fatty. This means that the membrane serves as a barrier to the passage of water, ions and polar molecules, thus preventing such species from entering or leaving the cell.

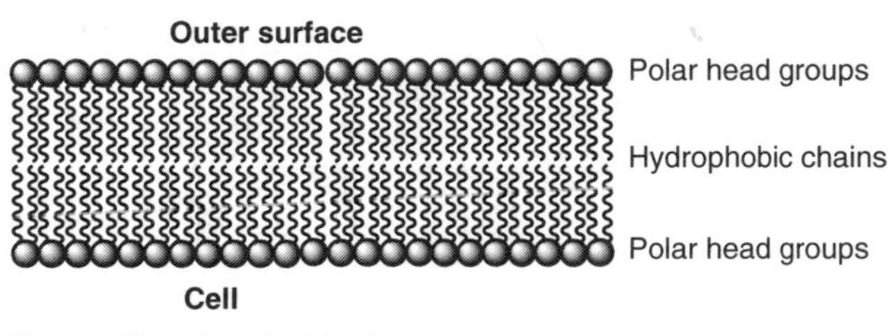

Fig. 2. The phospholipid bilayer.

Drug solubility

If a drug is to pass through the cells lining the gut wall, it must be able to dissolve in the hydrophobic center of the cell membranes. This means that the drug has to be lipid-soluble. However, it also has to be water soluble if it is to dissolve in the gut and the blood. Therefore, drugs must have a balance of hydrophilic and hydrophobic character. Drugs that are too polar and too hydrophilic will not cross the cell membranes and will not be absorbed. Equally, drugs that are not polar enough will be poorly soluble in aqueous environments. This too can lead to poor absorption since such drugs will coagulate as fatty globules in the gut and make poor surface contact with the gut wall.

The hydrophilic/hydrophobic character of the drug is the crucial factor affecting absorption, and, in theory, the molecular weight of the drug is unimportant. However, in practise, most orally active drugs have a molecular weight less than 500. This is because high molecular weight compounds are more likely to have a large number of polar functional groups, and it is the presence of these that limits absorption. As a rule of thumb, drugs should have no more than five hydrogen bonding donor groups and no more than 10 hydrogen bonding acceptor groups. Polar drugs which break these rules are not usually orally active and have to be administered by injection.

Having said that, some highly polar drugs *are* absorbed from the digestive system if they can 'hijack' specific **carrier proteins** in the cell membrane. Carrier proteins are essential to a cell's survival since they smuggle the highly polar building blocks required for various biosynthetic pathways (e.g. amino acids and nucleic acid bases). If the drug bears a structural resemblance to one of these building blocks, then it too may be smuggled into the cell by the carrier protein. Other highly polar drugs can be absorbed into the blood supply if they have a low molecular weight (less than 200) as that means they can pass between the cells lining the gut wall, rather than through them. Occasionally, polar drugs with high molecular weight can cross the cells of the gut wall without actually passing through the membrane. This involves a process known as **pinocytosis** where the drug is engulfed by the cell membrane and a membrane bound vesicle is pinched off to carry the drug across the cell (*Fig. 3*). The vesicle then fuses with the membrane to release the drug on the other side of the cell.

Sometimes, drugs are deliberately designed to be highly polar so that they are not absorbed from the gastrointestinal tract. These are usually antibacterial agents that are targeted against gut infections. By making them highly polar, more of the drug will reach the site of infection.

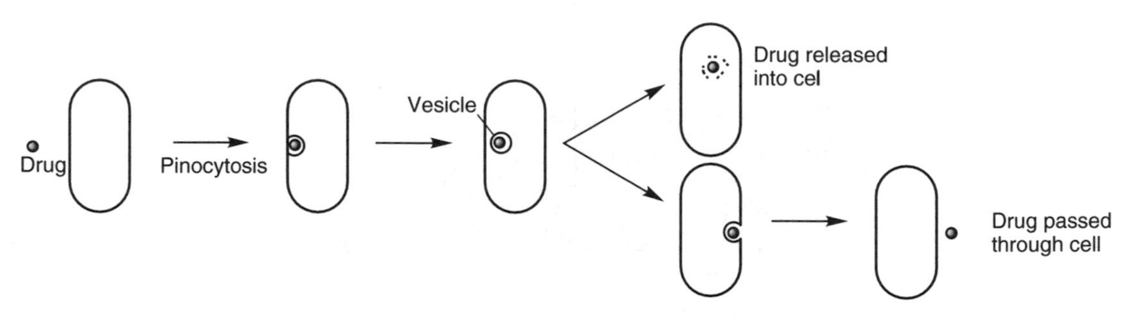

Fig. 3. Pinocytosis. Reproduced from An Introduction to Medicinal Chemistry, *G.L. Patrick, 2001, by permission of Oxford University Press.*

C3 DRUG DISTRIBUTION

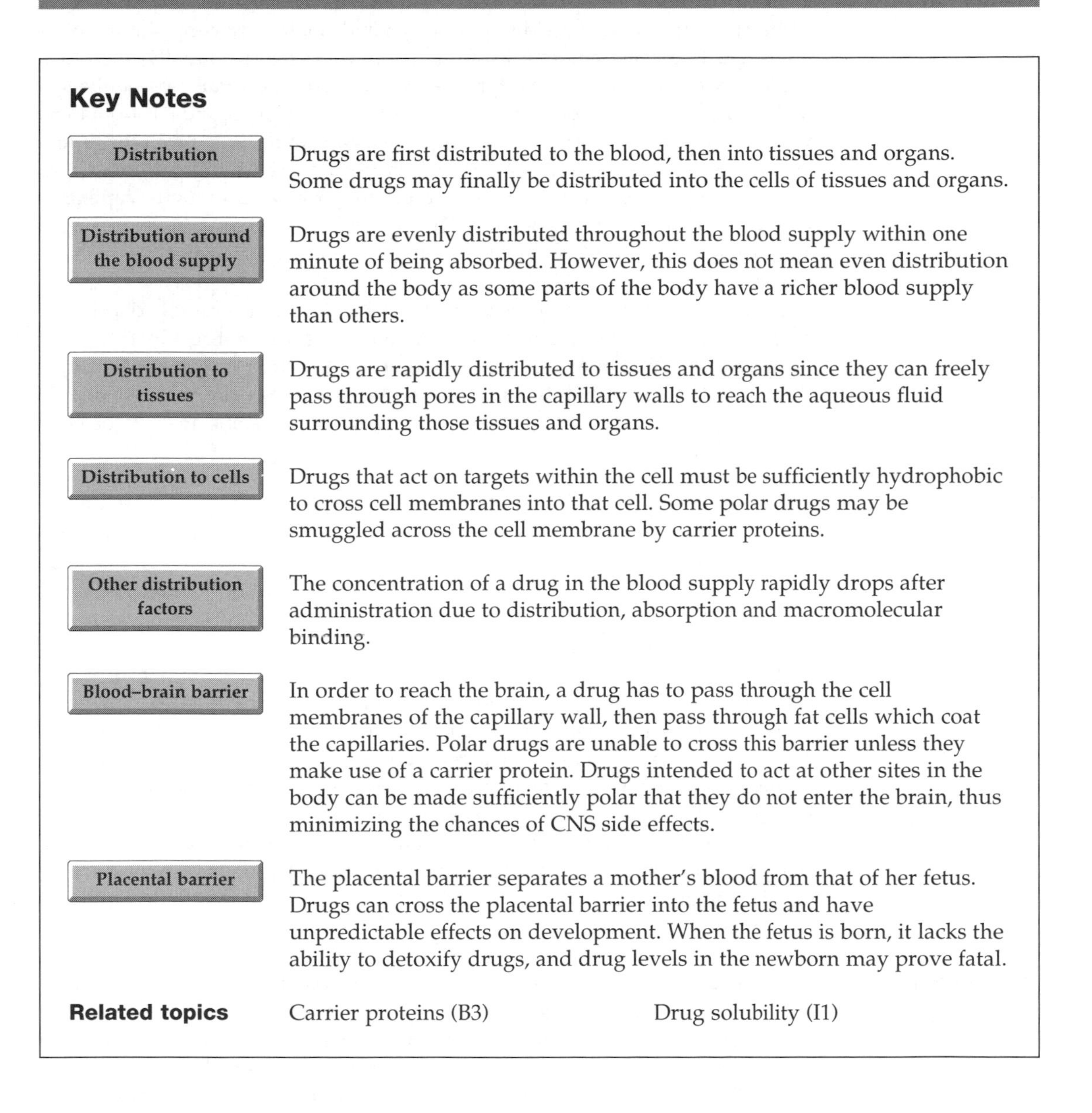

Key Notes

Distribution	Drugs are first distributed to the blood, then into tissues and organs. Some drugs may finally be distributed into the cells of tissues and organs.
Distribution around the blood supply	Drugs are evenly distributed throughout the blood supply within one minute of being absorbed. However, this does not mean even distribution around the body as some parts of the body have a richer blood supply than others.
Distribution to tissues	Drugs are rapidly distributed to tissues and organs since they can freely pass through pores in the capillary walls to reach the aqueous fluid surrounding those tissues and organs.
Distribution to cells	Drugs that act on targets within the cell must be sufficiently hydrophobic to cross cell membranes into that cell. Some polar drugs may be smuggled across the cell membrane by carrier proteins.
Other distribution factors	The concentration of a drug in the blood supply rapidly drops after administration due to distribution, absorption and macromolecular binding.
Blood–brain barrier	In order to reach the brain, a drug has to pass through the cell membranes of the capillary wall, then pass through fat cells which coat the capillaries. Polar drugs are unable to cross this barrier unless they make use of a carrier protein. Drugs intended to act at other sites in the body can be made sufficiently polar that they do not enter the brain, thus minimizing the chances of CNS side effects.
Placental barrier	The placental barrier separates a mother's blood from that of her fetus. Drugs can cross the placental barrier into the fetus and have unpredictable effects on development. When the fetus is born, it lacks the ability to detoxify drugs, and drug levels in the newborn may prove fatal.

Related topics Carrier proteins (B3) Drug solubility (I1)

Distribution

Once a drug has been absorbed into the blood, it is distributed around the body. This initially involves a journey round the blood supply, with subsequent distribution to the various tissues and organs fed by that blood supply. In some cases, drugs will be distributed into the cells of various tissues or organs. The rate and extent of distribution depends on various factors, including the physical properties of the drug itself.

Distribution around the blood supply

The vessels carrying blood around the body are called **arteries**, **veins** and **capillaries**, and the heart is the pump which drives the blood through these vessels. The major artery carrying blood from the heart is called the **aorta**, and as it moves further from the heart, it divides into smaller and smaller arteries – similar to the limbs and branches radiating from the trunk of a tree. Eventually, the blood vessels divide to such an extent that they become extremely narrow – equivalent to the twigs of a tree. These blood vessels are called capillaries and it is from these vessels that oxygen, nutrients and drugs can escape in order to reach the tissues and organs of the body. At the same time, waste products, such as cell breakdown products and carbon dioxide, are transferred from the tissues into the capillaries to be carried away and 'disposed of'. The capillaries now start uniting into bigger and bigger vessels resulting in the formation of veins, which return the blood to the heart.

Once a drug has been absorbed into the blood supply, it is rapidly distributed round the body. This only takes a minute since the entire blood volume completes one circulation in that time. Although the drug is evenly distributed throughout the blood supply, this does not mean that the drug is evenly distributed around the body, since the blood supply is richer to some areas of the body than to others.

Distribution to tissues

Drugs do not stay confined to the blood supply. If they did, they would be of little use as their targets are the cells of various organs and tissues. Therefore, the drug has to 'escape' from the blood supply in order to reach those targets. The body has an estimated 10 billion capillaries, which have a total surface area of 200 m^2. They probe every part of the body, so no cell is more than 20–30 μm away from a capillary. Each capillary is very narrow, not much wider than the red blood cells that pass through it. Its walls are made up of a thin, single layer of tightly packed cells. However, there are pores between the cells which are 90–150 Å in diameter and these pores are large enough to allow most drug-sized molecules to escape, but not large enough to allow the **plasma proteins** present in the blood to escape. Therefore, drugs do not have to cross cell membranes in order to leave the blood system and can be freely and rapidly distributed into the aqueous fluid surrounding the various tissues and organs of the body. Some drugs, however, bind to plasma proteins in the blood, and as the plasma proteins cannot leave the capillaries, any drug bound to these proteins is also confined to the capillaries and cannot reach its target.

Distribution to cells

Once a drug has reached the tissues, it can immediately be effective if its target site is a receptor situated in a cell membrane. However, there are many drugs that have to enter the individual cells of tissues in order to be effective, since their molecular targets are contained within. These include drugs that act as enzyme inhibitors, and those acting on nucleic acids or steroid receptors within the cell. Such drugs must be hydrophobic enough to pass through the cell membrane, unless they are able to utilize carrier proteins.

Other distribution factors

The concentration level of free drug circulating in the blood rapidly falls after administration due to the distribution patterns described above, but there are other factors that reduce the amount of drug in the blood. Drugs that are excessively hydrophobic are often absorbed into fatty tissues and removed from the blood supply. Drugs that are charged may be bound to various macromolecules and also removed from the blood supply. Drugs may also be bound reversibly to blood plasma proteins such as **albumin**, thus lowering the level of free drug. Therefore, only a small percentage of the administered drug will actually reach the desired target.

Blood–brain barrier

The **blood–brain barrier** is an important barrier that drugs have to negotiate if they are to act in the brain. The blood capillaries feeding the brains are lined with tight fitting cells that do not contain pores (unlike capillaries elsewhere in the body). Moreover, the capillaries are covered on the outside by a fatty barrier formed from nearby cells. Therefore, drugs entering the brain have to dissolve through the cell membranes of the capillaries and also through the fatty cells coating the capillaries. As a result, polar drugs such as **penicillin** do not easily enter the brain.

The existence of the blood–brain barrier makes it possible to design drugs that will act at various parts of the body (e.g. the heart), but have no activity in the brain, thus reducing any central nervous system side effects. This is achieved by increasing the polarity of the drug so that it does not cross the blood brain barrier. On the other hand, drugs that are intended to act in the brain must be designed so that they *are* able to cross the blood brain barrier. This means that they must have a minimum of polar groups, or have these groups temporarily masked (see prodrugs, Topic I1). Some polar drugs can cross the blood–brain barrier with the aid of carrier proteins.

Placental barrier

The **placental membranes** separate a mother's blood from the blood of her fetus. The mother's blood provides the fetus with essential nutrients and carries away waste products, but these chemicals must pass through the **placental barrier**. Since food and waste products can pass through the placental barrier, it is perfectly feasible for drugs to pass through as well. Drugs such as **alcohol**, **nicotine** and **cocaine** can all pass into the fetal blood supply. Fat-soluble drugs will cross the barrier most easily and drugs such as **barbiturates** will reach the same levels in fetal blood as in maternal blood. Such levels may have unpredictable effects on fetal development. They may also prove particularly hazardous once the baby is born. Drugs and other toxins can be removed from fetal blood via the maternal blood and detoxified, but once the baby is born it may have the same levels of drugs in its blood as the mother, but it does not have the same ability to detoxify or eliminate them. As a result, drugs will be longer lasting and may have fatal effects.

C4 Drug metabolism

Key Notes

Definition

Drug metabolism refers to the reactions undergone by a drug in the body. Metabolic enzymes exist mainly in the liver and catalyze reactions that increase the polarity of the drug.

Phase I metabolism

Phase I reactions usually involve the introduction of polar groups to a drug so that the drug can be excreted more easily. Particular groups in a molecule are more prone to metabolism than others. The cytochrome P450 enzymes are important metabolic enzymes involved in oxidative reactions. Drugs that enhance or inhibit the activity of these enzymes will affect the levels of drugs that are normally metabolized by them.

Phase II metabolism

Phase II reactions involve the formation of polar conjugates. Highly polar molecules such as glucose are linked to polar functional groups that may have been placed there by phase I metabolism.

Related topics

Drug excretion (C5)
Drug solubility (I1)
Drug stability (I2)
Drug metabolism studies (K3)
From lead compound to dianilino-phthalimides (L3)

Definition

Drug metabolism defines the reactions undergone by a drug when it is in the body. These reactions are catalyzed by metabolic enzymes, some of which are present in the gut wall and the blood supply, but most of which are in the liver. The role of the **liver** is crucial in drug metabolism. Drugs taken orally are absorbed through the gut wall into the blood, then passed through the liver before they proceed round the body. Thus, drugs have to survive the metabolic enzymes if they are to reach their target and show useful activity. In general, metabolic enzymes catalyze reactions that make drugs more polar and easier to excrete. Metabolic reactions are classed as being Phase I or Phase II reactions.

Phase I metabolism

Phase I reactions usually involve the introduction of a polar functional group or the modification of an existing functional group such that it becomes more polar (*Fig. 1*). There are certain structures in a drug that are more prone to drug metabolism than others. For example, *N*-methyl groups are often demethylated, while aromatic rings are oxidized to phenols; the terminal methyl group of an alkyl chain is often oxidized to an alcohol, while esters and amides are frequently hydrolyzed.

Oxidative reactions are particularly important in phase I metabolism, and these are catalyzed by a group of enzymes called the **cytochrome P450 enzymes**, which are located in the liver. There are at least 12 families of these enzymes in humans and they are capable of metabolizing a chemically diverse range of substrates ranging from drugs and food toxins to environmental chemicals. Because they are diverse in nature, individuals vary in the types of cytochrome P450 enzymes they may have. As a result the rate at which specific

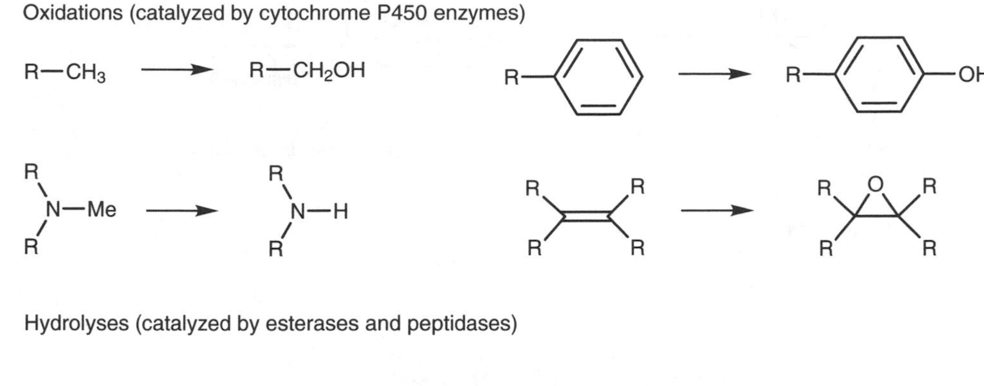

Fig. 1. Examples of Phase I metabolic reactions.

drugs are metabolized can vary from patient to patient, thus introducing an element of uncertainty into what the therapeutic dose of those drugs should be.

Another problem associated with the cytochrome P450 enzymes is the fact that some drugs can affect their activity. For example, **phenobarbitone** enhances the activity of cytochrome P450 enzymes, while the anti-ulcer agent **cimetidine** (Tagamet®) inhibits activity (*Fig. 2*). This can be a serious problem if the patient is taking another drug (e.g. **warfarin**) which is normally metabolized by the cytochrome P450 enzymes. Inhibition or enhancement of enzyme activity will mean that the blood concentration levels of the latter drug will be different from those when the drug is taken on its own. This could lead to sub-therapeutic levels of the drug if the activity of the cytochrome P450 enzymes is enhanced, or a drug overdose if the activity is inhibited. Certain foods can also influence the activity of cytochrome P450 enzymes (e.g. **brussel sprouts** and **grapefruit juice**) leading to the same types of problems. For this reason, many current medicinal chemistry projects aim to design drugs that are not metabolized by cytochrome P450 enzymes and which have no effect on cytochrome P450 activity.

Phase II metabolism

Phase II reactions are **conjugation reactions** where a polar molecule is linked to a suitable functional group on the molecule (*Fig. 3*). In many cases, the functional group that undergoes the phase II reaction may have been introduced by a phase I metabolic reaction. Once again, polarity is increased making the drug-conjugate

Phenobarbitone **Cimetidine**

Fig. 2. Drugs that influence the activity of cytochrome P450 enzymes.

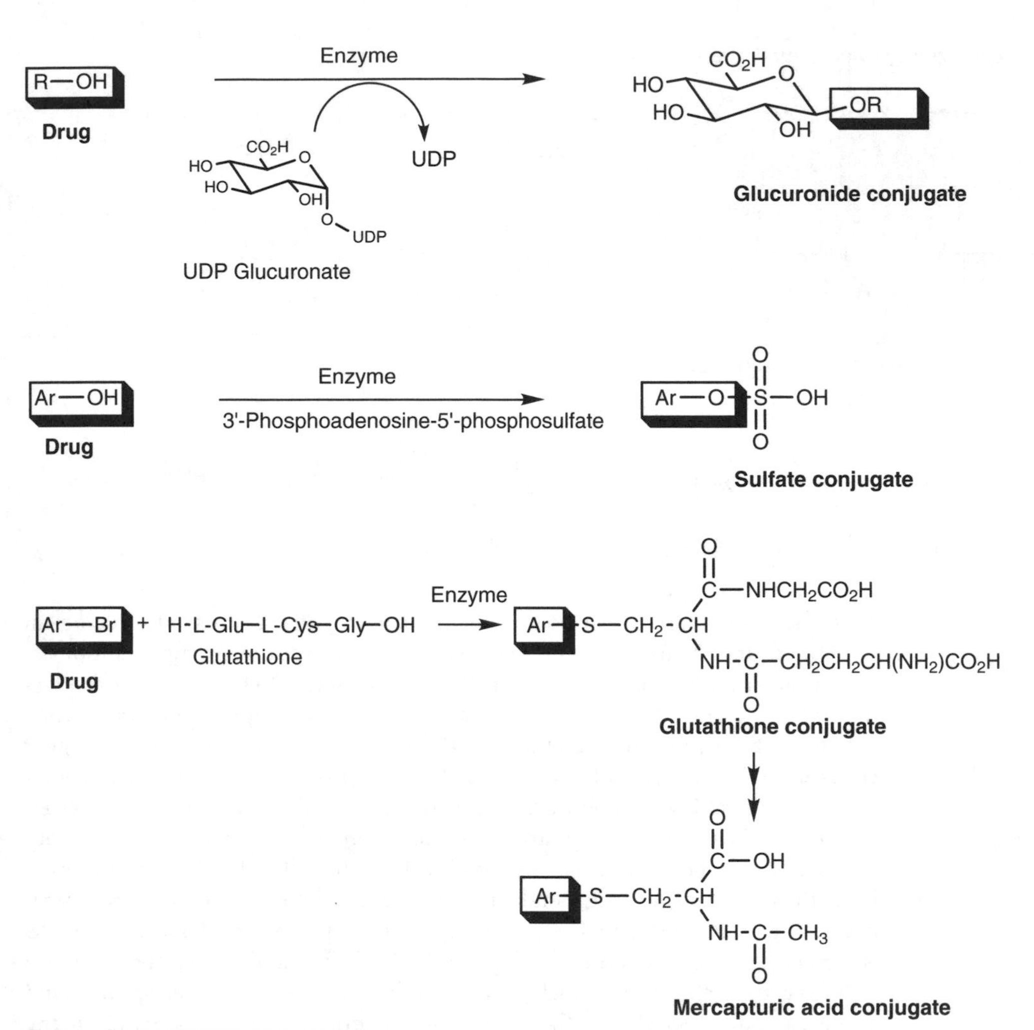

Fig. 3. Phase II metabolic reactions.

more easily excreted than the original drug. Phenols, alcohols, and amines react with UDP-glucose to form ***O*- or *N*-glucuronides**, where a highly polar glucose molecule is added to the drug. Some steroids form **sulfates**, while phenols, epoxides, and halides react with the tripeptide **glutathione** to give **mercapturic acid conjugates**.

C5 Drug excretion

Key Notes

Definition — Drugs and their metabolites are excreted from the body by a variety of routes, the most important being the kidneys.

Lungs — Gaseous and volatile drugs are excreted from the lungs by exhalation.

The bile duct — A small number of drugs are diverted from the blood supply into the gastrointestinal tract by passing down the bile duct. Such drugs can be reabsorbed from the gut.

Other routes — A certain proportion of a drug can be excreted through sweat, saliva and breast milk.

The kidneys — The most important route for excretion is through the kidneys. Blood is filtered under pressure to remove small molecules (such as drugs) and most of the water. However, water is quickly reabsorbed, setting up a concentration gradient, which leads to hydrophobic molecules being reabsorbed into the blood supply. Polar molecules remain in the nephrons and are excreted in urine.

Related topics — Drug solubility (I1) Drug metabolism (C4)

Definition

Drugs and their metabolites can be excreted from the body by a number of routes, such as via the lungs, the skin, the bile duct and the kidneys. In general, excretion via the kidneys is the most important route.

Lungs

Volatile or gaseous drugs can be excreted through the **lungs**. Such drugs pass out of the capillaries that supply the lungs then diffuse through cell membranes into the air sacs, from where they are exhaled. Gaseous **general anesthetics** are excreted in this way by moving down a concentration gradient from the blood supply into the lungs. They are also administered through the lungs, in which case the concentration gradient is in the opposite direction and the gas moves from the lungs to the blood supply.

The bile duct

The **bile duct** travels from the liver to the intestines and carries a yellow fluid called **bile**, which contains bile acids and salts that are important to the digestion process. It is known that a small number of drugs are diverted from the blood supply through the bile duct back into the intestines. Since this happens from the liver, the proportion of drug eliminated in this way has not had the opportunity to be distributed round the body. Therefore, the amount of drug distributed is less than that absorbed. However, once the drug has entered the intestine, it can be reabsorbed so that it can 'try again'.

Other routes

It is possible for as much as 10–15% of a drug to be lost through the skin in sweat. Drugs can also be excreted through saliva and breast milk, but these are minor excretion routes compared to the kidneys. There are concerns, however, that mothers may be passing on drugs such as **nicotine** to their babies through breast milk.

The kidneys

The **kidneys** are the principle route by which drugs and their metabolites are excreted (*Fig. 1*). The kidneys filter the blood of waste chemicals, which are subsequently removed in the urine. Drugs and their metabolites are excreted by the same mechanism.

Blood enters the kidneys via the **renal artery**. This divides into a large number of capillaries and each one of these forms a knotted structure called a **glomerulus**, which fits into the opening of a duct called a **nephron**. The blood entering these glomeruli is under pressure and so plasma is forced through the pores in the capillary walls into the nephron, carrying with it any drugs and metabolites that might be present. Any compounds that are too big to pass through the pores, such as plasma proteins and red blood cells, will remain in the capillaries with the remaining plasma. Note that this is a filtration process and it does not matter whether the drug is polar or hydrophobic. All drugs and drug metabolites will be passed equally efficiently into the nephron. However, this does not mean that every compound will be *excreted* with equal efficiency because there is more to the process than simple filtration.

The plasma and chemicals that have been filtered into the nephron then pass through the nephron on their route to the bladder and eventual excretion in the urine. However, only a small percentage of what begins that journey actually finishes it. This is because the nephron is surrounded by a rich network of blood vessels carrying the filtered blood away from the glomerulus. This permits much of the contents of the nephron to be reabsorbed into the blood supply. Much of the water that is filtered into the nephron is quickly reabsorbed, thus concentrating drugs and other agents in the nephron. A concentration gradient

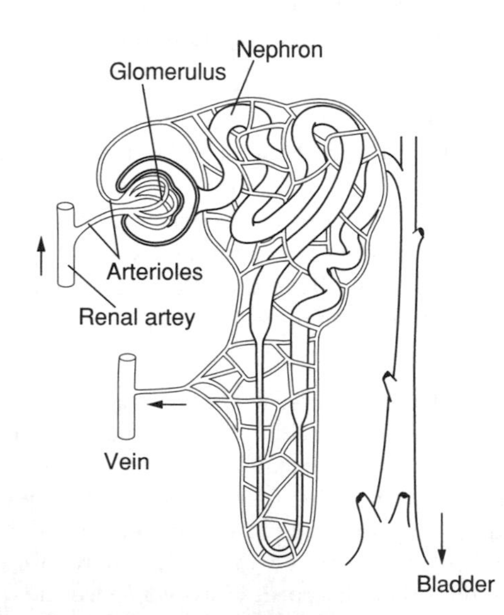

Fig. 1. Excretion by the kidneys.

is thus set up whereby the concentrations of the drug and its metabolites are greater in the nephron than in the blood supply, allowing these compounds to move back into the blood supply down the concentration gradient. However, this can only happen if the drug is sufficiently hydrophobic to pass through the cell membranes of the nephron. This means that hydrophobic compounds are efficiently reabsorbed back into the blood whereas polar compounds are not and are excreted.

This process of excretion explains the importance of drug metabolism to drug excretion. Drug metabolism makes a drug more polar so that it is less likely to be reabsorbed from the nephrons.

C6 Drug administration

Key Notes

Definition

Drug administration is the method by which a drug is introduced into the body. The method used depends on the target organ as well as the pharmacokinetic properties of the drug.

Oral administration

Oral administration exposes drugs to the greatest number of 'ordeals'. The drug must survive the acids of the stomach and the digestive enzymes of the gastrointestinal tract. It must dissolve through the fatty membranes of the cells lining the gut wall and survive metabolic enzymes.

Mucous membranes

Absorption through the mucous membranes of the mouth or nose avoids the hazards of the oral route. Drugs absorbed through the mouth can be kept under the tongue or sucked. Drugs absorbed through the nose can be sniffed as a powder or smoke.

Rectal

Drugs may be administered as suppositories by the rectal route when oral administration proves impractical. The drug is absorbed through the mucous membranes but absorption is less reliable than by the oral route.

Inhalation

Drugs such as the gaseous general anesthetics are inhaled and absorbed into the blood supply through the lungs. Anti-asthmatic drugs are designed to be poorly absorbed so they are concentrated in the airways.

Topical

Drug patches applied to the skin slowly release their drug so that it is absorbed through the skin into the blood supply, resulting in a steady release over a long time period.

Injection

Drugs can be injected into veins, muscles, under the skin, or into the cerebral spinal fluid. Injection allows the administration of accurate levels of drug and results in the fastest onset of action. However, the potential hazards are also greater. It is important to use sterile techniques to avoid the possibility of infection.

Subcutaneous implants

Minipumps can be implanted under the skin to monitor the chemistry of the blood and to release controlled amounts of a drug.

Related topics

Drug absorption (C2)
Drug solubility (I1)
Drug stability (I2)
Pharmacology and pharmaceutical chemistry (K2)

Definition

Drug administration is the method by which a drug is introduced into the body. There are various methods available and the method chosen will depend on the target organ and the pharmacokinetics of the drug. Anti-asthmatic drugs are targeted to the lungs and so it is clearly more sensible to give the drug by inhalation than by injection. The properties of the drug will often determine how it is administered. For example, drugs that are unstable when in contact with stomach acids cannot be given orally and should be administered by a different method. In some cases, the method of drug administration can solve many of the problems associated with pharmacokinetics.

Oral administration

The oral route is the easiest and most popular method of drug administration as far as the patient is concerned. There is also more chance that the patient will comply with the drug regime and complete the course. However, the oral route places the most demands on the chemical and physical properties of the drug. The drug has to be chemically stable when mixed with the acids of the stomach, as well as being metabolically stable when in contact with digestive enzymes in the stomach and intestines. Therefore, acid labile drugs, such as the **local anesthetics** and **insulin**, cannot be given orally. There are also the metabolic enzymes to contend with, mainly in the gut wall and the liver. Therefore, the amount of drug that eventually circulates round the blood supply will be less than that administered, and this is called the **first pass effect**.

The drug must have the correct hydrophilic/hydrophobic balance to be given orally. It must be polar enough to dissolve in the gut and the blood supply, but hydrophobic enough to dissolve through the cell membranes of the gut wall. Drugs given orally can be taken as pills, capsules or as solutions. Drugs taken in solution are absorbed more quickly and a certain percentage may even be absorbed through the stomach wall. For example, approximately 25–33% of **alcohol** is absorbed into the blood supply from the stomach, while the rest is absorbed from the upper intestine. Most drugs are taken as pills or capsules and are mostly absorbed in the upper intestine. The rate of absorption is partly determined by the rate at which the pills and capsules dissolve, which in turn depends on factors such as particle size and crystal form. In general, about 75% of an orally administered drug is absorbed into the body within 1–3 hours.

Specially designed pills and capsules can remain intact in the stomach to help protect acid labile drugs from stomach acids. The containers only degrade once they reach the intestine.

Mucous membranes

Some drugs can be absorbed through the mucous membranes of the mouth or nose, thus avoiding the digestive and metabolic enzymes encountered during oral administration. For example, **glyceryl trinitrate** (*Fig. 1*) taken by heart patients is placed under the tongue (sublingual administration). **Fentanyl** (*Fig. 1*) is an opiate analgesic, which can be given to children in the form of a lollipop and is absorbed through the mucous membranes of the mouth. Nasal deconges-

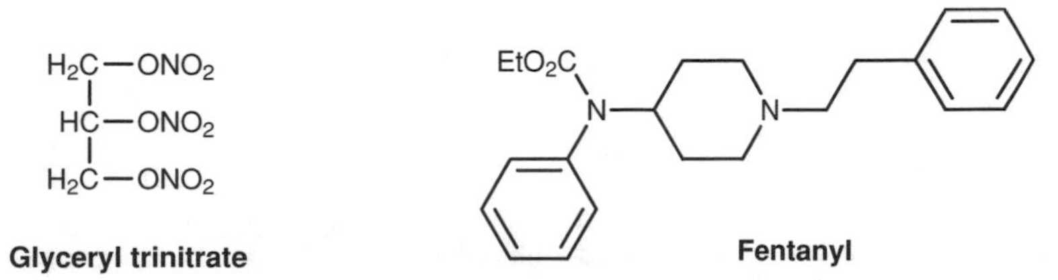

Fig. 1. Drugs absorbed through mucous membranes of the mouth.

tants are absorbed through the mucous membranes of the nose. **Cocaine** powder is absorbed through the mucous membranes of the nose when it is sniffed. **Nicotine** in the form of snuff can also be absorbed in this way. Drugs administered as eye drops or nose drops are also absorbed through mucous membranes.

Rectal

Some drugs are administered rectally as **suppositories**, especially if the patient is unconscious, vomiting or is unable to swallow. However, there are several problems associated with rectal administration, including membrane irritation. The extent of drug absorption can also be unpredictable.

Inhalation

Some drugs can be administered by inhalation, thus avoiding the digestive and metabolic enzymes of the gastrointestinal tract or liver. Once inhaled, the drugs are absorbed through the cell linings of the respiratory tract into the blood system. Assuming the drug is able to pass through the hydrophobic cell membranes, absorption is rapid and efficient, as the blood supply is in close contact with the cell membranes of the lungs. For example, **general anesthetic gases** are small, highly lipid soluble molecules which are absorbed, almost as fast as they are inhaled.

Nongaseous drugs can be administered as **aerosols**. Anti-asthmatic drugs are administered in this way, as it allows these drugs to be delivered to the lungs in far greater quantities than if they were given orally or by injection. In the case of anti-asthmatics, the drug is made sufficiently polar that it is poorly absorbed into the bloodstream. This localizes it in the airways and lowers the possibility of side effects elsewhere in the body (e.g. action on the heart). However, a certain percentage of an inhaled drug is inevitably swallowed and can reach the blood supply by the oral route.

Several drugs of abuse are absorbed through inhalation (e.g. **nicotine** (*Fig. 2*), **cocaine, marijuana, methamphetamine** (*Fig. 2*) and **heroin**). In the case of cigarettes or cannabis, smoking results in the production of carcinogenic tars, which are also inhaled. These are not absorbed into the blood supply but coat the lung tissue leading to long-term problems such as lung cancer.

Topical

Topical drugs are those that are applied to the skin. These are often administered as patches, such as the **nicotine patch**, which is used to help stop smoking. Other drugs that have been applied in this way include the analgesic **fentanyl**, the hypertensive agent **clonidine** (*Fig. 2*) and the hormone **estrogen**. Once applied, the drug is slowly released from the patch and absorbed through the skin into the blood supply over several days. As a result, the level of drug remains relatively constant over that period.

N CH3 N

Nicotine

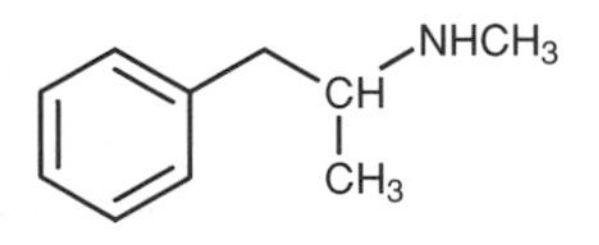

Methamphetamine

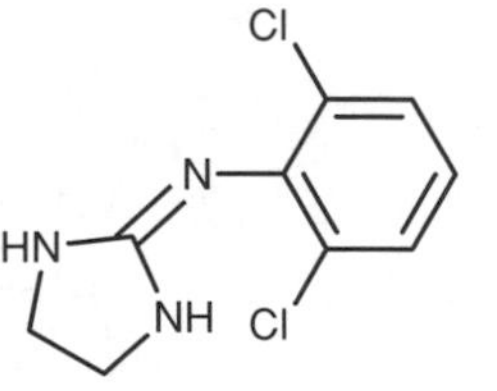

Clonidine

Fig. 2. Nicotine, methamphetamine and clonidine.

Injection

There are four methods by which drugs can be injected into the body – by intravenous, intramuscular, subcutaneous or intrathecal injection. Injection of a drug produces a much faster response than oral administration, as the drug reaches the blood supply more quickly. The levels of drug administered are also more accurate, as absorption by the oral route has a level of unpredictability due to the various problems previously described – the first pass effect. There are, however, potential problems with drug injection and the procedure can be more hazardous. For example, some patients may have an unexpected reaction to a drug and there is little one can do to reduce the level once the drug has been injected. Such side effects would be more gradual and treatable if the drug was given orally. Second, sterile techniques are essential when giving injections to avoid the risks of bacterial infection, or of transmitting hepatitis or AIDS from a previous patient. Third, there is a greater risk of receiving an overdose when injecting a drug.

The **intravenous** (i.v.) route involves injecting a solution of the drug directly into a vein. This method of administration is not particularly popular with patients, but it is a highly effective method of administering drugs in accurate doses and is the fastest of the injection methods. However, i.v. injection is also the most hazardous method of injection. Since its effects are rapid, the onset of any serious side effects or allergies are also rapid. It is therefore important to administer the drug as slowly as possible and to monitor the patient closely. Some drugs that can be safely given orally may prove hazardous if given by i.v. injection. Drugs that are dissolved in oily liquids cannot be given by i.v. injection since they may result in the formation of blood clots.

The **intramuscular** (i.m.) route involves injecting drugs directly into muscle. Usually, this is in the arm, thigh or buttocks. Drugs administered in this way do not pass round the body as rapidly as they would if given by i.v. injection, but they are still absorbed faster than by oral administration. The rate of absorption depends on various factors such as the blood supply to the muscle, the solubility of the drug, and the volume of the injection. Drugs are often administered by intramuscular injection when they are unsuitable for i.v. injection, so it is important to avoid injecting into a vein.

Subcutaneous injection involves injecting the drug just under the surface of the skin. Absorption depends on the blood supply to the skin and the ability of the drug to enter the blood vessels. Drugs that are irritant should not be administered in this way as they can cause severe pain and may damage local tissue.

Intrathecal injection is injection into the spinal cord. Antibacterial agents that do not normally cross the blood brain barrier are often administered in this way.

Subcutaneous implants

Minipumps have been developed that can be implanted under the skin and release drugs at a controlled rate. For example, a pump has been developed that monitors the level of **insulin** in the blood, and then releases insulin as required to keep the level constant. This avoids the problem of the large fluctuations in insulin levels that inevitably arise from regular injections.

C7 DRUG DOSING

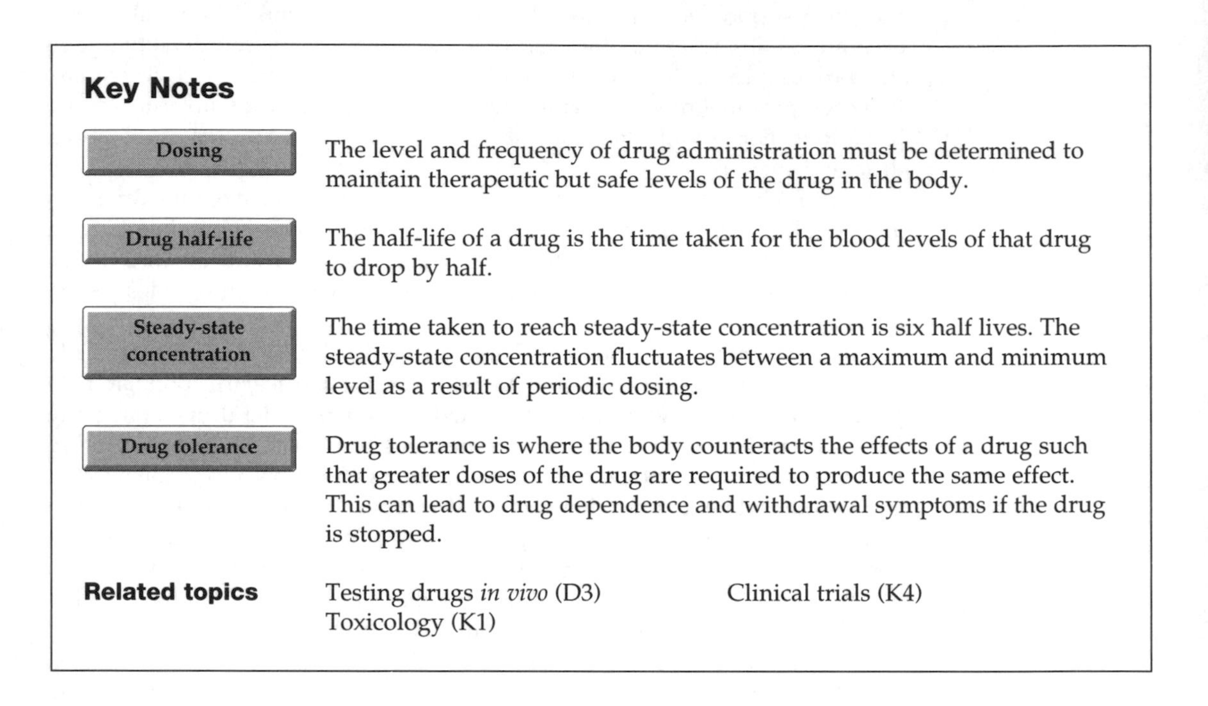

Key Notes

Dosing	The level and frequency of drug administration must be determined to maintain therapeutic but safe levels of the drug in the body.
Drug half-life	The half-life of a drug is the time taken for the blood levels of that drug to drop by half.
Steady-state concentration	The time taken to reach steady-state concentration is six half lives. The steady-state concentration fluctuates between a maximum and minimum level as a result of periodic dosing.
Drug tolerance	Drug tolerance is where the body counteracts the effects of a drug such that greater doses of the drug are required to produce the same effect. This can lead to drug dependence and withdrawal symptoms if the drug is stopped.

Related topics Testing drugs *in vivo* (D3) Toxicology (K1) Clinical trials (K4)

Dosing

Knowledge of a drugs pharmacokinetics is essential in order to predict the level and frequency of drug administration required to maintain a therapeutic, but safe level of the drug in the blood supply. In general, the concentration of free drug in the blood (i.e. not bound to plasma protein) is a good indication of the availability of that drug at its target site. This does not mean that blood concentration levels are the same as the concentration levels at the target site. However, any variations in blood concentration will result in similar fluctuations at the target site. Thus, blood concentration levels can be used to determine therapeutic and safe dosing levels for a drug.

Drug half-life

The **half-life** of a drug is the time taken for the concentration of a drug in blood to fall by half. The removal or elimination of a drug takes place both through excretion and drug metabolism and is not linear with time. Therefore, drugs can linger in the body for a significant time period. For example, if a drug has a half-life of 1 h, then 1 h after administration there would be 50% of it left. After 2 h there would be 25% of the original dose left and after 3 h, 12.5% would remain. Therefore, it would take 7 h for the level of that drug to fall below 1% of the original dose. Some drugs, such as the opiate analgesic **fentanyl**, have short half-lives (45 min), whereas others, such as **diazepam** (Valium) have a half-life measured in days. In the latter case, recovery from the drug may take a week or more.

Steady-state concentration

Because drugs start to be metabolized and eliminated as soon as they are administered, it is necessary to provide regular doses in order to maintain therapeutic levels in the body. Therefore, it is important to know the half-life of the drug in order to calculate the frequency of dosing required to reach and maintain these levels. In general, the time taken to reach a **steady-state concentration** is six times the drug's half-life. For example, the concentration level of a drug with a half-life of 4 hours, supplied at 4-hourly intervals is shown in *Table 1*.

Table 1. Fluctuation of drug concentration levels on regular dosing

Time of dosing (h)	Max level (μg/ml)	Min level (μg/ml)
0	1.0	0.5
4	1.5	0.75
8	1.75	0.87
12	1.87	0.94
16	1.94	0.97
20	1.97	0.98
24	1.98	0.99

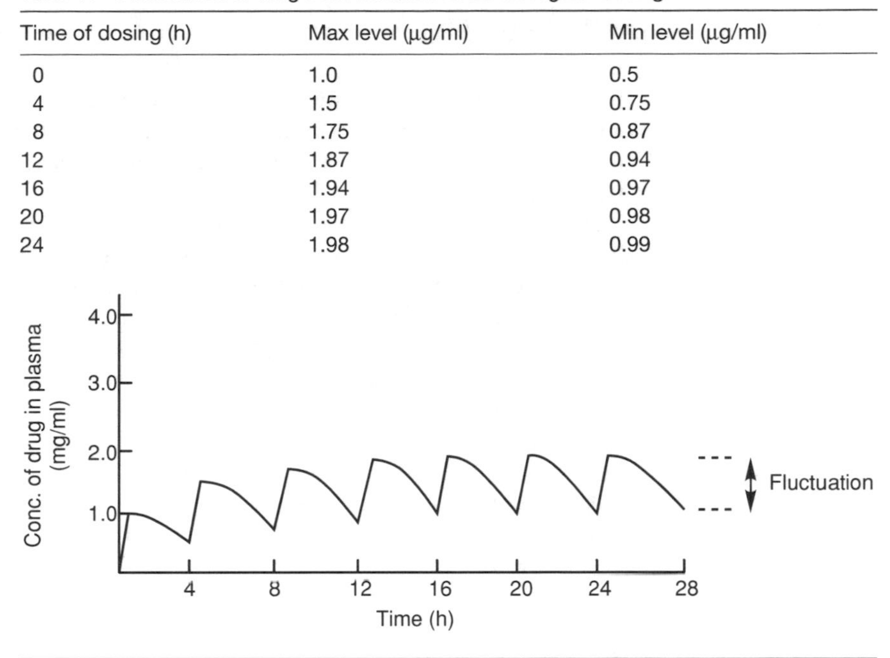

Note that there is a fluctuation in concentration level in the period between each dose. The level is at a maximum after each dose and falls to a minimum before the next dose is administered. It is important to ensure that the level does not drop below the therapeutic level, but does not rise to such a level that side effects are induced.

The time taken to reach steady-state concentration is not dependent on the size of the dose, but the concentration level of drug in the blood achieved at steady-state is. Therefore, the final dose level at steady-state concentration depends on the size of each dose given, as well as the frequency of the doses.

During clinical trials, blood samples will be taken from patients at regular intervals to determine the concentration of the drug in the blood. This will help determine the proper dosing regime in order to get the ideal concentration of drug in the blood.

Drug tolerance

With certain drugs, it is found that the effect of the drug diminishes after repeated doses, and it is necessary to increase the size of the dose in order to achieve the same results. This is known as **drug tolerance**. There are several mechanisms by which drug tolerance can occur.

The drug can induce synthesis of metabolic enzymes, which will result in increased metabolism of the drug. **Pentobarbital** (*Fig. 1*) is a barbiturate sedative that induces enzymes in this fashion.

Alternatively, the target receptors for the drug may adapt to the presence of

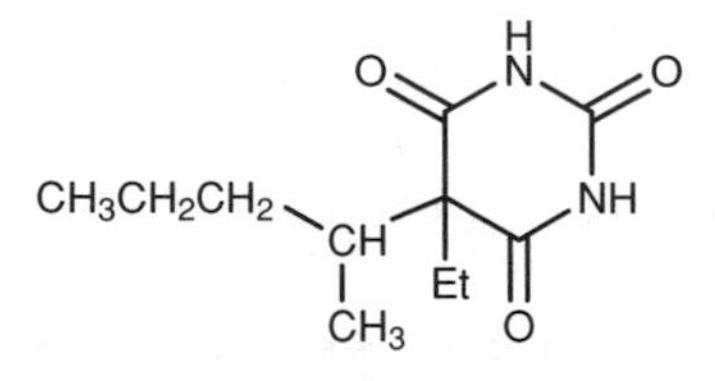

Fig. 1. Pentobarbital.

the drug. Occupancy of the target receptor by an antagonist may induce cellular effects that result in the synthesis of more receptors to counteract the antagonism. As a result, more drug will be needed in the next dose to occupy all the receptors.

Physical dependence is usually associated with drug tolerance. Physical dependence occurs when a patient needs the drug in order to feel normal. If the drug is withdrawn, uncomfortable **withdrawal symptoms** may arise, which can only be stopped by retaking the drug. These effects can be explained in part by the mechanism that leads to drug tolerance. For example, if cells have synthesized more receptors to counteract the presence of an antagonist, the removal of the antagonist will leave the body with too many receptors for its normal function. This results in a 'kick-back' effect where the cell becomes oversensitive to the normal neurotransmitters or hormones, and this is what produces the symptoms of withdrawal. These will continue until the excess receptors have been broken down by normal cellular mechanisms, a process which may take several days or weeks.

D1 TESTING DRUGS

Key Notes

Testing — Relevant tests are required to test the activity and selectivity of compounds against specific targets.

***In vitro* and *in vivo* tests** — *In vivo* tests involve the use of live animals, whereas *in vitro* tests do not. *In vitro* tests are more suitable for routine testing. However, they do not test how effectively a drug will reach its target in a living system. They also give negative results for prodrugs. *In vivo* tests are important in demonstrating the physiological effects of a drug.

Related topics

Testing drugs *in vitro* (D2)
Testing drugs *in vivo* (D3)
The lead compound (E1)
Definition of SAR (G1)
Toxicology (K1)
Drug metabolism studies (K3)
Clinical trials (K4)
Testing procedures (L2)

Testing

Before any medicinal chemistry project is started, it is important to establish what testing procedures are going to be used to test whether compounds have the desired activity and how strongly they interact with the desired target. Other procedures are required to test how selective the compounds are by testing their activity at other possible targets. Further tests are required to check whether the compounds enhance or inhibit metabolic enzymes, since this could lead to unwanted drug–drug interactions (see Topic C4).

Once a suitable drug has been established, preclinical tests are carried out involving toxicology and drug metabolism, and, finally, clinical trials are started. Preclinical and clinical trials will be covered in Section K. In this section, we concentrate on drug testing at the lead discovery and drug optimization stages.

In vitro and *in vivo* tests

***In vivo* tests** involve the use of live animals, whereas ***in vitro* tests** do not. It is generally cheaper and quicker to carry out *in vitro* tests, so most routine testing is carried out *in vitro*. *In vivo* tests are restricted to potentially useful compounds and are used to test whether active compounds *in vitro* are still active *in vivo*. Many drugs that are active *in vitro* turn out to be inactive *in vivo*. This is because a drug may not be able to reach its target in the living system. *In vivo* tests will also determine whether a drug has the physiological effect desired. For example, some research projects aim to develop compounds that show selectivity towards a specific receptor in the belief that this interaction may lead to a desired physiological effect (e.g. relief of depression). However, if a selective compound for that receptor has not been prepared before, there is no certainty that the desired response will occur.

Sometimes, compounds are found which are active *in vivo*, but are inactive *in vitro*. This is usually because the compounds involved are acting as **prodrugs**. In other words, they are not active themselves, but are metabolized in the body to the active compound. The dye **prontosil** (Topic M4) is an example of this. It is inactive *in vitro*, but when it is administered *in vivo*, it is metabolized to the antibacterial agent **sulfanilamide**.

D2 TESTING DRUGS *IN VITRO*

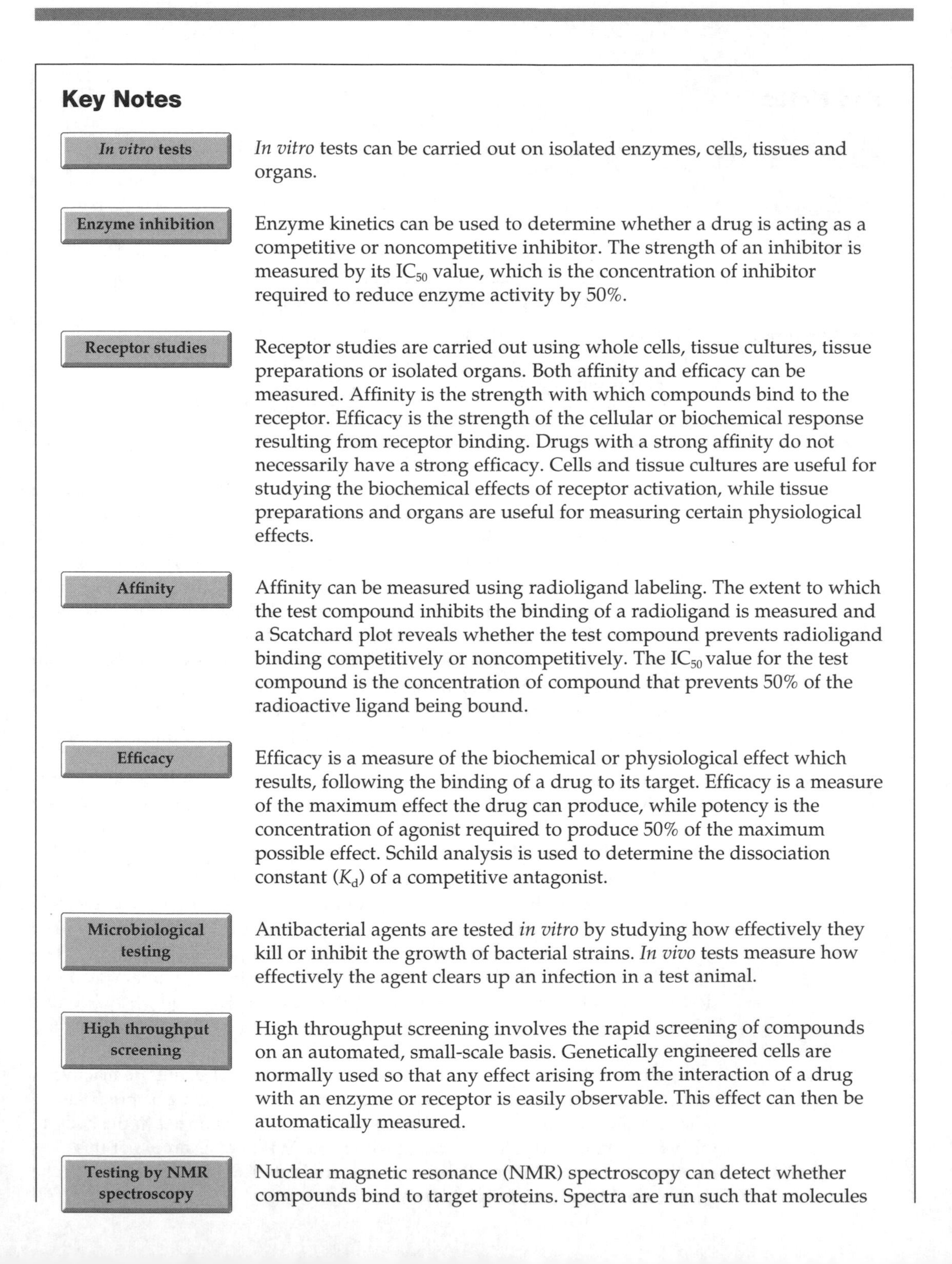

Key Notes

***In vitro* tests**	*In vitro* tests can be carried out on isolated enzymes, cells, tissues and organs.
Enzyme inhibition	Enzyme kinetics can be used to determine whether a drug is acting as a competitive or noncompetitive inhibitor. The strength of an inhibitor is measured by its IC_{50} value, which is the concentration of inhibitor required to reduce enzyme activity by 50%.
Receptor studies	Receptor studies are carried out using whole cells, tissue cultures, tissue preparations or isolated organs. Both affinity and efficacy can be measured. Affinity is the strength with which compounds bind to the receptor. Efficacy is the strength of the cellular or biochemical response resulting from receptor binding. Drugs with a strong affinity do not necessarily have a strong efficacy. Cells and tissue cultures are useful for studying the biochemical effects of receptor activation, while tissue preparations and organs are useful for measuring certain physiological effects.
Affinity	Affinity can be measured using radioligand labeling. The extent to which the test compound inhibits the binding of a radioligand is measured and a Scatchard plot reveals whether the test compound prevents radioligand binding competitively or noncompetitively. The IC_{50} value for the test compound is the concentration of compound that prevents 50% of the radioactive ligand being bound.
Efficacy	Efficacy is a measure of the biochemical or physiological effect which results, following the binding of a drug to its target. Efficacy is a measure of the maximum effect the drug can produce, while potency is the concentration of agonist required to produce 50% of the maximum possible effect. Schild analysis is used to determine the dissociation constant (K_d) of a competitive antagonist.
Microbiological testing	Antibacterial agents are tested *in vitro* by studying how effectively they kill or inhibit the growth of bacterial strains. *In vivo* tests measure how effectively the agent clears up an infection in a test animal.
High throughput screening	High throughput screening involves the rapid screening of compounds on an automated, small-scale basis. Genetically engineered cells are normally used so that any effect arising from the interaction of a drug with an enzyme or receptor is easily observable. This effect can then be automatically measured.
Testing by NMR spectroscopy	Nuclear magnetic resonance (NMR) spectroscopy can detect whether compounds bind to target proteins. Spectra are run such that molecules

with short relaxation times (i.e. large molecules) are not detected. If the test compound binds to the protein, its spectrum will not be observed.

Related topics

Testing drugs (D1)
Definition of structure–activity relationships (G1)
Testing procedures (L2)

In vitro tests

There are many different ways in which *in vitro* tests can be carried out, involving studies on isolated enzymes, whole cells, tissue preparations and organs.

Enzyme inhibition

Many enzymes can be isolated and purified, then studied in solution. In the past, this could be a difficult process, especially if the enzyme was present in minute concentrations. However, with the advent of **genetic engineering**, it is now possible to identify the gene responsible for a particular enzyme and to introduce it into the chromosomes of fast-growing cells such as yeast or bacteria. These cells then produce the enzyme in much greater quantities than would be normal in the natural cell. The enzyme can then be isolated and purified more easily, or can be studied within the cell.

An isolated enzyme can be used for **kinetic studies** to test what effects an inhibitor has on the reaction catalyzed by the enzyme. In a typical series of experiments, the rate of the enzyme-catalyzed reaction is measured with respect to different substrate concentrations when there is no inhibitor present, then again when an inhibitor *is* present. A graph can then be drawn as shown below relating the reciprocals of rate and substrate concentration for both experiments (*Fig. 1*). This is known as a **Lineweaver–Burk** plot and will show two straight lines which demonstrate whether the inhibitor is acting competitively or noncompetitively.

With a competitive inhibitor, the lines cross the y-axis at the same point. This point corresponds to the maximum rate of reaction where substrate concentration is vastly greater than inhibitor concentration (remember that reciprocal values are used here and a low value on the x- or y-axis represents a high value for the substrate concentration or rate respectively). At very high concentrations, the substrate is able to compete so effectively against the fixed amount of inhibitor present that it occupies all the active sites available, allowing the maximum rate to be achieved.

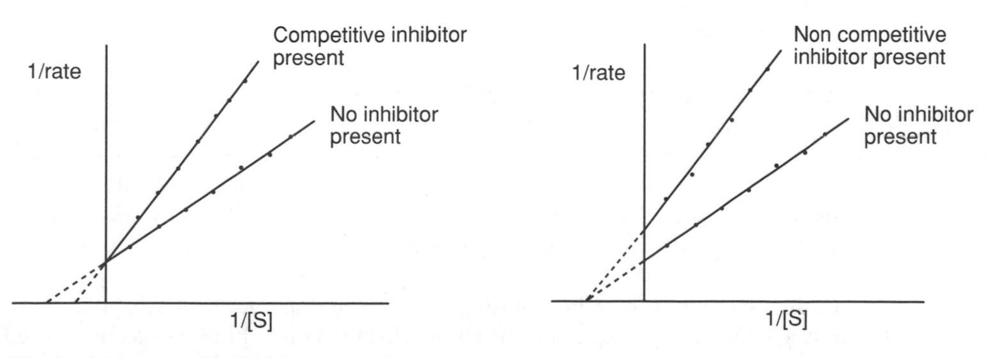

Fig. 1. Lineweaver–Burk plots.

With a non-competitive inhibitor the lines cross at different points on the y-axis, because the maximum rate achievable is less in the presence of the inhibitor than if the inhibitor is absent. This is because the inhibitor binds to a different part of the enzyme and is not displaced by the substrate, no matter how much substrate is present. Therefore, the percentage of enzyme occupied by the inhibitor is always unavailable and the maximum rate cannot be obtained.

The measure of how effectively a drug inhibits an enzyme is known as the **IC_{50} value**, which represents the concentration of inhibitor that is required to reduce enzyme activity by 50%.

Receptor studies

Since most receptors are located in cell membranes, it is not possible to isolate them like enzymes. Therefore, the use of whole cells, tissue cultures, tissue preparations or whole organs are suitable testing methods. Genetic engineering has been useful in the preparation of cells and tissue cultures that contain a high concentration of the receptor of interest. As with enzymes, the gene responsible for a receptor can be identified, cloned and inserted into the chromosomes of fast-growing cells such as yeast, bacterial or tumor cells. For example, **Chinese hamster ovarian cells** (CHO cells) are commonly used in such studies and express a large amount of the cloned receptor on their cell surface. Cell and tissue culture preparations offer several advantages over other forms of tests. Cells can be studied more easily and there are no complications arising in the drug having to cross barriers such as the gut wall in order to reach its target. The environment surrounding the cells can be easily controlled and both intra-cellular and intercellular events can be monitored. Primary cell cultures (i.e. cells that have not been modified) can be produced from embryonic tissues, while transformed cell lines are derived from tumoral tissue. Cells grown in this fashion are all identical.

Such cellular preparations can be used for a variety of purposes. They can be used to measure the **affinity** of drugs for a receptor (i.e. how strongly they bind), as well as measuring cellular responses resulting from receptor–ligand interactions. The latter measures the **efficacy** of a compound (i.e. how effective the compound activates the receptor to produce a cellular effect). It is important to appreciate that compounds with a high affinity for a receptor do not necessarily have a high efficacy. They may bind strongly, but the binding may be such that the receptor is poorly activated. Thus, an antagonist can have high affinity and zero efficacy.

Studies involving cellular and tissue culture preparations are useful in measuring both the affinity and efficacy of a receptor/drug interaction, but they are not able to measure the physiological effects that might arise from such an interaction. Such effects can be measured *in vivo*, but such studies are complicated by the fact that the drug has to reach the target tissue or organ. Furthermore, the target sites are subject to other influences, such as nerve input and hormonal control. To avoid these problems, tissue preparations and whole organs can be used to test for particular physiological effects. For example, muscle preparations can be tested to see whether a receptor agonist induces muscle contraction. An antagonist could also be tested by how efficiently it prevents an agonist from causing muscle contraction.

Affinity

Affinity is a measure of how strongly a drug (or ligand) binds to a receptor binding site, but it gives no indication of what effect that binding has on the cell or the organism. The affinity of a compound for a receptor can be measured using a process known as **radioligand labeling**. A known antagonist (or ligand) for the target receptor is labeled with radioactivity and is added to cells or tissue so that it can bind to the receptors present. Once equilibrium has been reached, the unbound ligand is removed by washing, filtration or centrifugation. The extent of binding can then be measured by detecting the amount of radioactivity present in the cells or tissue, and the amount of radioactivity that has been removed. The experiment is repeated several times using various concentrations of radioligand, but the same amount of cells or tissue. In the simplest situation, an equilibrium exists where the ligand is either bound or unbound to the receptor. The equilibrium constant for this process is defined as K_d – **the dissociation constant**.

$$[L] + [R] \rightleftharpoons \underset{\text{Receptor–ligand complex}}{[LR]} \qquad K_d = \frac{[L] + [R]}{[LR]}$$

The maximum number of receptors available in a test is the sum of the receptors that are occupied by the ligand ([LR]) and those that are unoccupied ([R]), i.e.

$$R_{tot} = [R] + [LR].$$

Rearranging this means that the number of receptors unoccupied by a ligand is

$$[R] = R_{tot} - [LR].$$

Substituting this into the first equation and rearranging leads to the **Scatchard equation**, where both [LR] and [L] are measurable:

$$\frac{[\text{Bound ligand}]}{[\text{Free ligand}]} = \frac{[LR]}{[L]} = \frac{R_{tot} - [LR]}{K_d}$$

A **Scatchard plot** (*Fig. 2*) compares the ratio of bound ligand to free ligand ([LR]/[L]) versus bound ligand [L] to give a straight line, where the slope is the reciprocal of the dissociation constant. This is known as the **affinity constant** ($K_a = 1/K_d$). The smaller the dissociation constant (typically 0.1–10 nM), the higher the affinity of the radioligand. The point where the line meets the x-axis represents the total number of receptors available (R_{tot}) (see line A in the graph below).

The radioligand experiment can be repeated in the presence of an unlabelled test compound. The test compound competes with the radioligand for the

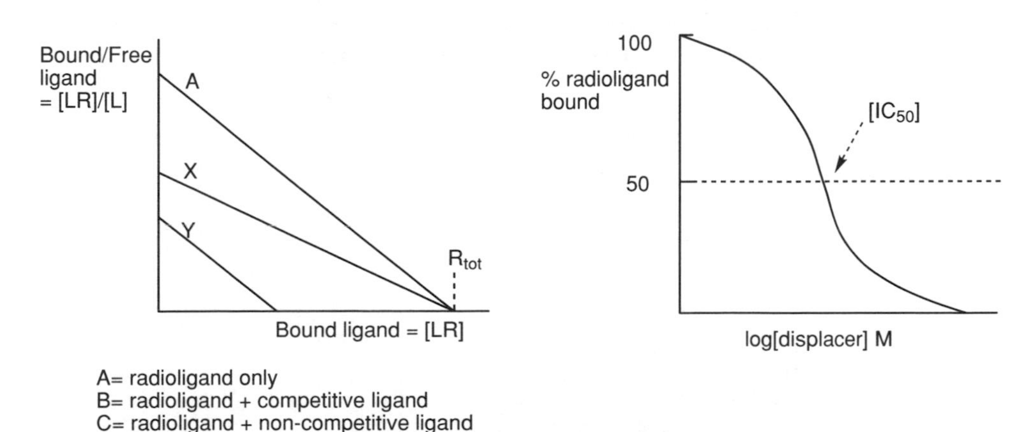

Fig. 2. Scatchard plot.

receptor's binding sites and is called a **displacer**. The stronger the affinity of the test compound, the more effectively it will compete for binding sites and the less radioactivity will be measured in the experiment for the cells or tissue. This will result in a different line in the Scatchard plot. If the test compound competes directly with the radiolabeled ligand for the same binding site on the receptor, the slope is decreased, but the intercept on the x-axis representing the total number of receptors (R_{tot}) remains the same (line X in the graph). In other words, if the radioligand concentration is much greater than the test compound it will bind to all the receptors available.

Agents that bind to the receptor at an allosteric binding site are not in competition with the radioligand for the same binding site and cannot be displaced by high levels of radioligand. This results in a line with an identical slope to line A, but which crosses the x-axis at a different point, thus indicating a lower total number of available receptors (line Y). Any receptor occupied by the test compound remains undetected no matter how much radioligand is used. The data from these displacement experiments can be used to plot a different graph, which compares the percentage of the radioligand that is bound to a receptor versus the concentration of the test compound. The result is a sigmoidal curve, termed the **displacement** or **inhibition curve**. This can be used to identify the **IC_{50} value** for the test compound (i.e. the concentration of compound that prevents 50% of the radioactive ligand being bound).

The **inhibitory** or **affinity constant (K_i or K_a)** for the test compound is the same as the IC_{50} value if noncompetitive interactions are involved. For compounds that *are* in competition with the radioligand for the binding site, the inhibitory constant depends on the level of radioligand present and is defined as shown below

$$K_i = \frac{IC_{50}}{1 + [L]_{tot}/K_d}$$

where K_d is the dissociation constant for the radioactive ligand, and $[L]_{tot}$ is the concentration of radioactive ligand used in the experiment.

Efficacy

Efficacy is a measure of the pharmacological effect resulting in a cell or tissue preparation as a result of a drug binding to its target receptor. The effect could be a biochemical event within the cell or a physiological event such as muscle contraction. The **EC_{50}** is the concentration of drug required to produce 50% of the maximum possible effect and reflects the **potency** of the drug (*Fig. 3*). In practice, **pD_2** is taken as the measure of potency where $pD_2 = -\log[EC]_{50}$.

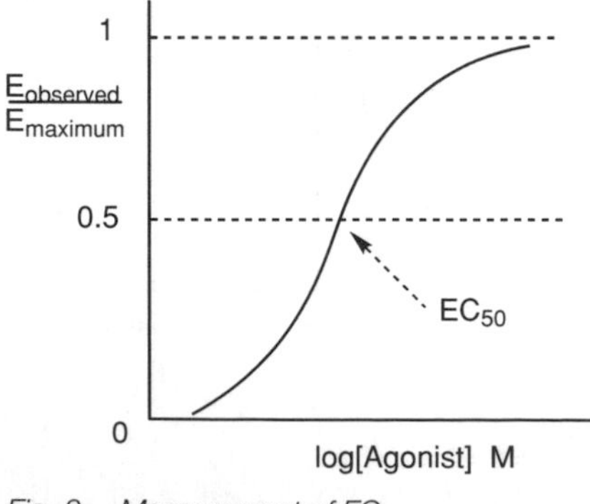

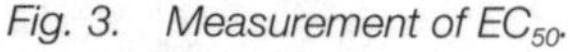
Fig. 3. Measurement of EC_{50}.

Schild analysis is used to determine the dissociation constant (K_d) of a competitive antagonist (*Fig. 4*). The effect resulting from an agonist activating its receptor is measured at different agonist concentrations. The experiment is then repeated in the presence of increasing concentrations of antagonist. Comparing the effect ratio ($E_{observed}/E_{maximum}$) versus the agonist concentration (log[agonist]) produces a series of sigmoidal curves where the EC_{50} of the agonist increases with increasing antagonist concentration. In other words, greater concentrations of agonist are required to counteract the antagonist. A **Schild plot** is then constructed which compares the reciprocal of the dose ratio versus the log of the antagonist concentration. (The **dose ratio** is the agonist concentration required to produce a specified level of effect when no antagonist is present, compared to the agonist concentration required to produce the same level in the presence of antagonist.) The line produced from these studies can be extended to the x-axis to find pA_2 which equals $-\log K_d$, where $\mathbf{pA_2}$ represents the affinity of the competitive antagonist.

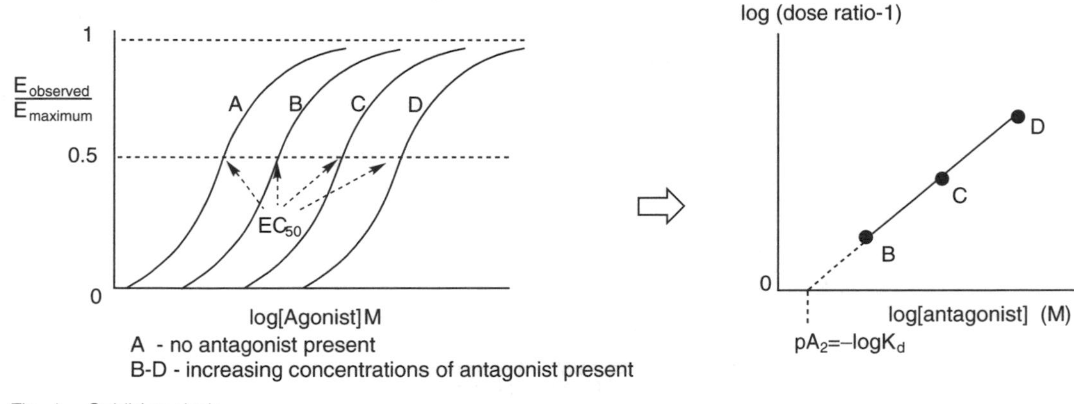

Fig. 4. Schild analysis.

Microbiological testing

Antibacterial drugs are tested *in vitro* by studying their effect on the growth of a range of bacterial strains. The drugs may be observed to inhibit growth, in which case they can be defined as **bacteriostatic**. Alternatively, they may actively kill bacterial cells, in which case they are defined as **bactericidal**. You may wonder why tests on bacteria are classed as *in vitro*, rather than *in vivo*, since live organisms are being tested. Here, the definition of *in vivo* or *in vitro* refers to the host animal infected by the microorganism. Thus, *in vivo* testing of an antibacterial agent tests how efficiently that agent clears up a particular infection in a host animal. *In vitro* testing is carried out on the organism responsible for the infection.

High throughput screening

High throughput screening involves the automated, small-scale testing of large numbers of compounds. Typically, the tests are carried out on a plate containing 96 wells. The tests used are designed to demonstrate an easily observable result should a compound interact with a target such as an enzyme or receptor. For example, a color change may result, which can be automatically measured spectroscopically. Other easily observable changes include cell growth, and the displacement of radioactively labeled ligands from receptors. Usually genetically modified cells are used for such tests since they have been designed to produce a readily measurable effect. Such mechanism-based bioassays using subcellular systems or cultured cells combine good reproducibility with the capacity for rapid sample throughput with high sensitivity and selectivity.

Enzyme inhibitors can be tested by using yeast cells which have been genetically engineered such that the yeast produces a human enzyme instead of an inherent enzyme. If the enzyme involved is crucial to cell growth, inhibitors will prevent the yeast cells from growing and this can be easily observed and measured.

Receptor antagonists can be studied by observing how effectively they inhibit the binding of a radioactively labeled ligand to the cell surface. Another approach is to use genetically modified cells where the activation of a target receptor switches on the production of an enzyme within the cell, which in turn catalyzes a reaction designed to release a dye molecule and produce a color change. For example, **estrogen agonists** and **antagonists** can be detected using genetically modified cells where a DNA sequence recognized by the **estrogen receptor** (see Topic B2) has been inserted into yeast DNA, along with the gene for an enzyme called **β-galactosidase** (*Fig. 5*).

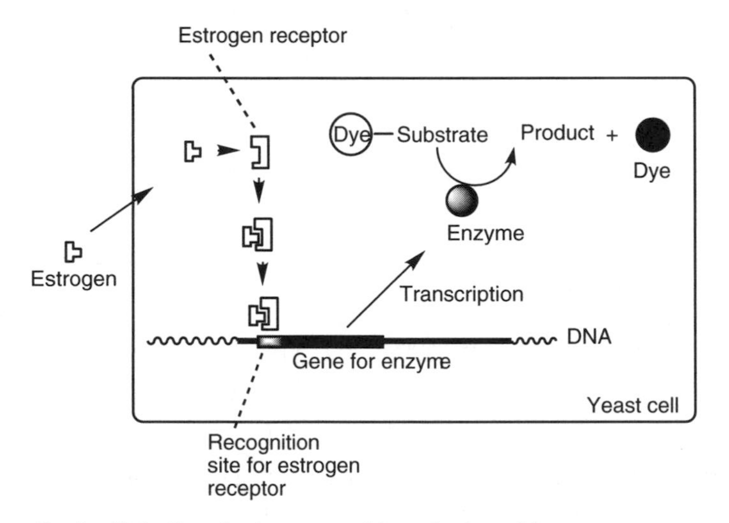

Fig. 5. Detection of estrogen agonists and antagonists.

When an estrogen agonist is tested, it binds to the estrogen receptor, and the receptor–agonist complex then binds to its recognition site on yeast DNA, switching on the transcription of the neighboring β-galactosidase gene. The enzyme is then synthesized in the cell and catalyzes the hydrolysis of a substrate that has been supplied to the cell, consisting of a sugar linked to a dye molecule. The enzyme hydrolyzes the link between the sugar and the dye, thus releasing the dye and producing a color change in the cell, which can be detected automatically using an **ELISA plate reader**. Therefore, estrogen agonists can be detected by the production of a color change. Estrogen antagonists can be detected if they are able to prevent estrogen from producing a color change.

Testing by NMR spectroscopy

Testing by **nuclear magnetic resonance (NMR) spectroscopy** is a recent innovation used to test whether compounds bind to target proteins. NMR spectroscopy works by irradiating a compound with a short pulse of energy to excite all the nuclei. The nuclei then slowly relax back to the ground state giving off energy as they do so to produce a spectrum. The time taken by different nuclei to relax back to their resting state is called the **relaxation time** and this varies depending on the size of the molecule. Small molecules have long relaxation times whereas large molecules have short relaxation times. Therefore, it is possible to delay the measurement of the energy emission so that only the spectra for small molecules are detected. This is the key to how protein binding is detected.

First of all, the NMR spectrum of the test compound is taken, then the target protein is added and the spectrum is re-run, introducing a delay in the measurement so that the protein signals are not detected. If the test compound fails to bind to the protein, then its NMR spectrum will still be detected. If, on the other hand, the compound *does* bind to the protein, it essentially becomes part of the protein. As a result, its nuclei will have a shorter relaxation time and no NMR spectrum will be detected.

This screening method can be applied to mixtures of compounds resulting from a natural extract or from a combinatorial synthesis. If any of the compounds present bind to the protein, signals due to that compound will disappear from the spectrum. This will show that a component of the mixture is active and identify whether it is worthwhile separating the mixture or not.

D3 TESTING DRUGS *IN VIVO*

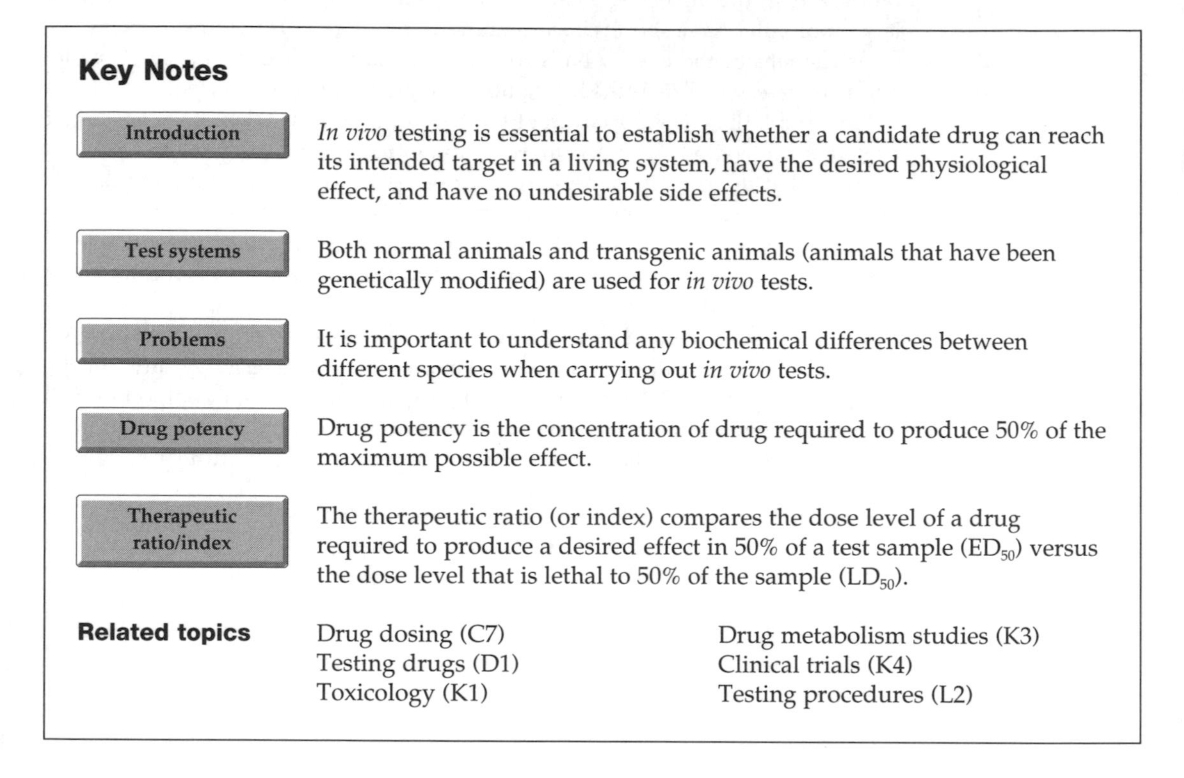

Key Notes

Introduction	*In vivo* testing is essential to establish whether a candidate drug can reach its intended target in a living system, have the desired physiological effect, and have no undesirable side effects.
Test systems	Both normal animals and transgenic animals (animals that have been genetically modified) are used for *in vivo* tests.
Problems	It is important to understand any biochemical differences between different species when carrying out *in vivo* tests.
Drug potency	Drug potency is the concentration of drug required to produce 50% of the maximum possible effect.
Therapeutic ratio/index	The therapeutic ratio (or index) compares the dose level of a drug required to produce a desired effect in 50% of a test sample (ED_{50}) versus the dose level that is lethal to 50% of the sample (LD_{50}).

Related topics

Drug dosing (C7)
Testing drugs (D1)
Toxicology (K1)
Drug metabolism studies (K3)
Clinical trials (K4)
Testing procedures (L2)

Introduction

***In vivo* testing** has always been a source of contention amongst animal rights protesters. As a result, the scientific community has gone a long way to reducing the number of such tests by introducing the *in vitro* tests described in the previous section. However, it is not possible to find a battery of *in vitro* tests that will replace *in vivo* testing altogether. Living systems are extremely complex with a large variety of control and feedback systems, all of which can have an influence on a drug's activity. The living system might find a way of counteracting the physiological effects of a drug, or unexpected physiological effects may occur. These can be unpredictable and dependent on the drug dose used. Moreover, it is difficult to find *in vitro* tests that model the journey a drug has to take through a living system. Finally, *in vitro* tests do not test for the unexpected. Unusual side effects in a living system will only be detected by *in vivo* testing. Therefore, *in vivo* testing is essential to establish whether a candidate drug can reach its intended target in a living system, have the desired physiological effect, and have no undesirable side effects.

Test systems

Some *in vivo* tests can be carried out on normal animals. For example, the effectiveness of **cocaine** as a local anesthetic was tested by dripping the compound into the eye of a frog then touching the cornea with a pin head. **Analgesics** are tested on animals by observing how long it takes them to move away from a heat source.

Not all drugs can be tested on normal animals. In order to test compounds for some disease states, it is necessary to induce that disease in the test animal. For example, in order to test anticancer drugs, tumors have been grafted onto mice. The drug is then injected into the mouse and the animal is studied to see whether the tumor stops growing or regresses (i.e. gets smaller).

Genetic engineering has been used to introduce human genes into test animals so that the animal produces a specific human enzyme or receptor, which will then interact with test compounds. Animals have also been genetically modified so that they are more susceptible to certain diseases (e.g. cancer). Drugs can then be tested to explore their preventative capability. Animals that have been genetically modified in these ways are called **transgenic animals**.

Problems

It is important to ensure that animal tests are as valid as possible to the human situation. This requires an understanding of any differences in biochemistry between animal cells and human cells. For example, **penicillin methyl ester** was found to be an effective antibacterial agent in mice or rats, but was inactive in rabbit, dog or humans. This was because the metabolic enzymes present in mice and rats were able to hydrolyze the methyl ester and release active penicillin (*Fig. 1*). In rabbit, dog and humans, no such metabolic reaction took place.

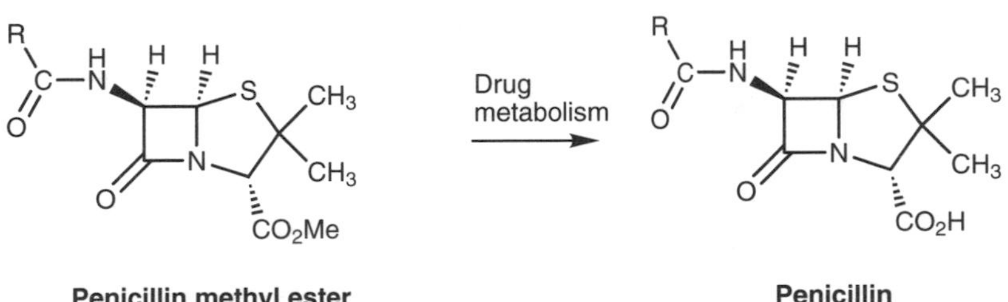

Fig. 1. Metabolism of penicillin methyl ester in mice and rats.

In vivo tests also demonstrate the more complex effects which may arise from administering a drug. For example, some **cholinergic agonists** (agents that mimic the neurotransmitter acetylcholine) have been subjected to *in vitro* tests on various tissues and organs. It can be shown that they dilate blood vessels by relaxing the smooth muscle of the vessel walls. It can also be shown that they decrease heart rate if they are tested on heart muscle. However, if the drugs are given intravenously in small doses to a test animal, they dilate the blood vessels as expected but cause an *increase* in heart rate. This is because the body has many feedback control systems which respond to physiological situations. Small intravenous doses of the cholinergic agonist result in the expected dilation of blood vessels, but this effect is recognized by the body's control systems, and nerve signals are sent to the heart to increase the heart rate.

Drug potency

The **potency** of a drug refers to the amount of drug required to achieve a defined biological effect. The smaller the dose required, the more potent the drug. **Efficacy** is a measure of the maximum biological effect that a drug can produce. It is possible for a drug to be potent (i.e. active in small doses), but have low efficacy. A measure of these properties is possible from dose–response curves, where the dose of a drug is related to the intensity of response (*Fig. 2*). The further to the left the curve is on the x-axis, the more potent the drug. A steep slope in the curve signifies that there is little difference between a dose

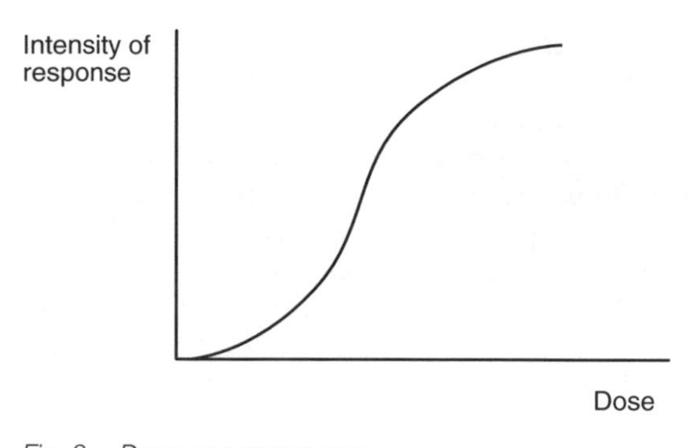

Fig. 2. Dose–response curve.

that produces a minimal effect and one that produces a maximum effect. The efficacy is the maximum point reached by the curve on the y-axis.

Different drugs will have different efficacies. For example, **caffeine** has a lower efficacy as a CNS stimulant than **amphetamine** (*Fig. 3*), while aspirin has a lower efficacy than **morphine** as a painkiller.

It should be noted that the dose level used for a particular drug may not be the dose required to produce the maximum effect, since that dose level may also introduce unacceptable side effects. In these situations, it is important that the drugs show gradual slopes in their dose–response curves to allow a range of therapeutically effective dose levels to be used.

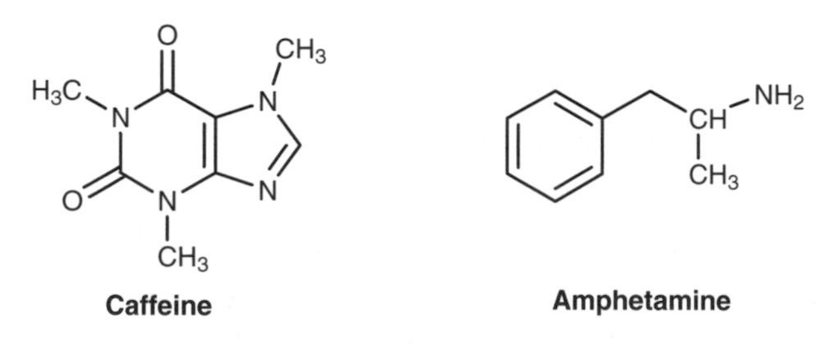

Fig. 3. Caffeine and amphetamine.

Therapeutic ratio/index

The $\mathbf{ED_{50}}$ of a drug is the dose required to produce the desired effect in 50% of test animals. The $\mathbf{LD_{50}}$ is the dose that is lethal to 50% of the animals tested. The ratio of LD_{50} to ED_{50} is known as the **therapeutic ratio** or index. A therapeutic ratio of 10 indicates an LD_{50}:ED_{50} ratio of 10:1. This means that a tenfold increase in the dose corresponding to the ED_{50} would result in a 50% death rate. The dose–response curves for a drug's therapeutic and lethal effects can be compared to determine whether the therapeutic ratio is safe or not (*Fig. 4*). Ideally, the curves should not overlap on the x-axis, which means that the more gradual the two slopes, the riskier the drug will be. The example in *Fig. 4* shows the therapeutic and lethal dose–response curves for a sedative. Here, a 50 mg dose of the drug will act as a sedative for 95% of the test animals, but will kill 5% of the animals. Such a drug would be unacceptable, even though it is effective in 95% of cases treated.

A better measure of a drug's safety is to measure the ratio of the lethal dose for 1% of the population to the effective dose for 99% of the population. A sedative drug with the ratio LD_1:ED_{99} of 1 would be safer than the one shown above.

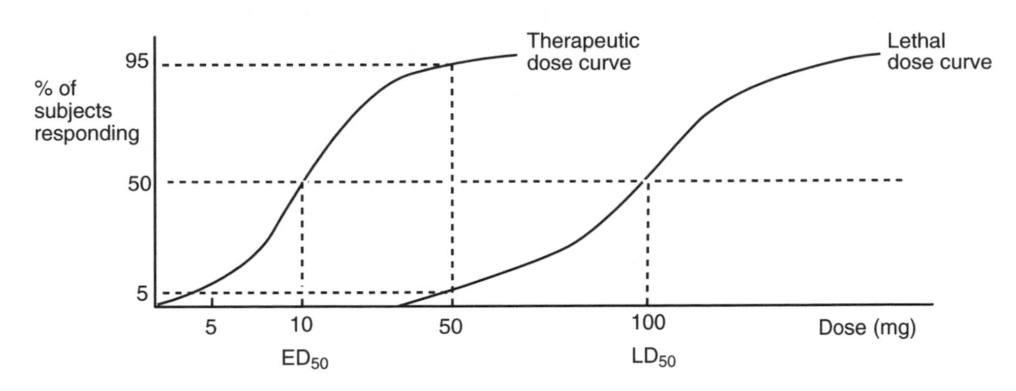

Fig. 4. Comparison of therapeutic and lethal dose curves.

Therapeutic ratios are of interest in animal testing, but they do not take into account any nonlethal toxic side effects. Therefore, the therapeutic dose levels that are eventually used on patients should have a minimum risk of side effects or long term toxicity.

E1 THE LEAD COMPOUND

Key Notes

Definition

A lead compound is the starting point when designing a new drug. The compound should have some desirable property that is likely to be therapeutically useful.

Sources

Lead compounds can be obtained from natural or synthetic sources. They may also be designed using computer modeling or NMR studies.

Searching for lead compounds

Suitable tests are required to search for lead compounds. The tests could be designed to detect physiological or cellular effects, or the binding of the compound with a macromolecular target such as a receptor.

Related topics

From concept to market (A2)
Testing drugs (D1)
Natural sources of lead compounds (E2)
Synthetic sources of lead compounds (E3)
Definition of structure–activity relationships (G1)
Aims of drug design (H1)
From lead compound to dianilino-phthalimides (L3)
Pyrazolopyrimidines (L6)
The antibiotic age (1945–1970s) (M5)
The age of reason (1970s to present) (M6)

Definition

Before any medicinal chemistry project can get underway, a **lead compound** is required. A lead compound will have some property considered therapeutically useful. The property sought will depend on the tests used to detect the lead compound, which in turn depends on the drug's target. The level of biological activity may not be particularly high, but that does not matter. The lead compound is not intended to be used as a clinical agent. It is the starting point from which a clinically useful compound can be developed. Similarly, it does not matter whether the lead compound is toxic or has undesirable side effects. Again, drug design aims to improve the desirable effects of the lead compound and to remove the undesirable effects.

Sources

Lead compounds can be obtained from a variety of different **sources** such as the flora and fauna of the natural world, or synthetic compounds made in the laboratory. There is also the potential of designing lead compounds using computer modeling or NMR spectroscopic studies.

Searching for lead compounds

In order to search for lead compounds, a suitable **test** is required. This could be a test that reveals a physiological effect in a tissue preparation, organ or test animal. Alternatively, it could be a cellular effect, resulting from the interaction of a lead compound with a particular target, such as a receptor or an enzyme; or a molecular effect, such as the binding of a compound with a receptor. In the last two situations, the molecular target is considered important to a particular disease state, and in such cases, the lead compound may not have the desired physiological activity at all! For example, there have been several examples where the natural agonist for a receptor was used as the lead compound in order to design a receptor antagonist. Here, the crucial property for the lead compound was that it should be recognized and bound to the binding site of the target receptor. The lead compound was then modified to bind as an antagonist rather than as an agonist. For example, the chemical messenger **histamine** was used as the lead compound in developing the anti-ulcer agent **cimetidine.** Histamine is an agonist that activates histamine receptors in the stomach wall to increase **gastric acid** release. Cimetidine acts as an antagonist at these receptors, thus reducing the levels of gastric acid released and allowing the body to heal the ulcer.

E2 Natural sources of lead compounds

Key Notes

Introduction

The natural world is a rich source of potential lead compounds. Evolution has resulted in the selection of potent biologically active structures, which serve a variety of purposes in nature. Many such compounds are secondary metabolites, which are produced in mature organisms.

Flora

Plants, trees and bushes are a traditional source of biologically active compounds which can serve as lead compounds, or as medicines in their own right. Historical records often provide clues as to which plants are worth studying. Many plant species in ecologically diverse areas have never been studied at all.

Animals

Venoms and toxins are a source of useful lead compounds since their potency indicates a strong interaction with important molecular targets in the body. A study of those interactions allows the design of useful drugs that will interact with these targets.

Microorganisms

Microorganisms have provided metabolites that are useful antibiotics in their own right, or have been used as lead compounds for the development of other antibacterial agents. Fungal metabolites have also been used as lead compounds in other fields of medicinal chemistry.

Marine chemistry

Marine chemistry is a promising source of new lead compounds for the future. Many biologically active compounds have been isolated from fish, coral, sponges, and marine microorganisms.

Human biochemistry

Neurotransmitters, hormones, and enzyme substrates are all potential lead compounds.

Advantages and disadvantages

Natural sources provide highly diverse and structurally unique lead compounds. However, it takes time to isolate lead compounds from natural sources, and the lead compounds are often complex in structure making them difficult to synthesize.

Related topics

From lead compound to dianilino-phthalimides (L3)
The age of herbs, potions and magic (M1)
The 19th century (M2)
A fledgling science (1900–1930) (M3)
The dawn of the antibacterial age (1930–1945) (M4)
The antibiotic age (1945–1970s) (M5)

Introduction

The **natural world** is particularly rich in potential lead compounds. For example, plants, trees, snakes, lizards, frogs, fungi, corals and fish have all yielded potent lead compounds which have either resulted in clinically useful drugs or have the potential to do so. There is a good reason why nature should be so rich in potential lead compounds. Years of evolution have resulted in the 'selection' of biologically potent natural compounds that have proved useful to the natural host for a variety of reasons. For example, a fungus that produces a toxin can kill off its microbiological competitors and take advantage of available nutrients.

Many of the pharmaceutically active compounds produced in nature are **secondary metabolites**. This means that they are not crucial to the early growth and development of the organism, but are only produced once the organism is mature. Another feature of secondary metabolites is the fact that many of them belong to a class of compounds called **alkaloids** – so called because they contain amine functional groups and are basic (alkaline) in character. This has led to the proposal that secondary metabolites in plants are produced to 'wrap up' excess nitrogen. During active growth, the plant requires a high level of nitrogen to synthesize proteins and nucleic acids. However, at maturity, this requirement for nitrogen drops dramatically. If nitrogen levels started to rise as a result, this could prove potentially toxic. Since plants cannot excrete chemicals in the same way as animals, their method of removing excess nitrogen might be to use up the nitrogen by synthesizing secondary metabolites. On the other hand, many secondary metabolites have remarkably complex structures and the fact that they are often highly potent lends support to the theory that they have a defensive or survival role for the host. Many plants, amphibians, reptiles, microorganisms and fish contain highly potent and toxic chemicals that are used in the chemical warfare waged between organisms.

Flora

Plants, bushes and trees have long been a source of biologically active compounds which can either be used directly in medicine or as lead compounds for the development of other drugs. Indeed, many of the oldest drugs used in medicine have been extracted from plants and trees. For example, the analgesic **morphine** comes from poppies and the anti-malarial compound **quinine** comes from the bark of the cinchona tree. In more recent times, the anticancer drug, **taxol**, was extracted from the yew tree, while the antimalarial drug, **artemisinin** (*Fig. 1*), was extracted from Chinese plants. Although these compounds are useful medicines in their own right, they have also been used as lead compounds in the design of other pharmaceutically useful compounds. One of the most recent discoveries has been a compound called **combretastatin** (*Fig. 1*),

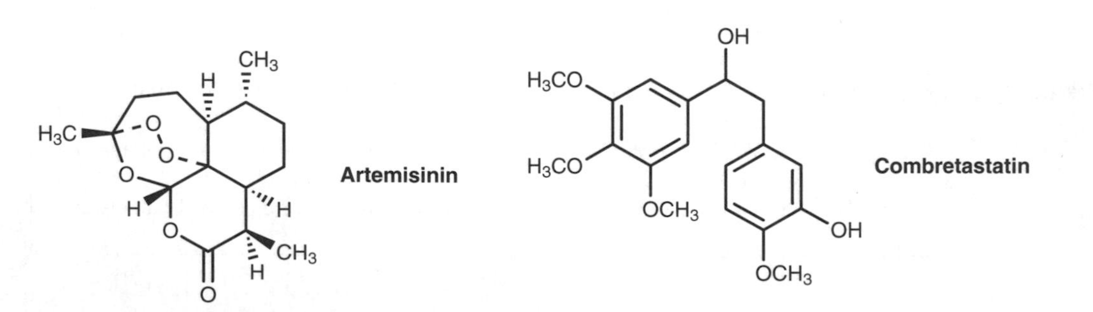

Fig. 1. Lead compounds from flora.

which was isolated from extracts of the African willow tree and has been used as a lead compound for novel anticancer agents that work by inhibiting the blood supply to tumors, thus starving them of nutrients.

The world has a huge variety of different plant species, so it is an awesome task trying to study each one. However, **folk medicine** and the records of ancient civilizations often give useful hints about which plants are most likely to contain useful lead compounds. Nevertheless, other plant species should not be ignored. There are thousands of plant species that have yet to be discovered in ecologically diverse areas such as the rain forests. Therefore, the world's flora still provides huge potential for the discovery of new lead compounds.

Animals

Several interesting drugs have been developed following the realization that **venoms** and **toxins** can act as lead compounds. It may seem strange to consider such lethal poisons as lead compounds for medicines, but the fact that they are so lethal demonstrates that they have a strong interaction with receptors or enzymes present in the body. Thus, they provide excellent lead compounds for the design of drugs that act on those target molecules.

The venoms of snakes and spiders have been particularly useful as lead compounds. The venoms themselves are not particularly useful in medicine since they are highly potent polypeptide structures that are difficult to administer due to their susceptibility to hydrolysis. Nevertheless, an understanding of how they work allows medicinal chemists to design simpler molecules that are easier to synthesize, are more stable in the presence of digestive and metabolic enzymes, and can be administered at dose levels that will have a beneficial effect. For example, the antihypertensive agent, **captopril**, was developed from **teprotide**, which is found in snake venom (*Fig. 2*).

Glu-Trp-Pro-Arg-Pro-Gln-Ile-Pro-Pro

Teprotide

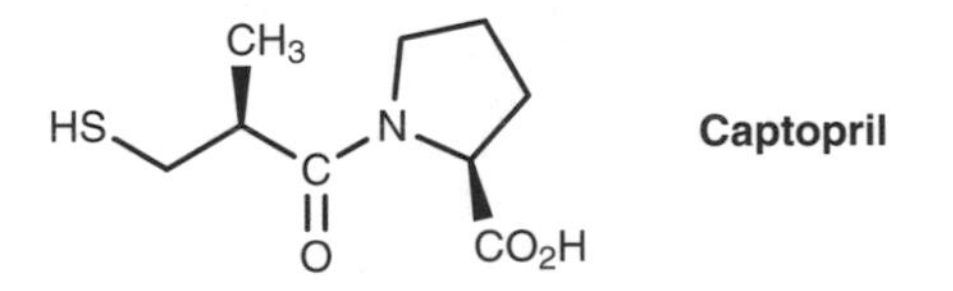

Captopril

Fig. 2. Teprotide and captopril.

Microorganisms

Microorganisms have been a popular source of antibiotics such as **penicillin**, **streptomycin**, **chloramphenicol**, and the **tetracyclines**. These have been the lead compounds for further antibacterial agents, such as the **semi-synthetic penicillins** and **cephalosporins**. Fungi have been a particularly good source of antibiotics. There is a good reason for this. Fungi tend to grow more slowly than bacteria and are therefore at a disadvantage when competing for nutrients. By releasing antibiotics, fungi can inhibit the growth of bacteria so they compete more effectively.

Lead compounds obtained from microorganisms have also been useful in other fields of medicine. For example, **asperlicin** (*Fig. 3*) isolated from *Aspergillus alliaceus* is a lead compound for projects aimed at developing novel anti-anxiety agents. This is because asperlicin has been observed to block a neurotransmitter thought to be involved in panic attacks. Another example is the fungal metabolite **lovastatin** (*Fig. 3*), which was the lead compound for a series of drugs that lower cholesterol levels.

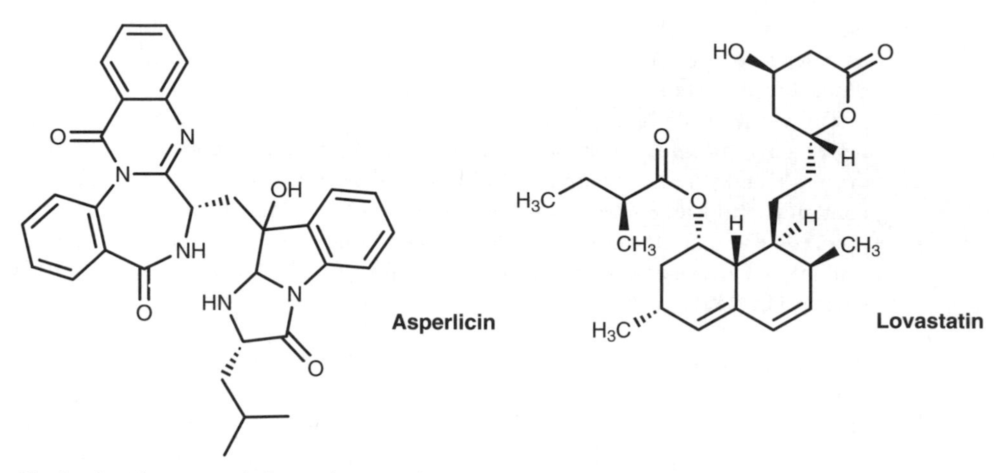

Fig. 3. Lead compounds from microorganisms.

CC-1065 is a metabolite isolated from *Streptomyces zelensis,* which has potent antitumor and antimicrobial activity (*Fig. 4*). It is too toxic to be used itself, but it has been used as a lead compound for other anticancer agents.

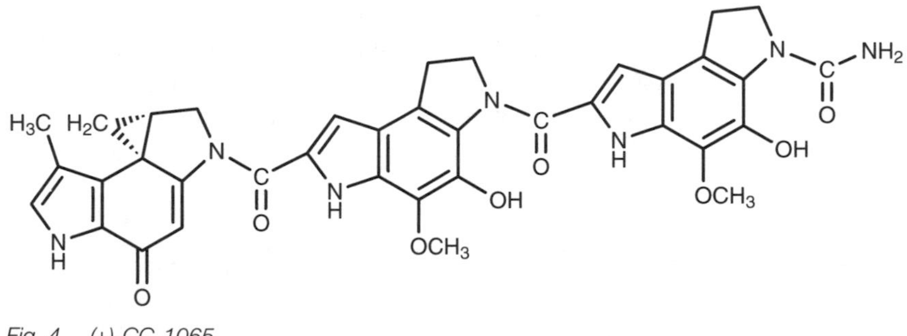

Fig. 4. (+)-CC-1065.

Marine chemistry

In recent years, marine chemistry has yielded a selection of highly potent compounds from a variety of sources, including coral, sponges, fish, jellyfish and marine microorganisms. Many of these have been used as lead compounds in the search for novel antiviral or antitumor drugs. For example, **curacin A** is obtained from a marine cyanobacterium and shows potent antitumor activity (*Fig. 5*).

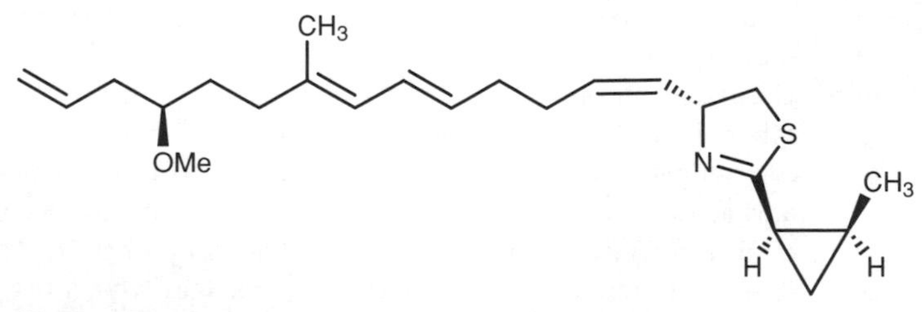

Fig. 5. Curacin A.

To date, no clinical drugs have been developed from marine lead compounds, but this is certain to change in the years to come. It has been estimated that only 5% of the world's oceans have been explored and there is a huge diversity of marine fauna and flora yet to be investigated.

Human biochemistry

An understanding of how the body works at the molecular level can help in identifying lead compounds from natural **biochemicals.** The natural agonists for receptors can be used as lead compounds for synthetic agonists that may demonstrate greater selectivity between receptor types and subtypes than the natural ligand. The natural agonist can also be used as the lead compound for receptor antagonists. As described previously, **epinephrine** and **histamine** were the lead compounds for the development of adrenergic and histamine antagonists respectively. The natural substrates of enzymes can be used as the lead compounds for the design of competitive inhibitors, while the natural compounds responsible for the allosteric control of enzymes could be lead compounds for the design of noncompetitive inhibitors.

Advantages and disadvantages

Through evolution, nature has selected chemicals that are pharmacologically active. Therefore, it is more likely that a search of natural extracts will produce a lead compound than a search of randomly synthesized compounds. It is also more likely that completely novel structures will be thrown up by nature, some of which would never occur to a synthetic chemist.

The disadvantage of looking to natural sources for lead compounds is that it can be a slow process: collecting the natural material, extracting it, then separating and purifying the active compounds. Furthermore, the active compounds are often highly complex in structure making them difficult to synthesize. As a result, it may be necessary to rely on the natural source for the lead compound, which could be wasteful and environmentally unfriendly. Difficulties in synthesis also mean that one is restricted in the number of analogs that can easily be prepared when trying to develop the lead compound.

E3 SYNTHETIC SOURCES OF LEAD COMPOUNDS

Key Notes

Synthesis

Synthetic compounds may prove to be useful lead compounds, regardless of whether they were synthesized with a medicinal aim in mind or not.

Combinatorial synthesis

Combinatorial synthesis involves the automated or semi-automated synthesis of compounds using solid phase synthetic techniques. Far more compounds can be synthesized in a particular time period than by conventional synthetic methods and so the odds of finding a lead compound are increased.

Compound data banks/libraries

Pharmaceutical companies synthesize many thousands of compounds each year. These compounds are stored and can be tested as lead compounds for novel targets. The disadvantage of using such banks is their lack of structural diversity.

Pharmacophore and substructure searches

The pharmacophore defines the relative positions of important atoms or functional groups required for binding to a particular target. Computer searches of existing compounds can identify those that contain the desired pharmacophore and might act as new lead compounds.

Related topics

Combinatorial synthesis (F3)
The pharmacophore (G4)
Pyrazolopyrimidines (L6)
The dawn of the antibacterial age (1930–1945) (M4)

Synthesis

Large numbers of novel structures are synthesized in research laboratories across the world for a diverse range of synthetic projects. These are a potential source of lead compounds, and pharmaceutical companies will often enter into agreements with research teams in order to test their compounds. Many of these structures may have been synthesized in research topics unrelated to medicinal chemistry, but are still potential lead compounds. The history of medicinal chemistry has many examples of lead compounds that were discovered from synthetic projects that had no medicinal objective in mind. For example, **prontosil** was manufactured as a dye, but was the lead compound for the development of the **sulfonamides**.

Combinatorial synthesis

Pharmaceutical companies have their own teams of organic synthetic chemists synthesizing large numbers of structures in their quest for lead compounds. In the past, this could be a long expensive business with no guarantee of success. Compounds were synthesized one at a time by conventional synthetic methods, and the odds of finding a lead compound were increased by synthesizing as many compounds as possible. However, the only way to do that was to employ more 'pairs of hands'. That all changed with the advent of **combinatorial synthesis**. In essence, combinatorial synthesis is the automated or semi-automated synthesis of compounds using **solid phase** synthetic techniques. The reactions are carried out in small scale and so it is possible for one chemist to produce far more compounds in a month than a large team of chemists could do in a year. With the potential to produce hundreds to thousands of new structures in any year, there is now a good chance that combinatorial synthesis will throw up at least one lead compound.

Compound data banks/libraries

Pharmaceutical companies have synthesized vast numbers of compounds and synthetic intermediates over the years. These are stored and are referred to as **compound banks** or **libraries**. Such storage is not just for historical purposes – pharmaceutical companies are constantly coming up with potentially new targets as a result of the various **genome projects** that are mapping the DNA of humans and other organisms. Once a new target is discovered, a search has to be made for a lead compound that will have some affinity for it. Many compounds that have been synthesized in the past may prove to have affinity for these new targets and are worth testing. The advantage in using compound banks is that they are a source of readily available structures that can be quickly tested on a new target. The disadvantage is that they often lack structural diversity since companies often synthesize hundreds and thousands of compounds based on the structure of one lead compound.

Pharmacophore and substructure searches

Novel lead compounds are always in demand, whether they are for a drug target that has just been discovered, or for an established target that has already had several drugs developed to interact with it. In the latter situation, it is common for a pharmaceutical company and its competitors to make literally thousands of analogs of an established lead compound and to investigate that particular structural class of compounds 'to death'. Eventually, there comes a time when further progress becomes difficult. Alternatively, a competitor may have 'wrapped up' that field through its various patents. Under those circumstances, it becomes necessary to find a lead compound having a totally different kind of structure, which will interact with the same target. If such a new structure could be found, it would allow the company to devise novel structures that avoid existing patent restrictions, and that may also have improved properties over the existing groups of compounds.

One way of searching for new structural groups is to carry out a computerized search of chemical data banks. This is done by comparing known drugs that act at a certain receptor and identifying the common **pharmacophore** (i.e. the atoms and functional groups required for binding and activity – see Topic G4). The pharmacophore specifies the relative position of these groups in three dimensions and can be used by a computer to search data banks of known compounds to see if they contain the same pharmacophore. If the compounds identified are structurally different from those drugs currently used in the field, they can be tested to see if they have any activity. If they do, they represent novel lead compounds.

Pharmacophore searching can also be used to search existing compounds for lead compounds against novel targets. If the structure of the target's binding site is known or hypothesized, it is possible to identify potential binding regions based on the types of amino acids present in the binding site. Computer modeling can then be used to create a pharmacophore that would interact with those regions. The pharmacophore can then be used as previously described to find potential lead compounds.

It has to be emphasized that there is no guarantee that this approach will work. Compounds may be discovered that have the correct pharmacophore, but which have no affinity for the target. This is because the molecular skeleton itself may prevent the compound fitting into the intended binding site.

F1 Synthetic considerations

Key Notes

Introduction

Synthetic analogs of a lead compound are prepared to study structure–activity relationships, and to find compounds with improved activity and reduced side effects.

Modifications of the lead compound

Reactions can be carried out on functional groups present in the lead compound. The resulting analogs are useful in identifying which functional groups are important for binding and activity.

Full synthesis

A full synthesis should have the minimum number of steps and allow the use of a large variety of reactants. Full syntheses are not feasible for complex natural products.

Semi-synthetic syntheses

A semi-synthetic synthesis involves the use of a naturally occurring starting material in order to synthesize the lead compound and its analogs. The starting material may be a biosynthetic intermediate or the lead compound itself.

Biosynthesis

An analog of a biosynthetic intermediate may be converted by the biosynthetic pathway to an analog of the lead compound. Alternatively, the growth conditions may be altered to produce analogs. The range of analogs that is possible by biosynthesis is restricted by the substrates accepted by the biosynthetic enzymes.

Related topics

Stereochemistry (F2)
Combinatorial synthesis (F3)
Simplification of complex molecules (H3)
Extra binding interactions (H5)
Patenting and chemical development (J1)
Drug metabolism studies (K3)
From lead compound to dianilino-phthalimides (L3)
The antibiotic age (1945–1970s) (M5)

Introduction

Once a lead compound has been discovered, it is important to synthesize as many analogs of it as possible. This is important for two reasons. First, studying the activity of a large range of analogs allows the identification of the functional groups that are important in binding the lead compound to its target and for activity. Such a study is known as a **structure–activity relationship** (SAR) study (Section G). Second, analogs of the lead compound may show improved activity or reduced side effects pointing the way for further drug development.

Modifications of the lead compound

It is usually possible to carry out reactions on the lead compound itself in order to make a series of analogs. This has been a popular strategy for over a century, and was the method by which **aspirin** and **heroin** were synthesized (*Fig. 1*).

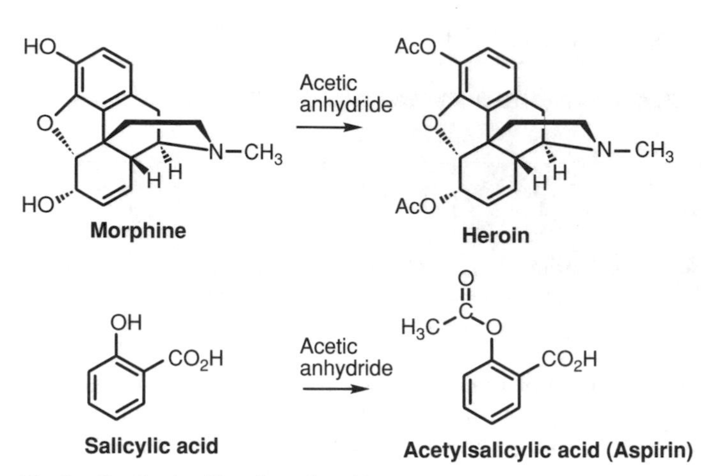

Fig. 1. Synthesis of heroin and aspirin.

The advantages in modifying the lead compound are that it can give access to analogs relatively quickly. The fact that these reactions are carried out on functional groups present in the lead compound also establishes whether these groups are important for binding and activity. The disadvantage is that the range of analogs is limited by the functional groups that are present. Furthermore, the lead compound itself may not be readily available, especially if it is from a natural source.

The sort of **functional groups** that are commonly found in lead compounds include alcohols, phenols, amines, aromatic/heteroaromatic rings, carboxylic acids, alkenes, and ketones. Knowledge of the typical reactions that can be carried out on such functional groups allows the synthesis of various analogs (*Fig. 2*). Some of the most common reactions carried out on lead compounds are

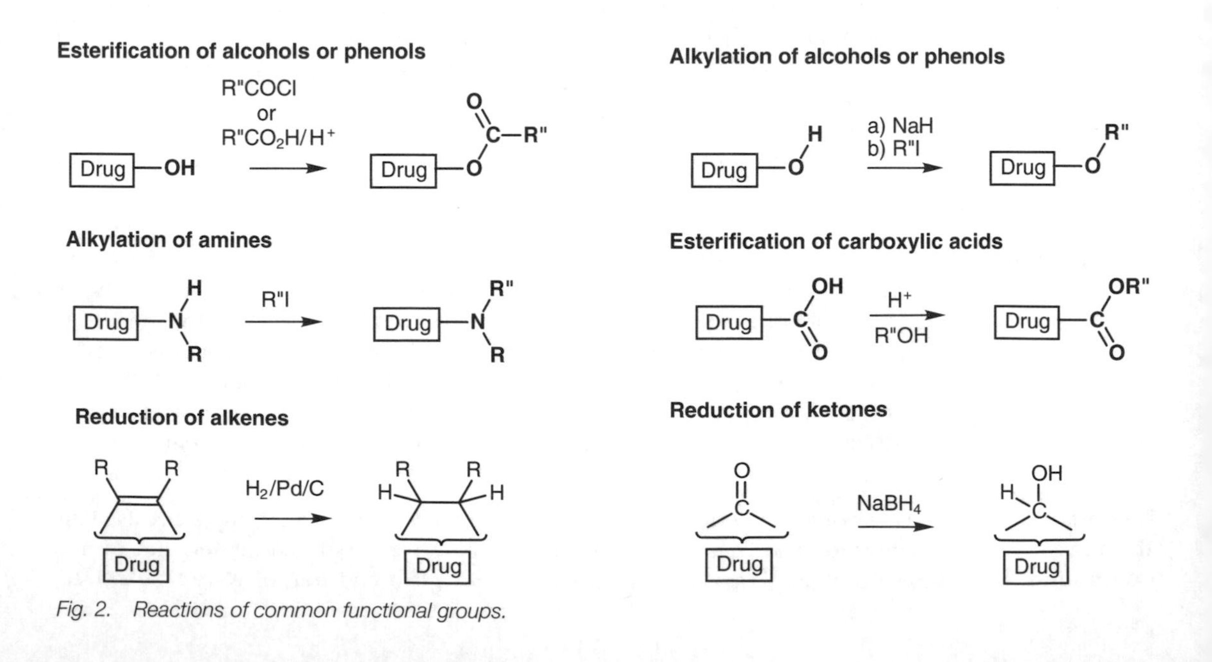

Fig. 2. Reactions of common functional groups.

the **esterification** of alcohols or phenols, and the **alkylation** of amines. When attempting these reactions it is important to consider the other functional groups present. It may be necessary to protect such groups to prevent them reacting as well. In some cases, it may not be possible to carry out a reaction if the reaction conditions are too vigorous and are likely to degrade the molecule. For example, the conditions required to reduce an aromatic ring may be too vigorous.

Full synthesis

A **full synthesis** involves synthesizing the lead compound and its analogs from simple starting materials. Ideally, the synthesis should have a minimum number of steps. It should also allow the use of a wide variety of reactants, so as many different analogs can be synthesized as possible. For example, the synthesis of the β-blocker **propranolol** is ideal in the sense that it only involves two synthetic steps from simple starting materials (*Fig. 3*).

The two-step synthesis also allows the use of virtually any phenol or amine to produce a massive range of analogs (*Fig. 4*).

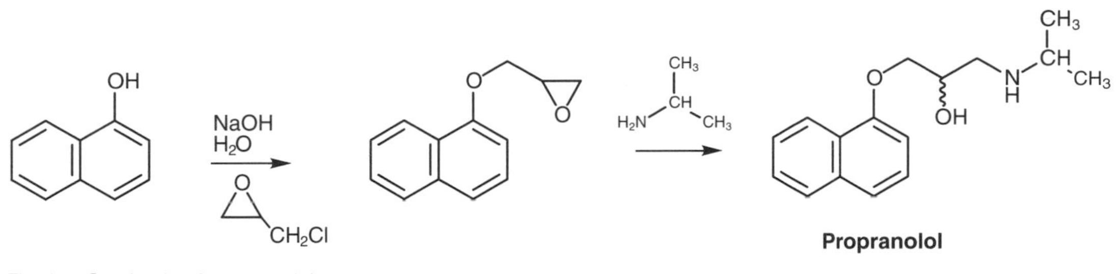

Fig. 3. Synthesis of propranolol.

Full syntheses are feasible if the structure of the lead compound is relatively simple. Indeed, a full synthesis may have been used to obtain the lead compound in the first place. However, full syntheses are often impractical for the complex structures which are often extracted from natural sources, and which may contain several functional groups and complex ring systems. Therefore, the full synthesis of a complex natural product may prove too time-consuming, too costly or low yielding to be practical. In such cases, the methods below may need to be employed.

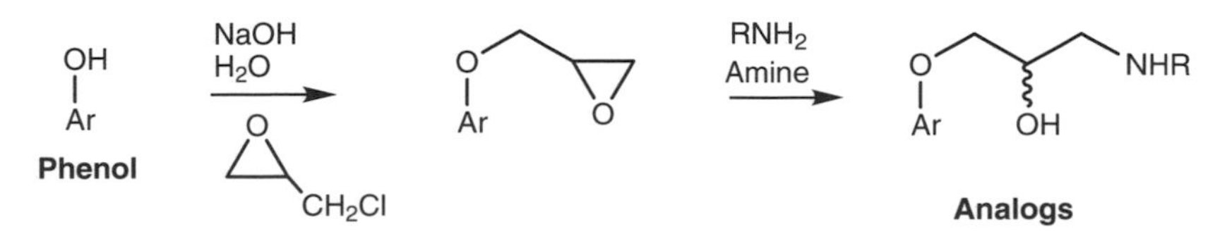

Fig. 4. Synthesis of propranolol analogs.

Semi-synthetic syntheses

Semi-synthetic syntheses involve the synthesis of the lead compound and/or its analogs from a naturally occurring compound. There are two approaches to this. First, it may be possible to split a group off the lead compound and then add different groups in its place. For example, **morphine** can be demethylated using **vinyloxycarbonyl chloride** (VOC-Cl) (*Fig. 5*). The resulting secondary amine can then by alkylated with a variety of different alkyl halides.

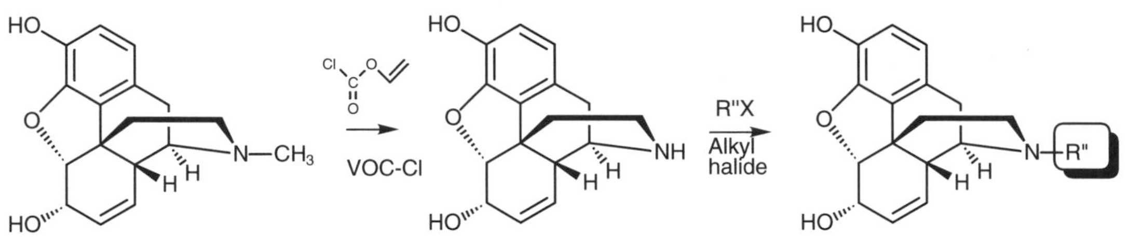

Fig. 5. Synthesis of semi-synthetic morphine analogs.

The acyl side chain of **penicillin G** can be cleaved using enzymatic methods (*Fig. 6*). The product obtained can then be esterified with different acid chlorides to produce a large range of **semi-synthetic penicillins**.

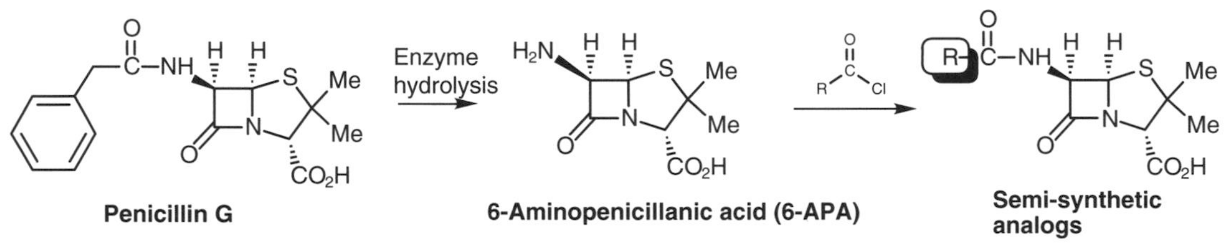

Fig. 6. Synthesis of semi-synthetic penicillins.

The second approach is to harvest and isolate a biosynthetic precursor of the lead compound from the original natural source. For example, **6-aminopenicillanic acid** (6-APA) is a biosynthetic intermediate for penicillin. It is possible to harvest this compound by fermentation and then react it with different acid chlorides as described above. A more recent example is that of **taxol** (*Fig. 7*). It is prepared by extracting **10-deacetylbaccatin** III from the needles of the yew tree,

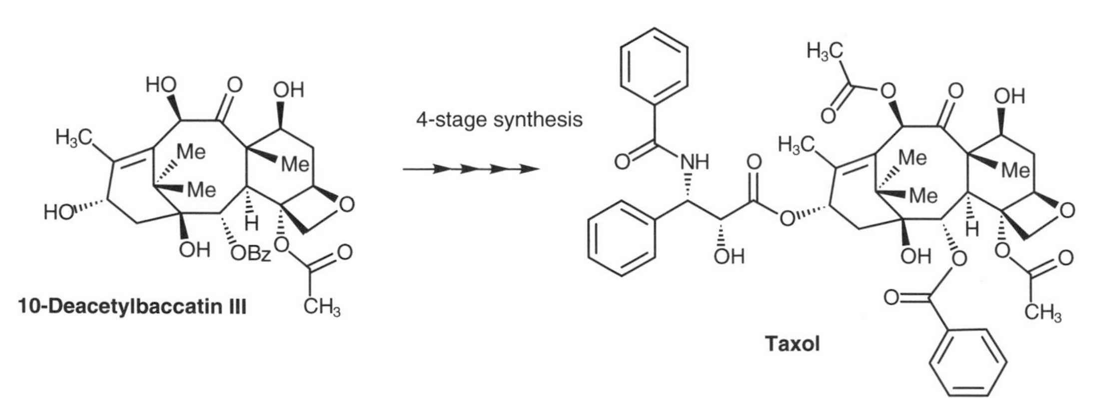

Fig. 7. Semisynthetic synthesis of taxol.

then carrying out a four-stage synthesis. Different analogs can be synthesized by varying the reagents used in the four stages.

Biosynthesis

In certain situations it may be possible to prepare analogs of a lead compound by letting an organism do all the work. If the biosynthetic route to the natural compound is known, an analog of a biosynthetic intermediate can be synthesized and fed to the organism in the hope that it will be converted to an analog. For example, **gliotoxin** (*Fig. 8*) is a fungal metabolite with antitumor activity. One of its biosynthetic intermediates is a cyclic dipeptide involving the amino acids L-phenylalanine and L-serine. By feeding the fungus with a cyclic dipeptide consisting of L-phenylalanine and L-alanine, it was possible to isolate an analog of gliotoxin where the hydroxymethylene group was replaced with a methyl group.

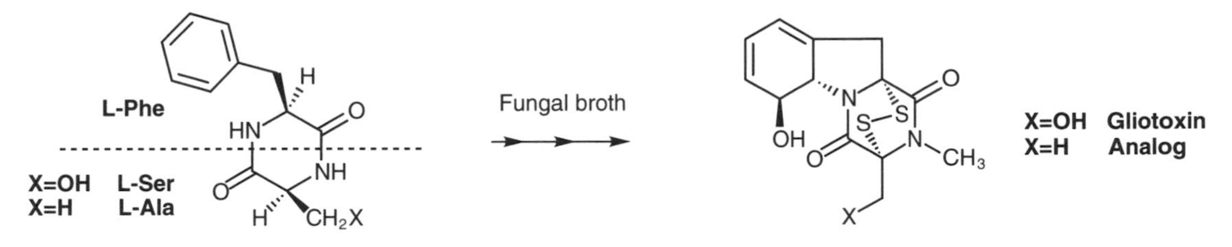

Fig. 8. Biosynthetic synthesis of gliotoxin analog.

Alternatively, changing the growth conditions can result in production of analogs. For example, the first analogs of penicillin were prepared by adding different carboxylic acids to the fermentation medium. Adding phenylacetic acid ($PhCH_2CO_2H$) resulted in the biosynthesis of benzyl penicillin (**Penicillin G**), whereas the addition of phenoxyacetic acid ($PhOCH_2CO_2H$) resulted in the formation of phenoxymethyl penicillin (**Penicillin V**) (*Fig. 9*).

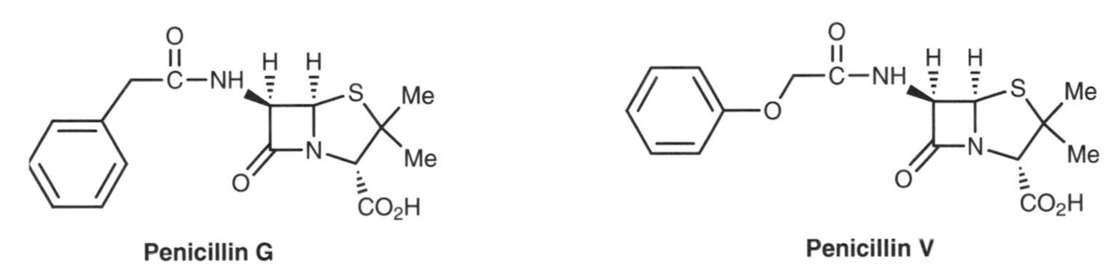

Fig. 9. Penicillins G and V.

The disadvantage in using an organism to carry out the biosynthesis of analogs is that the range of analogs is severely restricted by whatever substrates the biosynthetic enzymes will accept. For example, in penicillin biosynthesis only carboxylic acids of the general formula $ArCH_2CO_2H$ are accepted. Therefore, it is impossible to produce penicillins with a branched side chain using carboxylic acids of general formula $ArCH(X)CO_2H$. Since many of the most effective penicillins used today have branched side chains (e.g. **ampicillin** and **amoxycillin**), they would never have been discovered by this method of synthesis.

F2 STEREOCHEMISTRY

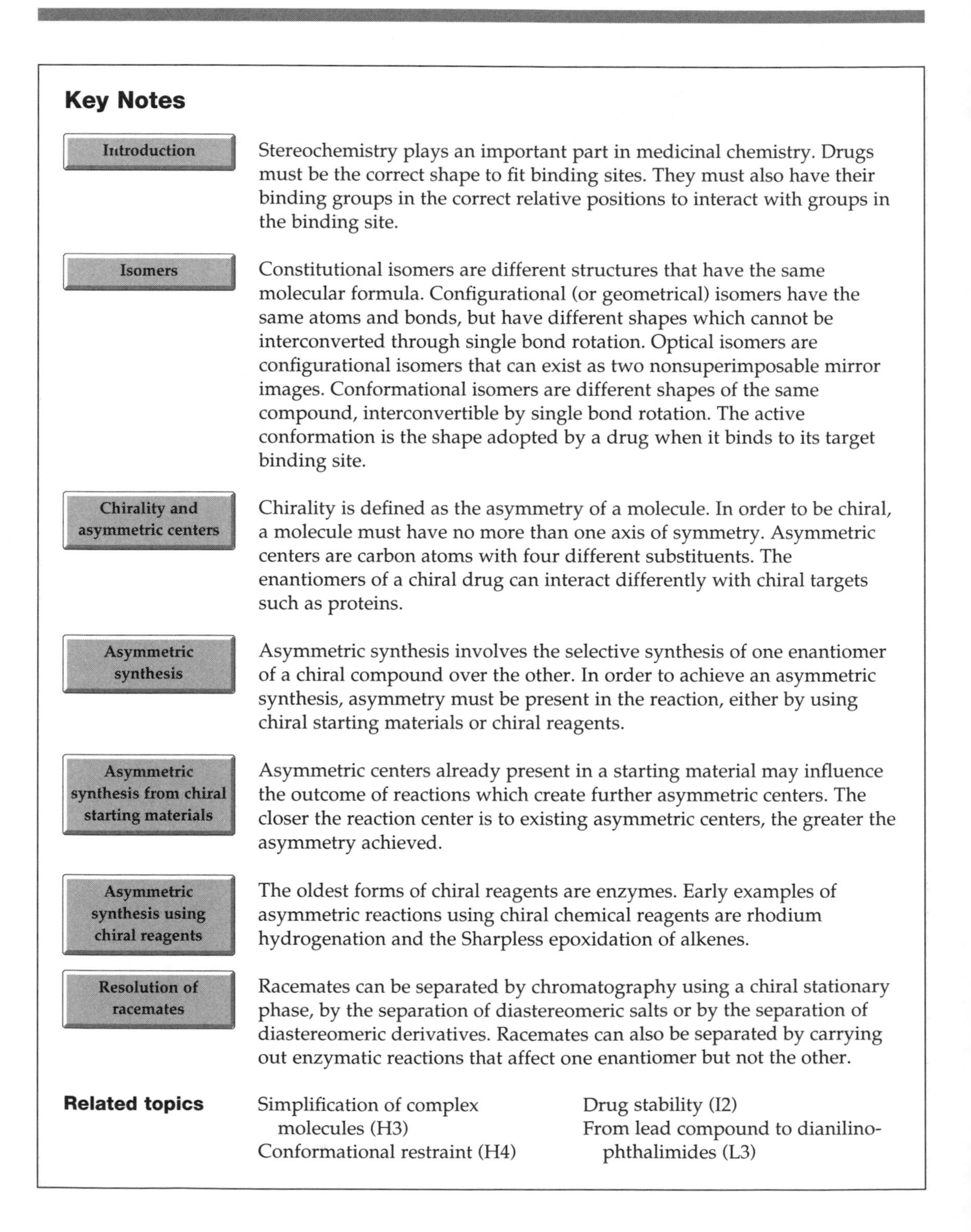

Key Notes

Introduction

Stereochemistry plays an important part in medicinal chemistry. Drugs must be the correct shape to fit binding sites. They must also have their binding groups in the correct relative positions to interact with groups in the binding site.

Isomers

Constitutional isomers are different structures that have the same molecular formula. Configurational (or geometrical) isomers have the same atoms and bonds, but have different shapes which cannot be interconverted through single bond rotation. Optical isomers are configurational isomers that can exist as two nonsuperimposable mirror images. Conformational isomers are different shapes of the same compound, interconvertible by single bond rotation. The active conformation is the shape adopted by a drug when it binds to its target binding site.

Chirality and asymmetric centers

Chirality is defined as the asymmetry of a molecule. In order to be chiral, a molecule must have no more than one axis of symmetry. Asymmetric centers are carbon atoms with four different substituents. The enantiomers of a chiral drug can interact differently with chiral targets such as proteins.

Asymmetric synthesis

Asymmetric synthesis involves the selective synthesis of one enantiomer of a chiral compound over the other. In order to achieve an asymmetric synthesis, asymmetry must be present in the reaction, either by using chiral starting materials or chiral reagents.

Asymmetric synthesis from chiral starting materials

Asymmetric centers already present in a starting material may influence the outcome of reactions which create further asymmetric centers. The closer the reaction center is to existing asymmetric centers, the greater the asymmetry achieved.

Asymmetric synthesis using chiral reagents

The oldest forms of chiral reagents are enzymes. Early examples of asymmetric reactions using chiral chemical reagents are rhodium hydrogenation and the Sharpless epoxidation of alkenes.

Resolution of racemates

Racemates can be separated by chromatography using a chiral stationary phase, by the separation of diastereomeric salts or by the separation of diastereomeric derivatives. Racemates can also be separated by carrying out enzymatic reactions that affect one enantiomer but not the other.

Related topics

Simplification of complex molecules (H3)
Conformational restraint (H4)
Drug stability (I2)
From lead compound to dianilinophthalimides (L3)

Introduction

Stereochemistry refers to the shape of molecules and is important in understanding how drugs interact with their molecular targets. Clearly, potential drugs must have the correct shape if they are to fit into binding sites. The functional groups involved in binding the drug to the binding site must also be positioned correctly relative to each other, so that they can bind simultaneously.

Isomers

Isomers are classed as being constitutional, configurational and conformational. **Constitutional isomers** have identical molecular formulae, but the atoms are linked together by different bonds. For example, cyclohexane is a constitutional isomer of 1-hexene (*Fig. 1*).

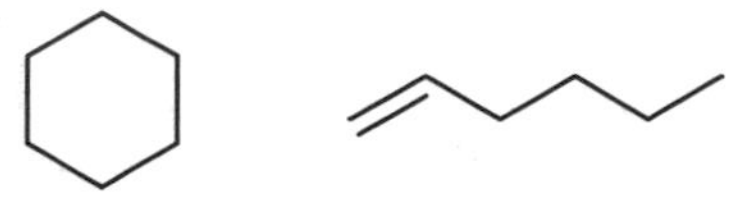

Fig. 1. Constitutional isomers.

Configurational isomers are structures that have the same atoms and bonds. However, the shapes of the molecules are different and cannot be interconverted without breaking bonds. Such molecules are also called **geometric isomers**. The *cis* and *trans* isomers of disubstituted alkenes or cycloalkanes are configurational isomers having different chemical and physical properties (*Fig. 2*).

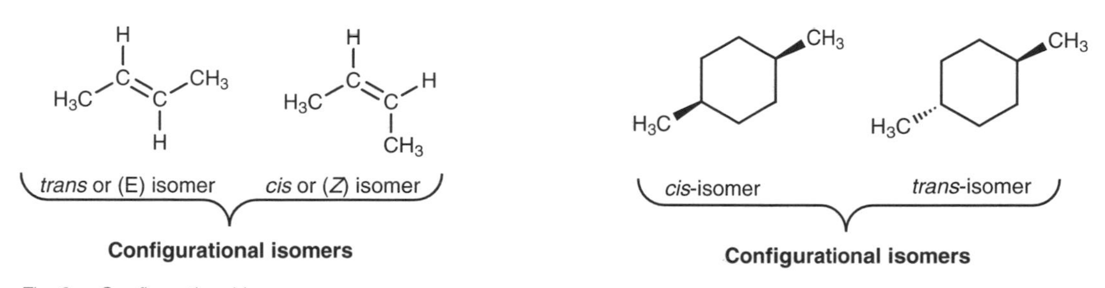

Fig. 2. Configurational isomers.

Another kind of configurational isomerism is **optical isomerism**. Optical isomers are compounds that can exist as two nonsuperimposable mirror images (e.g. lactic acid) (*Fig. 3*).

Conformational isomers are different shapes of the same compound arising from rotation around single bonds. Conformational isomers cannot usually be

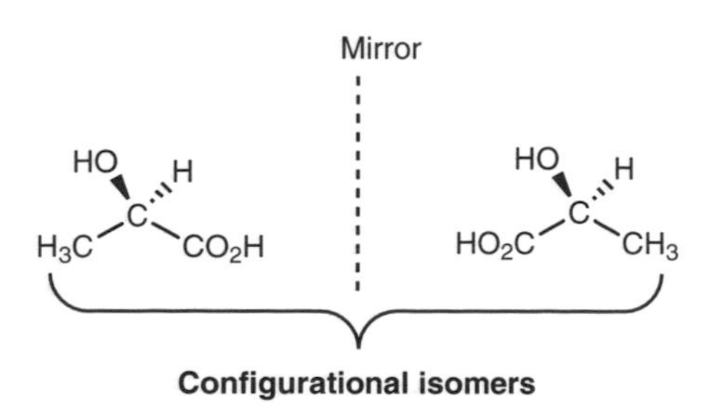

Fig. 3. Optical isomerism.

isolated. However, the percentage of molecules present in one conformation relative to another will depend on the relative stabilities of these conformations. This in turn depends on the various steric and electronic interactions that take place between groups. For example, two possible conformations for methylcyclohexane are where the methyl group is **axial** or **equatorial** (*Fig. 4*). The more stable conformation is the equatorial one, since the methyl group experiences steric and electronic repulsions when it is in the axial position. It can be calculated that 95% of methylcyclohexane molecules adopt the equatorial conformation at any one time, compared with 5% for the axial conformation.

Drugs have to adopt a particular shape or conformation in order to bind to their target binding sites. This is known as the **active conformation**, where all the drug's binding groups are in the correct relative positions to interact with the binding site. It is important to appreciate that the active conformation is not necessarily the most stable conformation. The energy lost in having to adopt a less favorable conformation can be offset by the energy gained by the binding interactions with the target binding site.

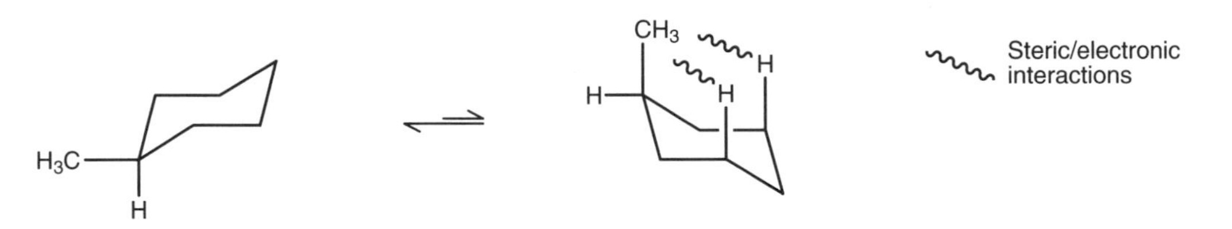

Equitorial conformation **Axial conformation**

Fig. 4. Conformational isomers of cyclohexane.

Chirality and asymmetric centers

Optical isomers are compounds that can exist as two nonsuperimposable mirror images. This property is related to the lack of symmetry (**asymmetry**) present in the molecule, and is known as **chirality**. Thus, asymmetric molecules are defined as **chiral** molecules, whereas symmetrical molecules are defined as **achiral** molecules. To be precise, a molecule must have no more than one axis of symmetry in order to be chiral.

Identifying the overall symmetry or asymmetry of a molecule is not an easy task. However, in most cases, the task is made easier by identifying what is known as an **asymmetric carbon center**. This is a carbon atom having four different substituents. For example, the asymmetric center for lactic acid is defined in *Fig. 5*. The presence of an asymmetric center usually means that the molecule is chiral.

The two mirror images of a chiral molecule such as lactic acid are defined as **enantiomers**. They have identical chemical and physical properties with two

HO H C H3C CO2H — Asymmetric center

Lactic acid

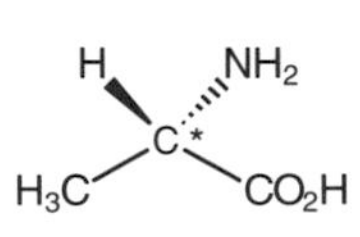

L-alanine

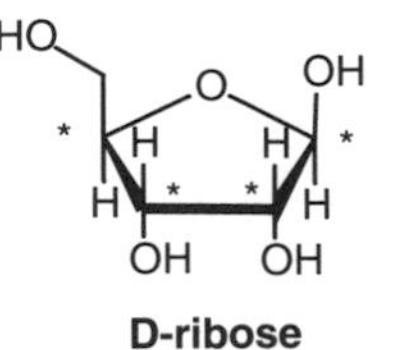

D-ribose

* Asymmetric center

Fig. 5. Asymmetric centers.

exceptions. First of all, solutions of each enantiomer will rotate plane polarized light in equal but opposite directions. Second, and more importantly, each enantiomer can interact differently with other chiral molecules. This is relevant to drugs since they interact with chiral targets. Enzymes and receptors are chiral because they are made up of chiral amino acids (e.g. L-alanine) (*Fig. 5*) (glycine is the exception to the rule). Nucleic acids are chiral as the sugars in the sugar–phosphate backbone are chiral (e.g. D-ribose) (*Fig. 5*).

The reason why enantiomers interact differently with chiral targets can be explained by considering how **lactic acid** binds to the binding site for **lactate dehydrogenase** (*Fig. 6*). Assuming that three interactions are responsible for binding, it can be seen that one enantiomer is able to form all three binding interactions simultaneously, whereas the other enantiomer cannot. As a result, one enantiomer would undergo the enzymatic reaction and the other would not. Similarly, one enantiomer of a chiral enzyme inhibitor could bind and inhibit the enzyme, whereas the other would not. The same principle holds true for the interactions of drugs with receptor binding sites.

The term **eutomer** is used to define the more active enantiomer of a chiral drug, while the term **distomer** is used for the less active enantiomer. The **eudismic ratio** is a measure of how the eutomer and distomer differ in their activities. It has often been observed that chiral drugs that are effective in low doses have the greatest eudismic ratio (**Pfeiffer's rule**).

$$\text{Eudismic ratio} = \left[\frac{\text{Activity of eutomer}}{\text{Activity of distomer}}\right]$$

Since different enantiomers can have different activities, it makes sense to synthesize only the more active enantiomer. However, the conventional synthetic reagents used in organic chemistry are not able to distinguish between the two enantiomers of a chiral compound, and so a conventional synthesis will give a mixture of both enantiomers – a **racemate**. For example, the two-stage synthesis of **propranolol** (Topic F1), gives a racemic product. At best, this is wasteful of reagents and chemicals as half the product may be inactive. However, there is a more serious matter. The wrong enantiomer may not be able to interact with the intended target, but it may interact with a totally different target leading to undesirable or toxic side effects. It is therefore better to use the single enantiomer if at all possible. (Having said that it is not always beneficial to use the more active enantiomer. For example, the more active enantiomer of Prozac was found to be less beneficial than the racemic mixture.)

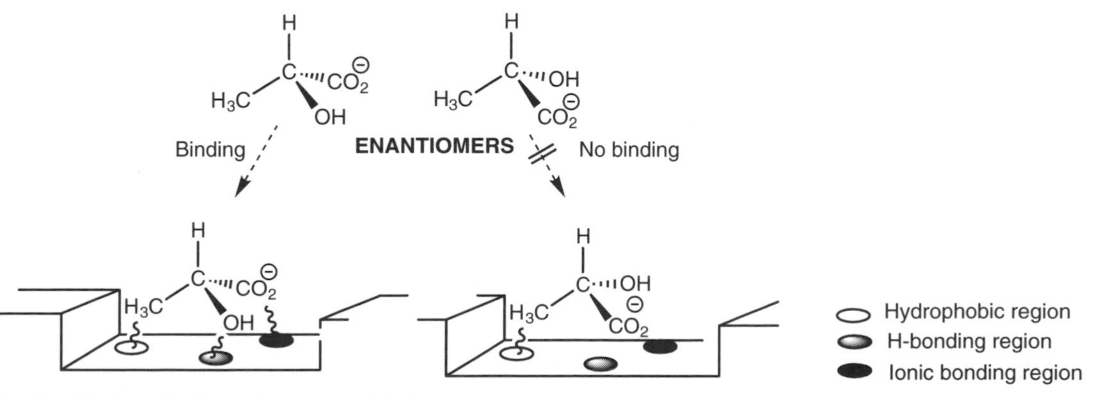

Fig. 6. Binding of lactic acid to lactate dehydrogenase.

There are two approaches to obtaining single enantiomers. One can either synthesize a racemic mixture of the drug using conventional methods, then separate the enantiomers, or one can use reagents that will produce one enantiomer in preference to the other – a process known as asymmetric synthesis.

Asymmetric synthesis

Asymmetric synthesis involves the selective synthesis of one enantiomer of a chiral product. In order to achieve this, there must be an element of asymmetry present in the reaction. This means that either the starting material or the reagents used in the reaction must be chiral. Asymmetric syntheses are usually longer than conventional syntheses and are more demanding since it is necessary to avoid conditions that are likely to cause racemization (e.g. strong heat or the presence of strong base). The criteria for a good asymmetric synthesis is that the final product should be obtained in both a high chemical yield and a high **optical yield**. The latter is a measure of how **enantioselective** the reaction has been and is also called the **enantiomeric excess** (e.e.). For example, an optical yield of 90% means that 90% of the product is a single enantiomer, while 10% is a racemate. This corresponds to an enantiomeric ratio of 95:5 for each enantiomer. Ideally, reactions should have an enantiomeric excess which is greater than 98%. Another requirement of a good asymmetric synthesis is that any expensive chiral reagents used in the reaction must be present in catalytic amounts, or else recoverable with high yield and high purity so that they can be recycled.

Asymmetric synthesis from chiral starting materials

If the starting material is chiral, then the asymmetry already present can often influence a reaction which creates a further asymmetric center, such that one configuration is preferred over the other. A case in point is the Grignard reaction shown in *Fig. 7*, which is carried out as part of a synthesis leading to a

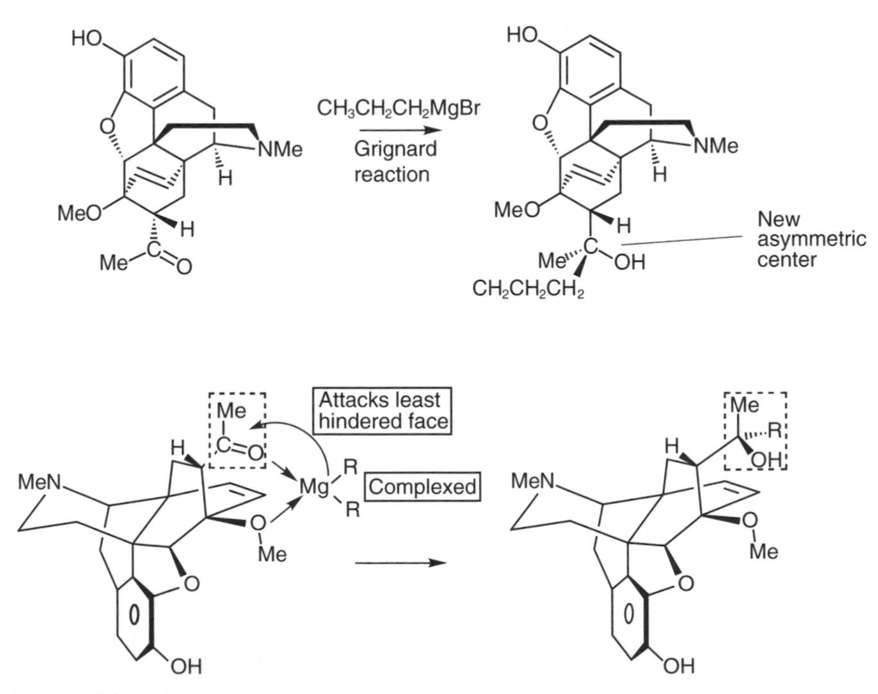

Fig. 7. Grignard reaction.

sedative called **etorphine**. The new asymmetric center resulting from this reaction has the configuration shown. This can be explained by the asymmetric magnesium complex, which is formed during the Grignard reaction prior to the alkyl group being transferred to the carbonyl center.

Successful asymmetric syntheses have been devised by starting with an easily available chiral starting material such as a carbohydrate. However, there are disadvantages in this approach since such starting materials are rarely obvious starting materials for the target compound and extra synthetic steps are required. For example, the asymmetric synthesis of **propranolol** is possible using a chiral sugar, but requires more steps than the two-stage racemic synthesis mentioned in Topic F1 (*Fig. 8*).

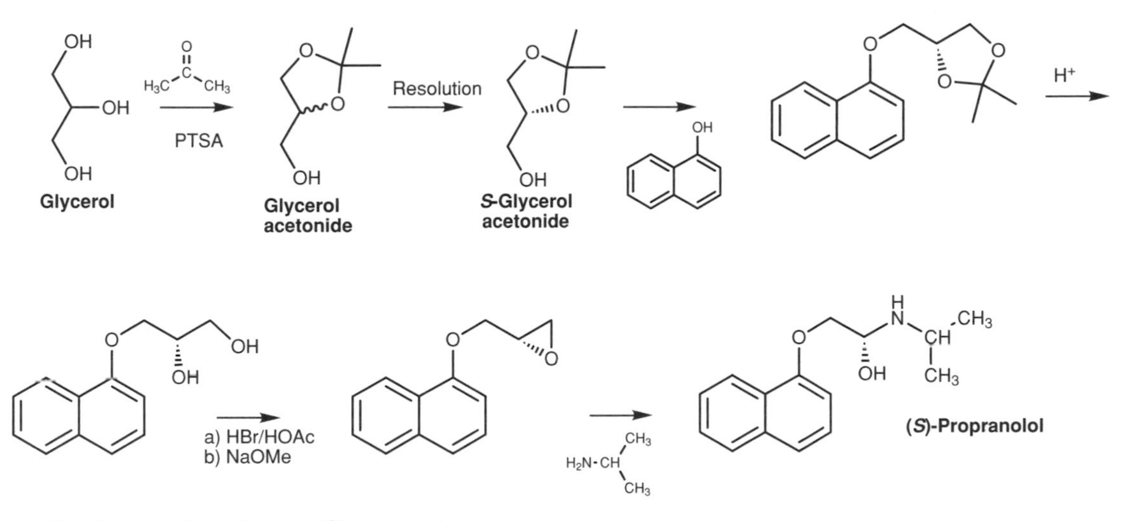

Fig. 8. Asymmetric synthesis of (S)-propranolol.

The presence of chirality in a molecule does not necessarily result in an asymmetric reaction elsewhere in the structure. In general, asymmetry is induced when a new asymmetric center is created close to the one that is already present. If the new center is some distance from the existing asymmetric center, then the influence will be slight.

Asymmetric synthesis using chiral reagents

Chiral reagents can be used to carry out asymmetric reactions on achiral molecules. The oldest chiral reagents in organic synthesis are **enzymes**. The reactions catalyzed by enzymes are likely to give high optical purities, in many cases 100%. More recently, asymmetric chemical reagents have been developed which favor the formation of one enantiomer over another. Two early examples are the **rhodium-catalyzed hydrogenation** of alkenes and the **Sharpless epoxidation** of alkenes (*Fig. 9*).

In both cases, a **metal template** is involved (Rh and Ti respectively) which binds the starting material and the various reagents into an organometallic complex. Chiral molecules present in each reaction also bind to the metal

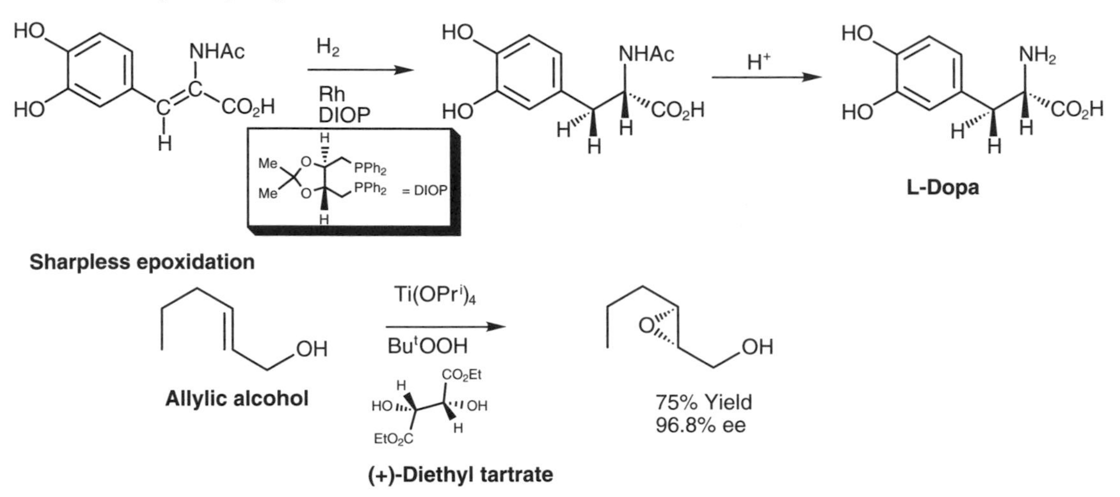

Fig. 9. Asymmetric syntheses with chiral reagents.

template, making the complex chiral. In the hydrogenation reaction, the chirality is introduced by a bidentate chiral ligand (**DIOP**). In the epoxidation reaction, the chiral molecule is (+)-**diethyl tartrate**. Neither the bidentate ligand in the hydrogenation reaction nor the diethyl tartrate in the epoxidation reaction undergo any form of reaction and are purely there to introduce the asymmetry required. Since they are unaffected by the reaction, they are known as **chiral auxiliaries**.

Resolution of racemates

The synthesis of a chiral product using achiral starting materials and reagents will produce a **racemic mixture**. However, it is possible to separate the enantiomers by a process known as **resolution**.

Preferential crystallization involves the crystallization of one enantiomer in preference to the other by seeding a supersaturated solution of the racemate with the desired enantiomer. Sometimes seeds of each enantiomer can be put in different locations on the same crystallizing dish allowing the crystallization of both enantiomers. By using seeds where one enantiomer forms small crystals and the other large crystals, the final crystals can be sifted to separate the enantiomers as the two will differ in crystal size. This method works only for racemates that form **conglomerates** (i.e. where the racemate normally crystallizes to give crystals made up solely of one enantiomer or the other). Only about 10–20% of organic compounds form conglomerates.

Chromatography can be used to separate enantiomers if a chiral compound is linked to the silica support (*Fig. 10*). For example, one could covalently link L-amino acids to the silica or impregnate the silica with the single enantiomers of chiral acids such as tartaric, malic or camphorsulfonic acid. The racemate is passed down the chiral column and each enantiomer forms a reversible complex with the chiral compound on the stationary phase. This complex is **diastereomeric** rather than enantiomeric since there is more than one asymmetric center present. Since the complexes are not mirror images, they will have different chemical and physical properties. This means that the complexation/decomplexation rate of each enantiomer will differ as the racemate passes down the column, resulting in one enantiomer traveling faster than the other.

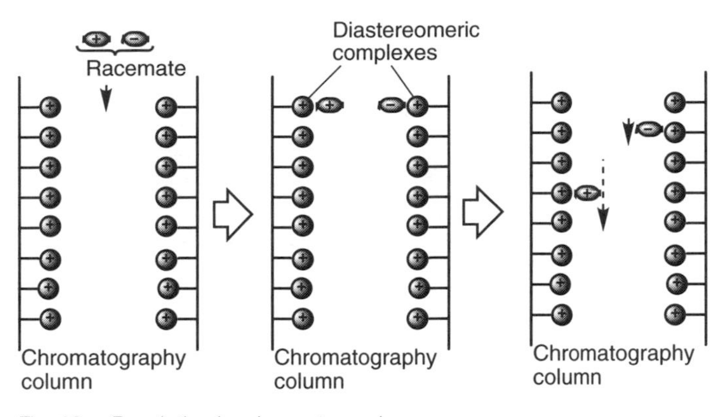

Fig. 10. Resolution by chromatography.

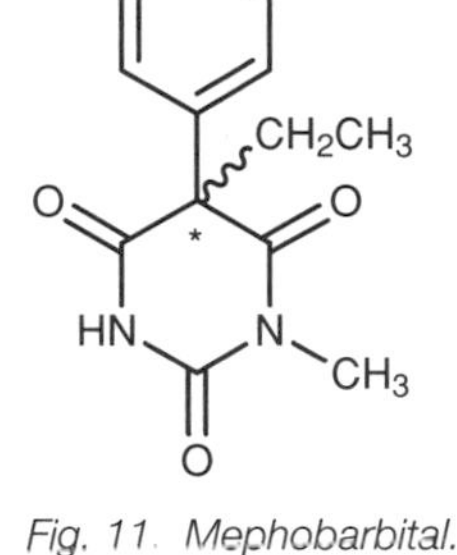

Fig. 11. Mephobarbital.

Another method is to use a naturally occurring chiral polymer as the stationary phase. For example, a racemic mixture of **mephobarbital** (*Fig. 11*) was resolved by passing it through a cellulose column.

The classical method of resolving racemates is to form **diastereomeric derivatives** of each enantiomer. Since diastereomers have different properties, they can be separated by conventional methods such as crystallization or distillation. After separation, the derivative can be hydrolyzed or neutralized to recover the pure enantiomer desired.

For example, a chiral carboxylic acid can be treated with the single enantiomer of an optically active amine such as quinine (*Fig. 12*). A diastereomeric salt is formed which can be separated by crystallization. Once the diastereomers have been formed, the pure enantiomers are recovered by treating the salt with dilute hydrochloric acid to remove the basic quinine.

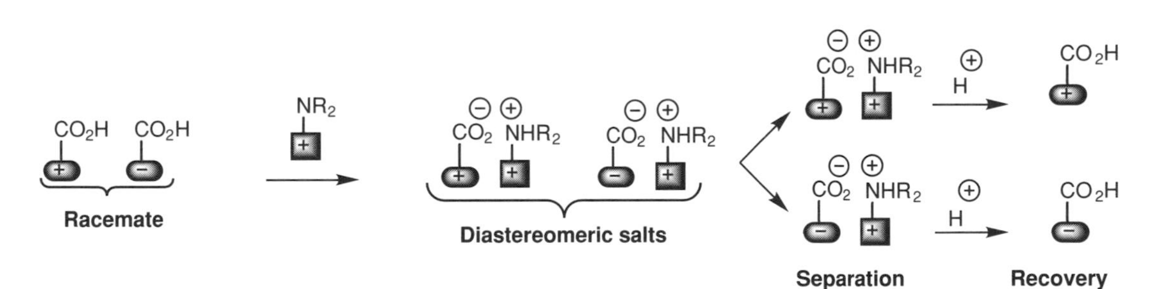

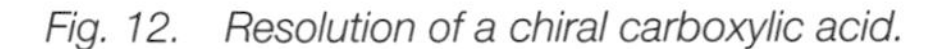
Fig. 12. Resolution of a chiral carboxylic acid.

A chiral amine can be treated with the single enantiomer of a chiral carboxylic acid, such as tartaric acid or malic acid, to form a diastereomeric salt, and similarly separated. Treatment with base then removes the acid to recover the desired amine.

If the chiral compound lacks a carboxylic acid or amine group, it is possible to form a diastereomeric derivative by temporarily linking a chiral compound to another functional group. For example, if the racemate has an alcohol present it could be treated with the single enantiomer of camphorsulfonyl chloride to give a camphorsulfonate derivative (*Fig. 13*). The diastereomeric derivative is separated then hydrolyzed.

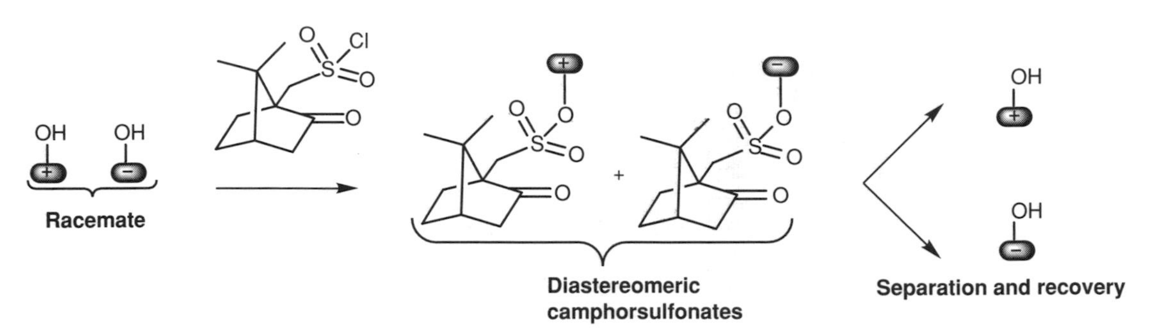

Fig. 13. Resolution of a chiral alcohol.

Kinetic resolution involves carrying out a reaction where only one of the enantiomers is affected. In order to do this, it is necessary to react the racemic mixture with an enantiopure reagent. The method relies on the formation of **diastereomeric transition states**, which will have different stabilities. If there is a large difference in stability, then the reaction will favor one enantiomer over the other. **Enzymes** have been particularly useful in resolving racemates. For example, chiral esters can be resolved using **lipase enzymes** that catalyze the hydrolysis of one enantiomeric ester rather than the other. It is then a simple procedure to separate the product carboxylic acid from the unreacted ester.

Inevitably, any resolution process results in half of the racemic product being wasted. This could be avoided if the 'unwanted' enantiomer could be racemized and the resolution process repeated. For example, **D-phenylglycine** is used in the synthesis of **ampicillin** but is synthesized as a racemate. In order to separate the enantiomers, the racemate can be converted to a primary amide (**D,L-phenylglycinamide**) which is then treated with an aminopeptidase enzyme to hydrolyze the L-enantiomer but not the D-enantiomer (*Fig. 14*). The D-enantiomeric amide can then be separated from L-phenylglycine and hydrolyzed to the desired D-phenylglycine. The unwanted L-phenylglycine is then racemized with sulfuric acid and converted back to racemic D,L-phenylglycinamide with ammonia. The resolution/separation/racemization cycle can be repeated until virtually all the original racemate has been converted to the desired enantiomer.

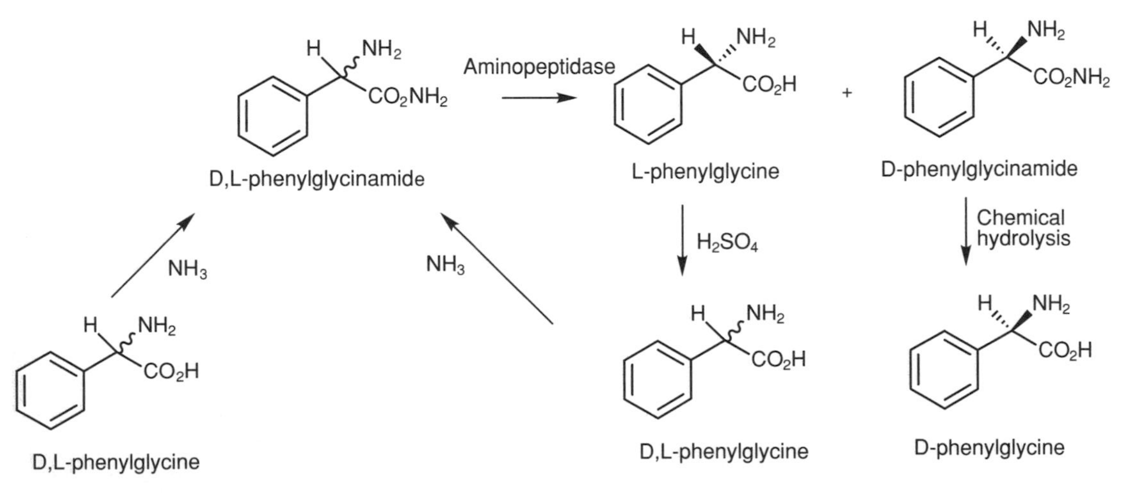

Fig. 14. Kinetic resolution of D,L-phenylglycine.

An alternative approach to the resolution/racemization cycle described above is to carry out a reaction which inverts the asymmetric center of the 'wrong' enantiomer once the resolution has been carried out. For example, the enantiomeric alcohol (II) was required as an intermediate for a drug synthesis (*Fig. 15*). The racemic ester (I) was resolved by treating it with a microorganism called *Athrobacter* whose enzymes hydrolyzed the 'unwanted' enantiomer. The products could then be separated and the desired enantiomer (II) isolated. The 'unwanted' enantiomer was converted to a sulfonate, which was then treated with water. An S_N2 nucleophilic substitution took place resulting in inversion of the asymmetric center and formation of the desired enantiomer (II). Thus, the desired enantiomer was obtained in 99% yield and 99% optical purity from the original racemic mixture.

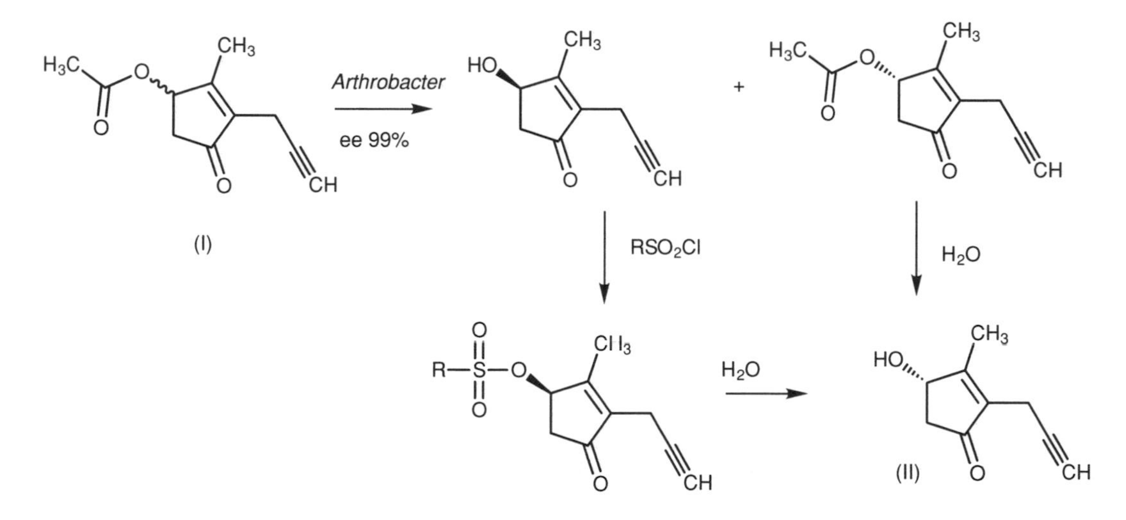

Fig. 15. Kinetic resolution and inversion of unwanted enantiomer.

F3 COMBINATORIAL SYNTHESIS

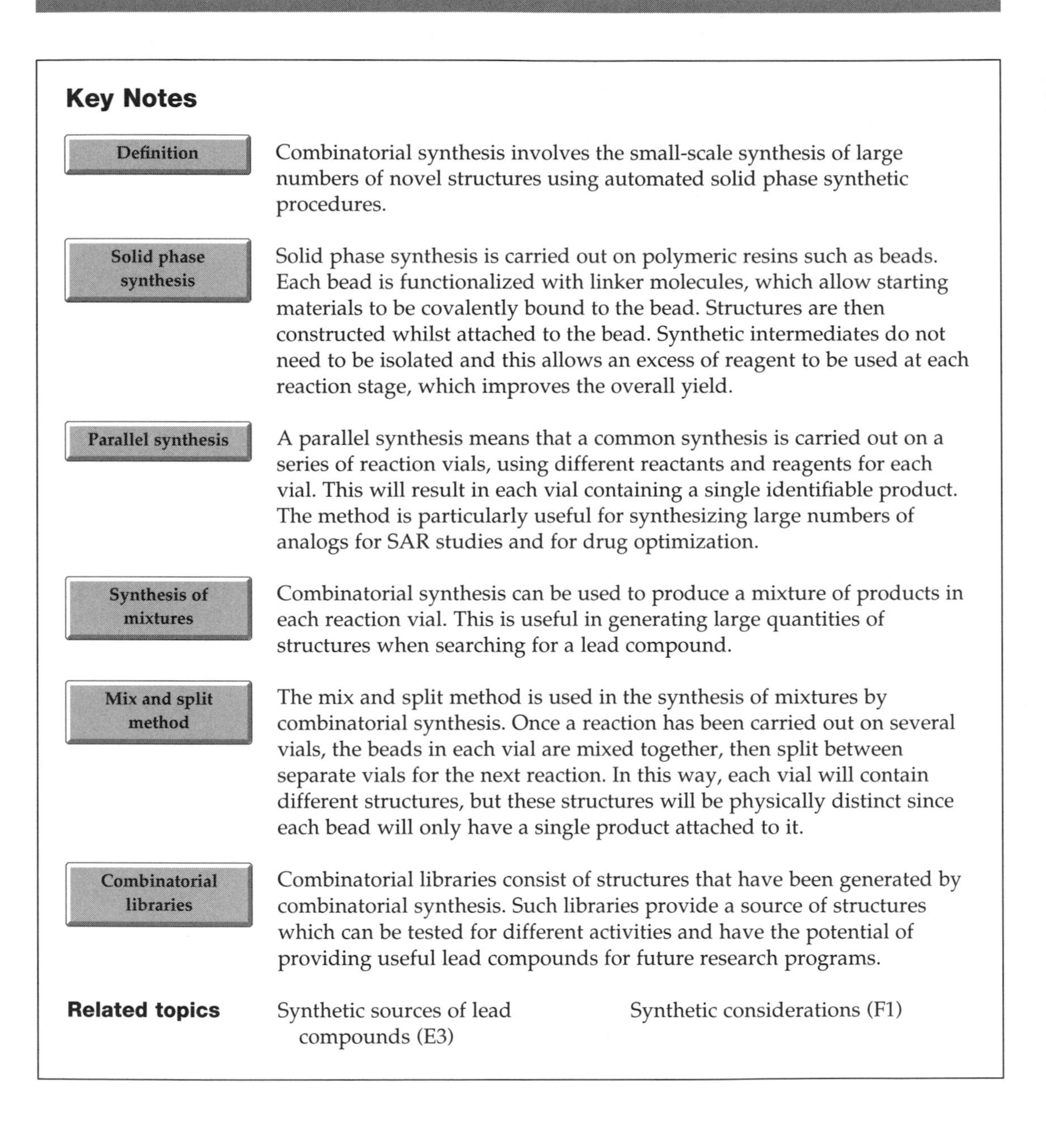

Key Notes

Definition — Combinatorial synthesis involves the small-scale synthesis of large numbers of novel structures using automated solid phase synthetic procedures.

Solid phase synthesis — Solid phase synthesis is carried out on polymeric resins such as beads. Each bead is functionalized with linker molecules, which allow starting materials to be covalently bound to the bead. Structures are then constructed whilst attached to the bead. Synthetic intermediates do not need to be isolated and this allows an excess of reagent to be used at each reaction stage, which improves the overall yield.

Parallel synthesis — A parallel synthesis means that a common synthesis is carried out on a series of reaction vials, using different reactants and reagents for each vial. This will result in each vial containing a single identifiable product. The method is particularly useful for synthesizing large numbers of analogs for SAR studies and for drug optimization.

Synthesis of mixtures — Combinatorial synthesis can be used to produce a mixture of products in each reaction vial. This is useful in generating large quantities of structures when searching for a lead compound.

Mix and split method — The mix and split method is used in the synthesis of mixtures by combinatorial synthesis. Once a reaction has been carried out on several vials, the beads in each vial are mixed together, then split between separate vials for the next reaction. In this way, each vial will contain different structures, but these structures will be physically distinct since each bead will only have a single product attached to it.

Combinatorial libraries — Combinatorial libraries consist of structures that have been generated by combinatorial synthesis. Such libraries provide a source of structures which can be tested for different activities and have the potential of providing useful lead compounds for future research programs.

Related topics Synthetic sources of lead compounds (E3); Synthetic considerations (F1)

Definition

Combinatorial synthesis is a relatively recent innovation in medicinal chemistry. However, it has had a major impact and it is now used by all the major pharmaceutical industries. Essentially, combinatorial synthesis is an automated process by which large numbers of novel structures are synthesized on a small scale using **solid phase** chemistry. There are two main approaches: parallel synthesis and the synthesis of mixtures.

Solid phase synthesis

Solid phase synthesis is used for combinatorial synthesis since it lends itself to automation. In order to carry out a solid phase synthesis there has to be a polymeric support or **resin** which is inert to the reaction conditions used in the synthesis. This often takes the form of very small beads. There also has to be a functional group present on the resin (a **linker**) which will allow molecules to be covalently linked to the solid support. A solid phase synthesis is carried out by linking the first molecule in the synthesis to the solid phase, then carrying out the rest of the synthesis on the polymer-bound structure (*Fig. 1*). Obviously, the link holding the structure to the polymer must be stable to all the reaction steps. Finally, the product has to be released from the polymer using suitable reaction conditions that do not degrade the product.

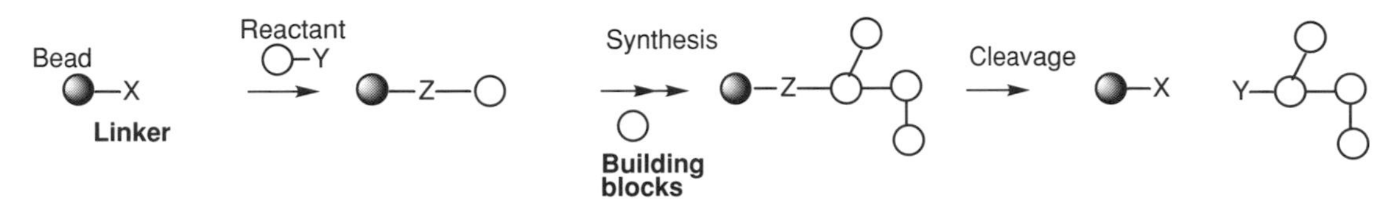

Fig. 1. Solid phase synthesis (X, Y, Z = functional groups).

There are several advantages to using solid phase synthesis, the chief one being increased yields. Since the reaction sequence is carried out on a polymer-bound structure, there is no need to isolate and purify reaction intermediates. Large excesses of reagents can be used to force reactions to completion and the excess reagents can be easily removed by washing the resin with suitable solvents.

There are a large variety of solid supports with different linkers, which allow the attachment of molecules through different functional groups. One of the most commonly used solid supports is the **Wang resin**, which contains an alcohol functional group as part of the linker unit (*Fig. 2*). It is therefore possible to attach carboxylic acids or carboxylic acid derivatives to this resin.

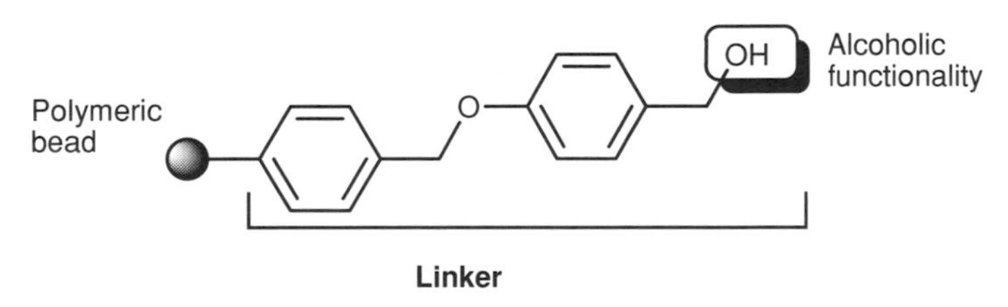

Fig. 2. Wang resin.

Figure 3 shows the combinatorial synthesis of antibacterial structures called **fluoroquinolones**. The first structure in the synthesis is covalently linked to the Wang resin via an ester bond. Once attached, the rest of the synthesis can be carried out and the final product is released by treating the resin with trifluoroacetic acid.

Parallel synthesis

A **parallel combinatorial synthesis** involves carrying out the same reaction sequence on a series of reaction vials, but using different reactants and reagents for each vial (*Fig. 4*). The beads in any individual vial will all contain the same

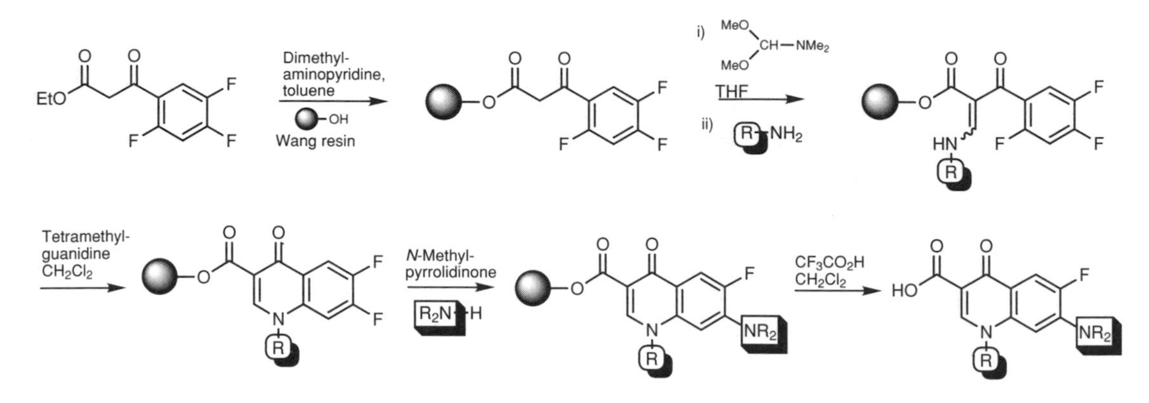

Fig. 3. Combinatorial synthesis of fluoroquinolones.

structure. However, different vials will contain different structures, depending on the reactants and reagents added to each vial. Therefore, in a parallel synthesis, each reaction vial contains a unique product and the structure of that compound will be known based on the reagents that were used. For example, in *Fig. 3*, different fluoroquinolones can be synthesized in separate vials depending on the amines (RNH_2 and R_2NH) added to each vial.

Parallel synthesis allows the rapid synthesis of a large number of analogs based on a common skeleton. This is useful in providing a series of compounds for studies into structure–activity relationships. It is also useful in fine-tuning or optimizing a lead compound in order to find a structure with improved activity or reduced side effects.

Synthesis of mixtures

Combinatorial synthesis is also used to deliberately synthesize a mixture of compounds in each vial. In this situation, the structures attached to each individual bead will be identical, but different beads in the same vial will contain

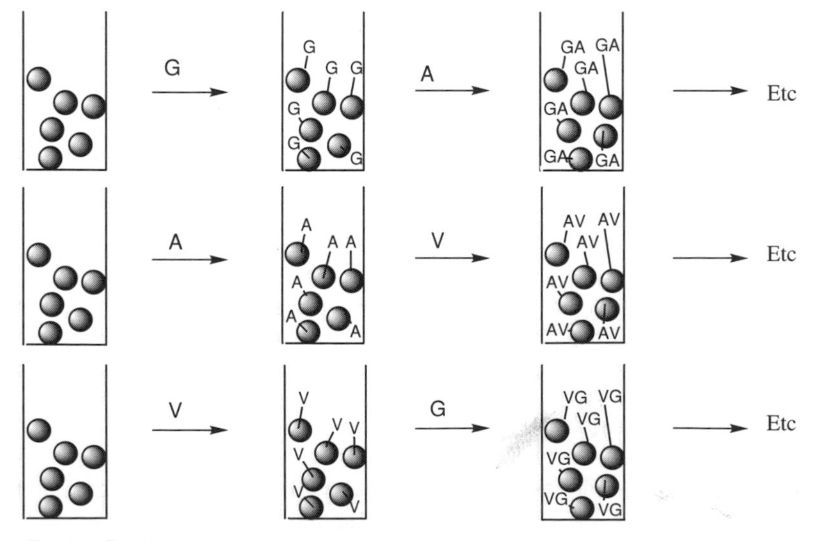

Fig. 4. Parallel synthesis of peptides.

different structures. This form of combinatorial synthesis is useful in drug discovery where one is searching for a lead compound. By synthesizing different structures in each vial, it is possible to make extremely large quantities of different compounds in a short period of time, thus increasing the chances of finding a structure that has the desired activity.

Often, the composition of compounds present in each mixture is not known with any certainty, but that is not important to begin with. The idea is to test each mixture for activity, rather than to test individual compounds. If the mixture proves to be inactive, then it is stored as part of a **combinatorial library**. If on the other hand, the mixture is active, the emphasis can then switch to identifying which compound or compounds in the mixture are responsible for the activity.

Mix and split method

The synthesis of mixtures by combinatorial synthesis is carried out by a **mix and split** method (*Fig. 5*). First of all, different starting materials are linked to beads in separate reaction vials. For example, a different amino acid is added to each vial so that all the beads in that vial contain that amino acid. The next stage is to pool all the beads together and to thoroughly mix them. Once that is done, samples of the mixture are placed into separate vials so each vial contains the same mixture. Therefore, in this example, each vial contains beads with glycine, valine and alanine attached. The next step in the reaction sequence can then be carried out using a different reactant for each vial. For example, in the case of the peptide synthesis shown, a different amino acid is added to each vial. Coupling can take place resulting in every bead containing a dipeptide which is unique to that bead. In the example shown, all the possible dipeptides containing glycine, alanine and valine have now been synthesized in three separate vials. It is known which vial contains a specific dipeptide, but it is not known which specific bead contains that dipeptide.

The process of mixing and splitting can now be repeated to synthesize all possible tripeptides containing these amino acids and so on. The important thing to note is the economy of effort involved. The mix and split method allows the synthesis of nine different dipeptides using three vials, compared with using nine vials in parallel synthesis.

The mix and split method is not restricted to peptide synthesis. Many hetero-

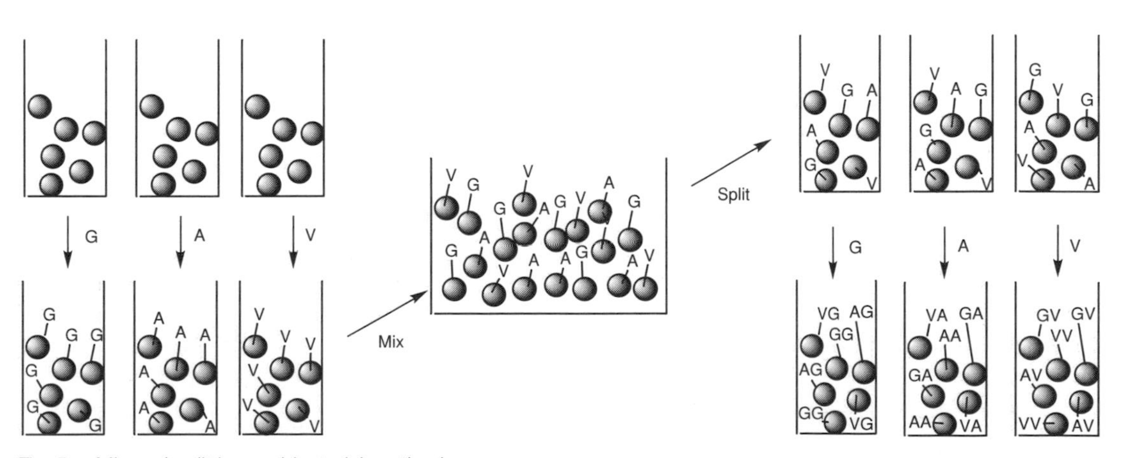

Fig. 5. Mix and split in combinatorial synthesis.

cyclic syntheses such as the one shown in *Fig. 3* have been developed and can be used to synthesize large mixtures of analogs.

Combinatorial libraries

A combinatorial synthesis will produce a large number of structures which may or may not be present as mixtures, depending on whether a mixed or a parallel synthesis has been carried out. These structures can be stored so that they provide a ready source of compounds for future reference. The stock of compounds produced by a particular combinatorial synthesis is called a **combinatorial library** and may be tested for lead compounds in any future drug research program that the company wishes to carry out.

G1 DEFINITION OF STRUCTURE–ACTIVITY RELATIONSHIPS

Key Notes

Structure–activity relationships

Analogs of a lead compound are synthesized and tested to see how structural variations affect activity (structure–activity relationships (SAR)). The results identify groups that are important to binding interactions and activity.

Rationale

The conclusions drawn from SAR studies depend on the test procedures used. *In vitro* testing can be used to determine the structural features that are important for binding interactions as well as the cellular effects arising from that binding. *In vivo* tests can be used to determine the structural features that are important to the overall physiological activity. The latter take pharmacokinetics into account as well as binding interactions.

Related topics

From concept to market (A2)
Testing drugs (D1)
The lead compound (E1)
Binding interactions (G2)
Functional groups as binding groups (G3)
The pharmacophore (G4)
Quantitative structure–activity relationships (G5)
From lead compound to dianilino-phthalimides (L3)
Pyrazolopyrimidines (L6)

Structure–activity relationships

Once a lead compound has been identified (Section E), a series of analogs is synthesized and tested to determine how structural variations affect the pharmacological activity. This is known as **structure–activity relationships** (SAR). The results can be used to identify which groups in the lead compound are important for binding interactions and activity. They can also be used to identify any trends in physical properties that might affect activity. Extra binding interactions might also be discovered.

Rationale

The conclusions that can be drawn from an SAR study depend on the method of testing used. *In vitro* tests can be carried out to test the **affinity** of a range of analogs for a target binding site and will indicate the groups that are important for binding. If a group is modified and the binding affinity drops, then this indicates that the original group is important for binding. If a structural change has little effect on affinity, then the portion of the molecule that was modified is unimportant. If a structural change results in significantly enhanced affinity, it suggests that an additional interaction with the target has been discovered.

Such studies are useful in determining the groups involved in important binding interactions, but they are not necessarily relevant to the **efficacy** of the compounds or their physiological activity. For example, testing the binding affinity of a range of compounds for a particular receptor gives no indication as to whether the molecules are acting as agonists or antagonists. Since antagonists generally bind more strongly than agonists, identifying the structural groups required for highest affinity may well identify the desirable features of an antagonist, but is of little use if one wants to know the desirable features of an agonist.

In vitro tests that measure a cellular or pharmacological effect are more relevant since the results obtained indicate the important groups required, not only for target binding but also for efficacy. However, it cannot be assumed that the structural features identified as important for activity *in vitro* will also be important for activity *in vivo*. This is because some structural features that are good for binding interactions to the target might be bad for the drug's pharmacokinetics (i.e. how effectively the drug reaches its target in the body).

In vivo tests will demonstrate whether structural variations are important to physiological activity in a test animal or human. However, it may be difficult to rationalize why such features are important since there are many more factors to consider. For example, differences in activity may be due to different receptor-binding interactions, or they may be due to altered pharmacokinetics (i.e. how easily the drug is absorbed, distributed and excreted by the test animal).

Normally, *in vitro* tests are carried out for SAR studies, but some *in vivo* tests may also be carried out to identify any features having pharmacokinetic importance.

G2 Binding interactions

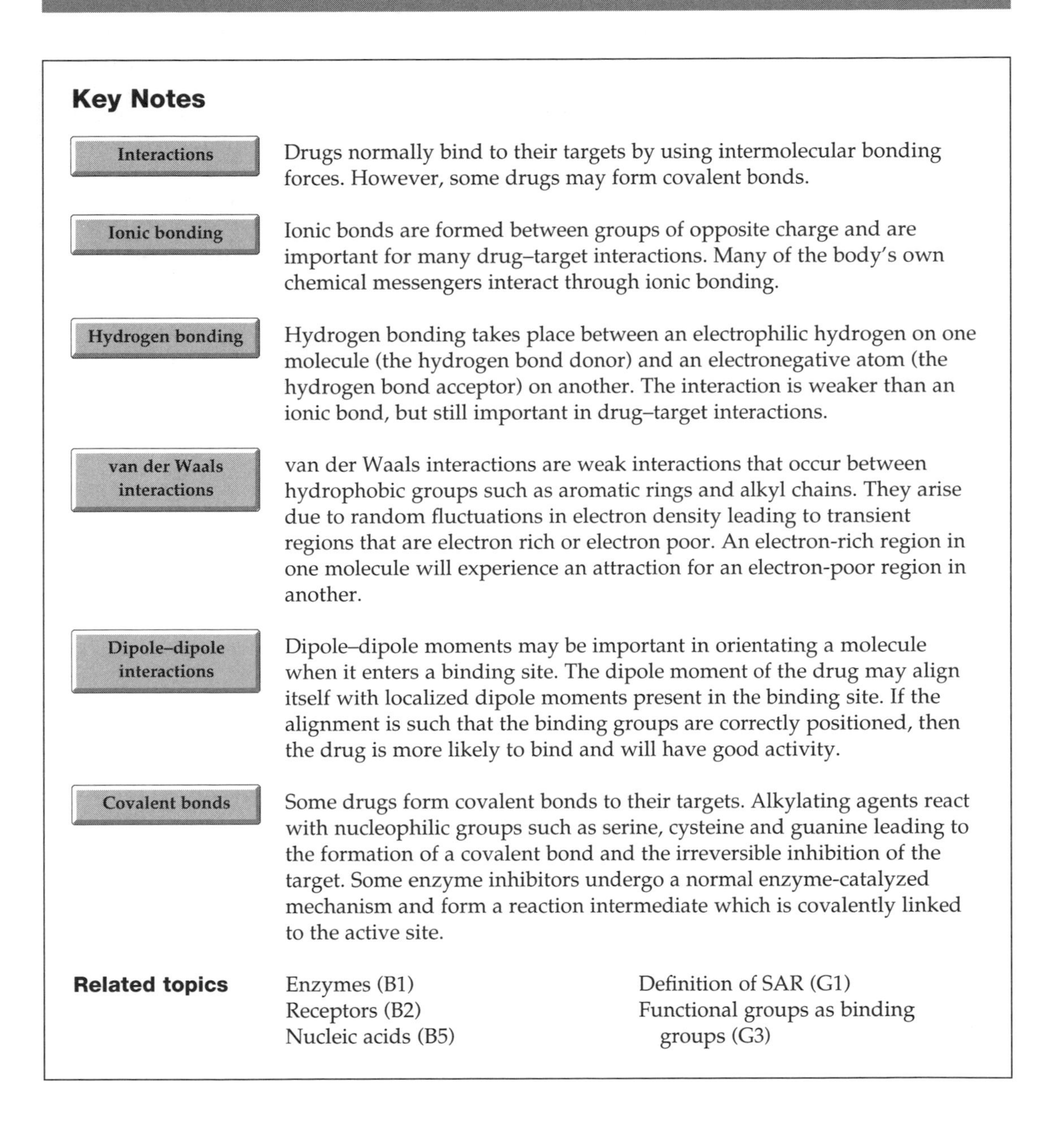

Key Notes

Interactions	Drugs normally bind to their targets by using intermolecular bonding forces. However, some drugs may form covalent bonds.
Ionic bonding	Ionic bonds are formed between groups of opposite charge and are important for many drug–target interactions. Many of the body's own chemical messengers interact through ionic bonding.
Hydrogen bonding	Hydrogen bonding takes place between an electrophilic hydrogen on one molecule (the hydrogen bond donor) and an electronegative atom (the hydrogen bond acceptor) on another. The interaction is weaker than an ionic bond, but still important in drug–target interactions.
van der Waals interactions	van der Waals interactions are weak interactions that occur between hydrophobic groups such as aromatic rings and alkyl chains. They arise due to random fluctuations in electron density leading to transient regions that are electron rich or electron poor. An electron-rich region in one molecule will experience an attraction for an electron-poor region in another.
Dipole–dipole interactions	Dipole–dipole moments may be important in orientating a molecule when it enters a binding site. The dipole moment of the drug may align itself with localized dipole moments present in the binding site. If the alignment is such that the binding groups are correctly positioned, then the drug is more likely to bind and will have good activity.
Covalent bonds	Some drugs form covalent bonds to their targets. Alkylating agents react with nucleophilic groups such as serine, cysteine and guanine leading to the formation of a covalent bond and the irreversible inhibition of the target. Some enzyme inhibitors undergo a normal enzyme-catalyzed mechanism and form a reaction intermediate which is covalently linked to the active site.

Related topics

Enzymes (B1)
Receptors (B2)
Nucleic acids (B5)
Definition of SAR (G1)
Functional groups as binding groups (G3)

Interactions

One of the most useful bits of information that can be obtained from SAR studies is the types of atoms and functional groups that are important in binding a drug to its target binding site. There are various forms of bonding that can take place. Usually these are intermolecular bonding interactions such as ionic bonds, hydrogen bonds, van der Waals interactions and dipole–dipole interactions. However, some drugs may form covalent bonds to their targets.

Ionic bonding

Ionic bonds are formed between groups of opposite charge. Many of the body's natural chemical messengers contain an amine functional group, which is ionized at body pH. These include neurotransmitters, such as **dopamine** and **norepinephrine** (noradrenaline), as well as hormones such as **epinephrine** (adrenaline) and **histamine**. The amine group is crucial to how these molecules bind to their target receptors since it is ionized and interacts with an ionized carboxylate group in the binding site. For example, the ionized amino group of epinephrine forms an ionic bond with an aspartate ion (Asp113) in the binding site of the adrenergic receptor (*Fig. 1*).

Drugs that mimic or antagonize such chemical messengers also require an ionized amine group in order to bind ionically, so ionic bonding is an important feature in many drug target interactions. Ionic interactions are also referred to as **electrostatic interactions** and are strong in character.

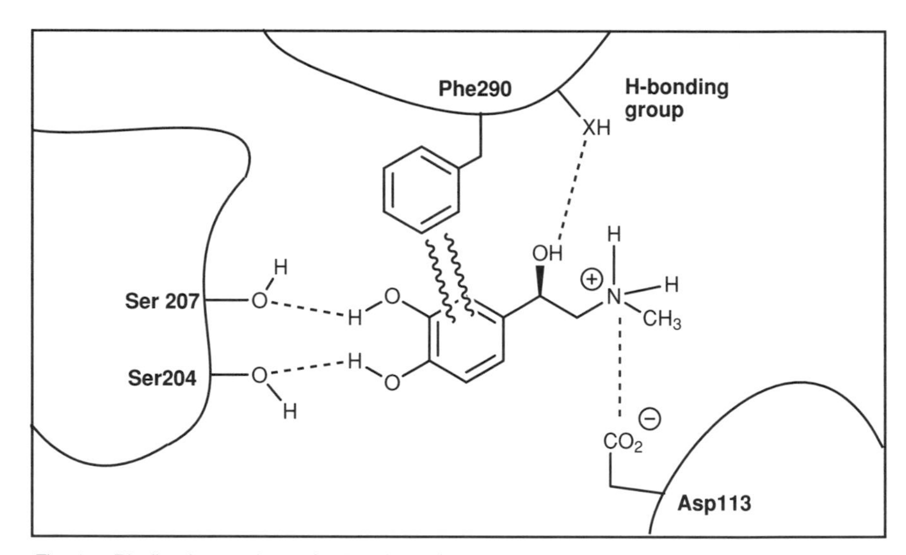

Fig. 1. Binding interactions of epinephrine (adrenaline).

Hydrogen bonding

Hydrogen bonding involves an interaction between two molecules, where one of the molecules acts as a hydrogen bond donor and the other acts as a **hydrogen bond acceptor**. The **hydrogen bond donor** contains a functional group which has a proton attached to an electronegative atom (i.e. oxygen or nitrogen). The electronegative atom has a greater share of electrons in the bond to hydrogen, and so this makes the hydrogen slightly positive and electrophilic. The hydrogen bond acceptor contains a functional group that includes an electronegative atom such as oxygen or nitrogen.

Hydrogen bonding takes place between the slightly positive hydrogen of the hydrogen bond donor and the slightly negative atom of the hydrogen bond acceptor. For example, in *Fig. 1* the two phenol groups of **epinephrine** form hydrogen bonds to the alcohol groups of two serine residues (Ser207 and Ser204). The phenols act as hydrogen bond donors, while the serine hydroxyl groups act as hydrogen bond acceptors.

The interaction can be viewed as a 'mild' form of ionic interaction since the oxygen is slightly negative and the hydrogen is slightly positive. Hydrogen bonds are consequently weaker than ionic bonds. Nevertheless, hydrogen bonding is important in many drug–target binding interactions.

van der Waals interactions

van der Waals interactions are weaker than both ionic and hydrogen bonding and involve neutral hydrocarbon regions of a molecule (i.e. alkyl groups or aromatic rings). Although these groups are nonpolar, the movement of electrons in the groups is not predictable and random fluctuations of electron density can result in transient regions where there is either a slight deficit or a slight excess of electrons. These areas only exist for a very short period of time, but they are significant enough to allow a mild interaction between molecules such that an area of transient electron deficiency in one molecule can interact with a transient electron-rich area of another molecule.

Such interactions explain why alkyl groups and aromatic rings are often important to the binding of a drug to its binding site. Alkyl groups may be capable of fitting into a hydrophobic pocket in the binding site and interacting with it by van der Waals interactions. Aromatic rings may be able to interact with planar hydrophobic regions or slots, again by van der Waals interactions. Many important chemical neurotransmitters and hormones contain aromatic and heteroaromatic rings. For example, **epinephrine** has an aromatic ring that can take part in van der Waals interactions with the aromatic ring of Phe290 in the adrenergic binding site.

Dipole–dipole interactions

The **dipole moment** of a molecule represents the relative orientation of electron density. Drugs having a dipole moment are likely to align with localized dipole moments in the binding site such that the dipole moments are parallel and in opposite directions. Compared to the previous interactions, very little has been written regarding the importance of dipole–dipole interactions to drug–target binding. However, it is likely that dipole–dipole interactions are important in orientating molecules when they enter a binding site so the interactions above can take place effectively. A dipole moment that is incorrectly orientated could well result in a drop in activity. This was observed with analogs of the anti-ulcer drug **cimetidine**.

Covalent bonds

Most drugs interact with their targets using intermolecular bonds. However, some drugs form **covalent bonds** to their target. Drugs containing good leaving groups (e.g. alkyl halides) can act as electrophiles and react with nucleophilic amino acid residues, such as serine and cysteine, in the target binding site, leading to an irreversible link between the drug and the target. This can lead to permanent antagonism of a receptor or irreversible inhibition of an enzyme. Some alkylating drugs react with nucleic acids resulting in the formation of irreversible covalent bonds to nucleophilic guanine groups, leading to the disruption of DNA function.

However, not all covalent bonds formed between drugs and their targets are irreversible. Some enzyme inhibitors are designed to act as substrates and to undergo the enzyme-catalyzed reaction. This inevitably leads to the formation of a covalent link between the drug and the enzyme. In the normal state of affairs, this bond would be quickly broken again. However, the drug is designed to stay attached much longer, resulting in inhibition of the enzyme's usual reaction. In theory, the formation of the covalent bond is reversible, although in many cases the rate of bond cleavage is so slow that it is effectively irreversible.

G3 FUNCTIONAL GROUPS AS BINDING GROUPS

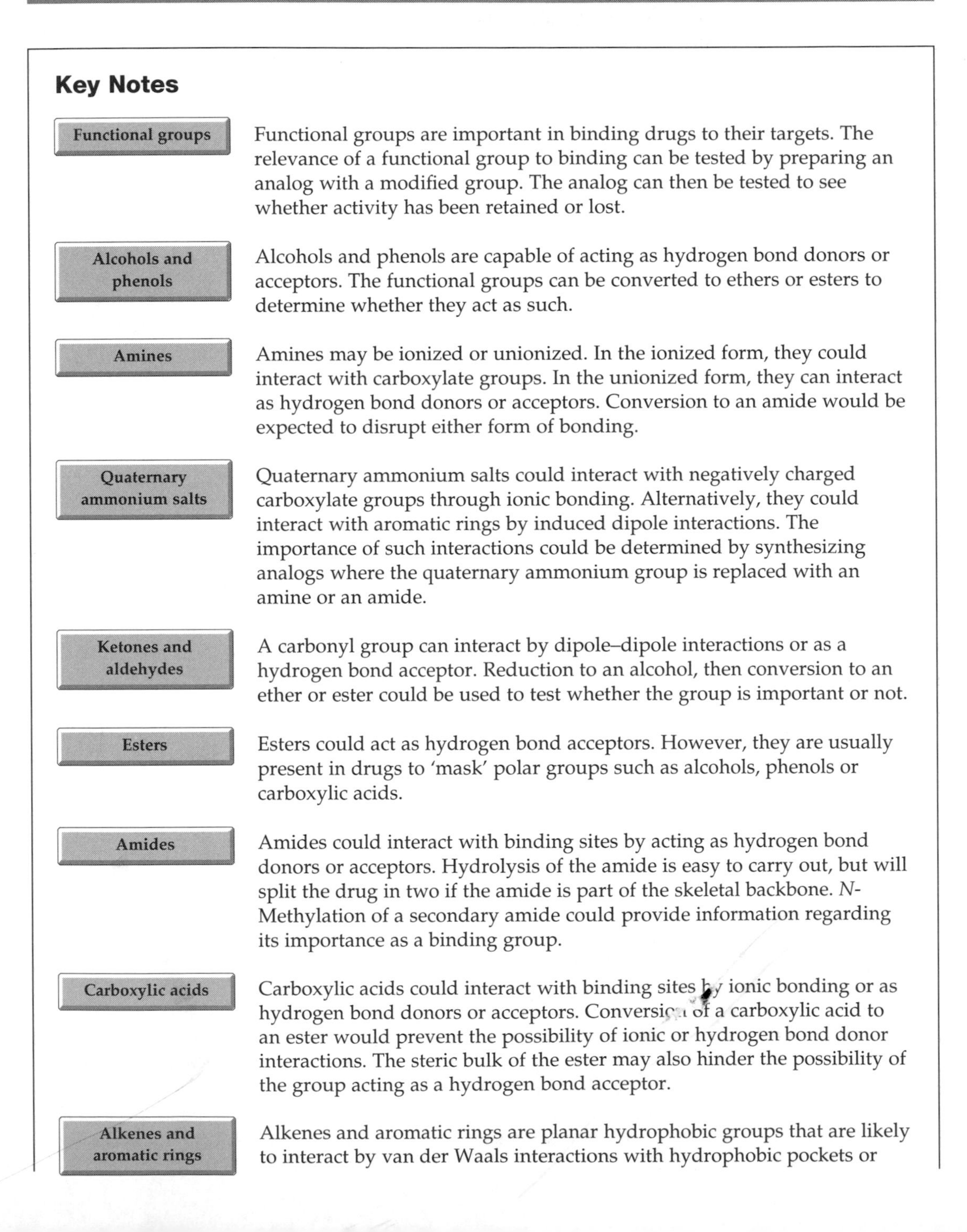

Key Notes

Functional groups	Functional groups are important in binding drugs to their targets. The relevance of a functional group to binding can be tested by preparing an analog with a modified group. The analog can then be tested to see whether activity has been retained or lost.
Alcohols and phenols	Alcohols and phenols are capable of acting as hydrogen bond donors or acceptors. The functional groups can be converted to ethers or esters to determine whether they act as such.
Amines	Amines may be ionized or unionized. In the ionized form, they could interact with carboxylate groups. In the unionized form, they can interact as hydrogen bond donors or acceptors. Conversion to an amide would be expected to disrupt either form of bonding.
Quaternary ammonium salts	Quaternary ammonium salts could interact with negatively charged carboxylate groups through ionic bonding. Alternatively, they could interact with aromatic rings by induced dipole interactions. The importance of such interactions could be determined by synthesizing analogs where the quaternary ammonium group is replaced with an amine or an amide.
Ketones and aldehydes	A carbonyl group can interact by dipole–dipole interactions or as a hydrogen bond acceptor. Reduction to an alcohol, then conversion to an ether or ester could be used to test whether the group is important or not.
Esters	Esters could act as hydrogen bond acceptors. However, they are usually present in drugs to 'mask' polar groups such as alcohols, phenols or carboxylic acids.
Amides	Amides could interact with binding sites by acting as hydrogen bond donors or acceptors. Hydrolysis of the amide is easy to carry out, but will split the drug in two if the amide is part of the skeletal backbone. *N*-Methylation of a secondary amide could provide information regarding its importance as a binding group.
Carboxylic acids	Carboxylic acids could interact with binding sites by ionic bonding or as hydrogen bond donors or acceptors. Conversion of a carboxylic acid to an ester would prevent the possibility of ionic or hydrogen bond donor interactions. The steric bulk of the ester may also hinder the possibility of the group acting as a hydrogen bond acceptor.
Alkenes and aromatic rings	Alkenes and aromatic rings are planar hydrophobic groups that are likely to interact by van der Waals interactions with hydrophobic pockets or

planar regions in the binding site. Reducing either group would result in a bulkier group which could fail to interact as efficiently.

Alkyl halides

Alkyl halides are reactive groups that can react with nucleophiles present on proteins and nucleic acids. Nucleophilic substitution results in a covalent bond between the macromolecule and the drug. Alkyl fluorides are unreactive. Fluorine is often introduced to affect the electronic properties of the drug, or as a metabolic blocker.

Thiols

Thiols are often found in drugs that interact with zinc metalloproteinases. The thiol group can form a strong bond to the zinc ion. Methylation or oxidation of the thiol group should cause a dramatic fall in activity.

Miscellaneous groups

Many other functional groups are present in drugs for reasons other than binding. Some functional groups are used to modify the electronic properties of a drug. Others are used as metabolic blockers or conformational restraints.

Alkyl groups

Alkyl groups can act as important binding groups if they form van der Waals interactions with hydrophobic regions of the binding site. Varying the size of the groups allows an exploration of the hydrophobic region.

Related topics

Enzymes (B1)
Receptors (B2)
Testing drugs *in vitro* (D2)
Definition of structure–activity relationships (G1)
Binding interactions (G2)
Enhancing existing binding interactions (H6)
Drug solubility (I1)

Functional groups

Functional groups can play an important part in binding drugs to their targets, and SAR studies can identify whether these groups are important or not. These studies can be carried out by preparing analogs from the lead compound itself. In order to determine whether a functional group is important, it is converted to another functional group. If the activity falls, the original functional group is important to binding. If the activity is unaffected, the original functional group is unimportant. If the experiment is to be valid, *in vitro* tests must be carried out that do not require the drug to cross any barriers such as cell membranes. The derivative must also be stable to the test conditions. In some cases, it may be necessary to carry out a full synthesis to prepare a suitable analog.

Alcohols and phenols

There are many examples of drugs that contain **alcohol** or **phenol** functional groups. Such groups can interact as hydrogen bond donors (HBD) and/or hydrogen bond acceptors (HBA) (*Fig. 1*).

In order to test whether an alcohol or phenol is an important binding group, it can be converted to an ether or ester (*Fig. 2*), then the analog is tested to see whether activity falls. The formation of the ester or ether will prevent the functional group acting as a hydrogen bond donor. It would also be expected to interfere with the group's capacity to act as a hydrogen bond acceptor, since the increased bulk of the substituent would sterically hinder the interaction.

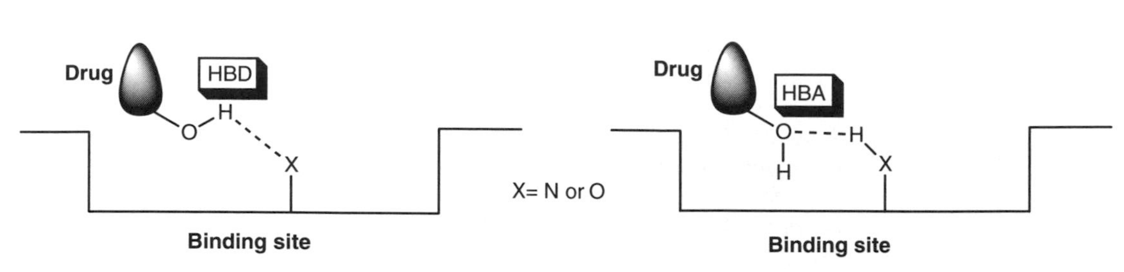

Fig. 1. Possible binding interactions of alcohols and phenols.

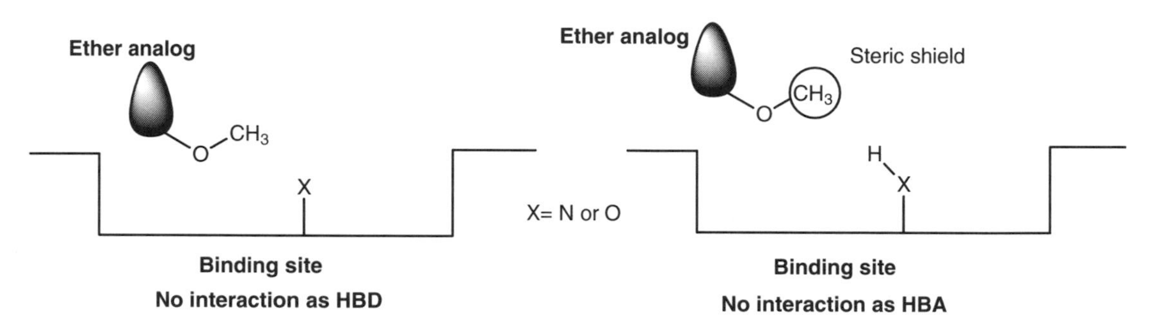

Fig. 2. Disruption of binding interactions by an ether.

Amines

Amines are extremely important functional groups in many drugs. Often the group is protonated at physiological pH, allowing the group to interact with a carboxylate group in the binding site (*Fig. 3*). The importance of such an interaction can be tested by converting the amine to an amide. The amide analog cannot ionize and so one would expect the activity to drop.

Some amines may not be significantly ionized at physiological pH but can interact with the target binding site by hydrogen bonding (*Fig. 4*). In this situation, primary and secondary amines can act as hydrogen bond donors, while primary, secondary and tertiary amines can act as hydrogen bond acceptors.

The importance of these interactions can also be tested by converting the amines to amides and testing the analogs for activity. Formation of a tertiary amide would prevent the possibility of the group acting as a hydrogen bond donor. Moreover, the bulky character of the amide group (whether secondary or tertiary) would be expected to have a steric effect, which would hinder either form of hydrogen bonding.

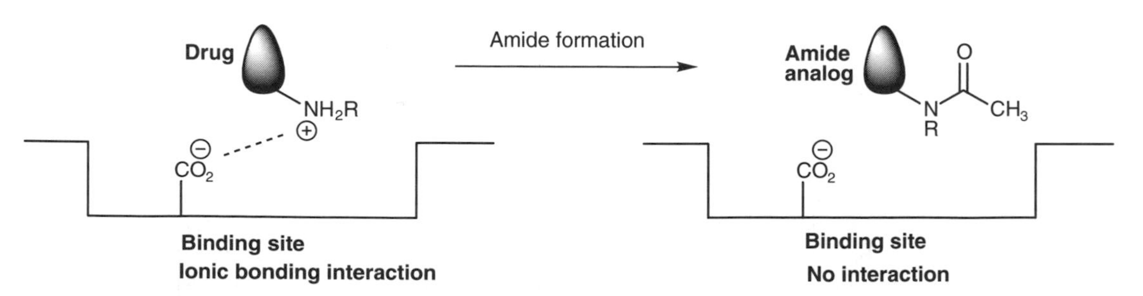

Fig. 3. Disruption of ionic bonding by conversion of an amine to an amide.

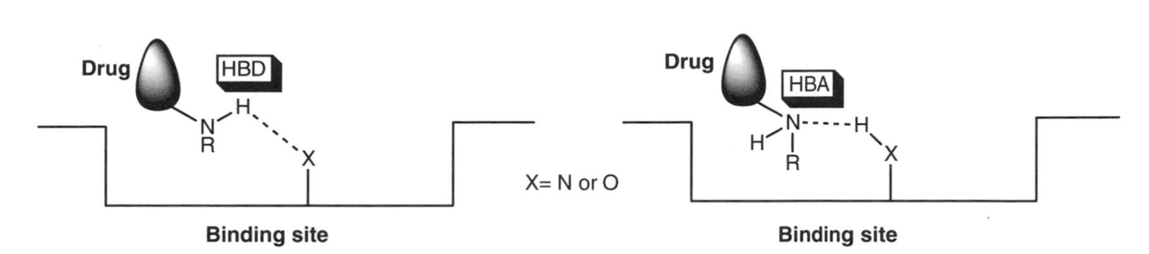

Fig. 4. Possible hydrogen bonding interactions of an amine.

Quaternary ammonium salts

Quaternary ammonium salts are ionized and can interact with carboxylate groups by ionic interactions (*Fig. 5*). Another possibility is an **induced dipole interaction** between the quaternary ammonium ion and aromatic rings. The positively charged nitrogen can distort the π electrons of the aromatic ring such that a dipole is induced, whereby the face of the ring is slightly negative and the edges are slightly positive. This allows an interaction between the slightly negative faces of the aromatic rings and the positive charge of the quaternary ammonium ion.

The importance of the above interactions could be tested by synthesizing an analog that has an amine group rather than a quaternary amine group. Of course, it is possible that such a group could become ionized by being protonated, and interact in the same way as the quaternary ammonium group. Converting the amine to an amide would prevent this possibility.

The neurotransmitter **acetylcholine** has a quaternary ammonium group which is thought to bind to its target receptor by ionic bonding and/or induced dipole interactions (*Fig. 6*). There is an aspartate group (Asp311) present in the binding site that is in the correct position for binding to the quaternary ammonium group by ionic bonding; but the quaternary group is also positioned in a hydrophobic pocket lined by several aromatic residues (Trp307, Tyr616, Trp613), which may allow induced dipole interactions to take place. Neuromuscular blocking agents are a group of drugs that require the presence of a quaternary ammonium group to be active. They act as cholinergic antagonists by preventing acetylcholine from binding to its receptor. **Suxamethonium** is an example of this group of drugs.

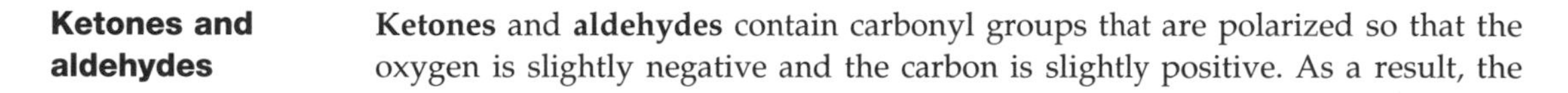

Ketones and aldehydes

Ketones and **aldehydes** contain carbonyl groups that are polarized so that the oxygen is slightly negative and the carbon is slightly positive. As a result, the

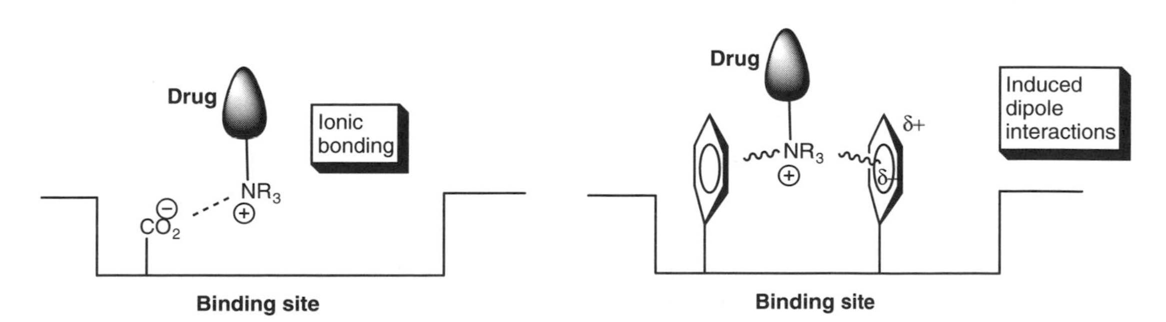

Fig. 5. Possible binding interactions of a quaternary ammonium ion.

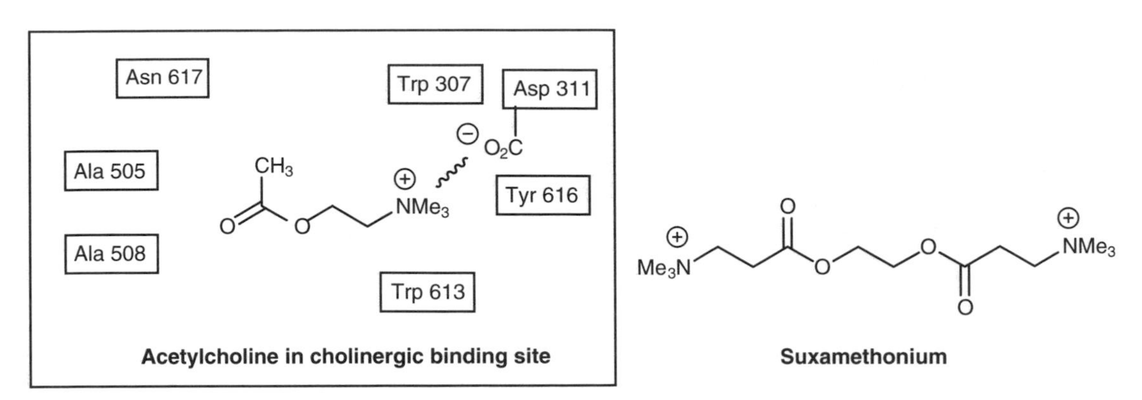

Fig. 6. Binding interactions of acetylcholine.

functional group has a dipole moment that may have some role to play in drug–target interactions (*Fig. 7*). However, it is more likely that the oxygen acts as a hydrogen bond acceptor. Ketones are relatively common in drugs, whereas aldehydes are not. In general, aldehydes are chemically more reactive than ketones and are easily oxidized to carboxylic acids, meaning that compounds containing such groups are less stable.

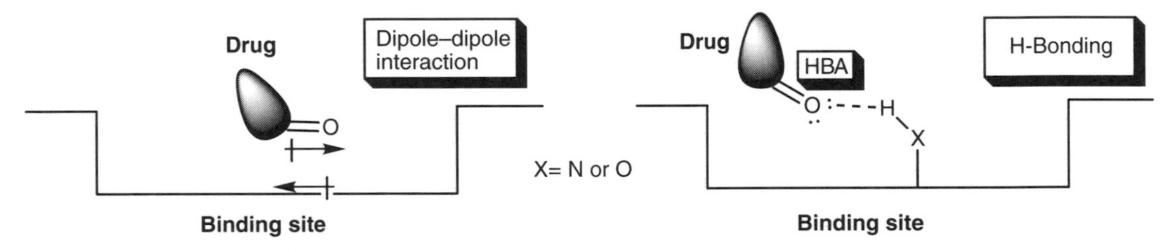

Fig. 7. Possible binding interactions for a ketone.

The importance of a carbonyl group to target binding could be determined by reducing it to an alcohol using sodium borohydride (*Fig. 8*). This may have an effect on activity since the geometry of the functional group is altered from planar to tetrahedral. This would certainly alter the orientation of the dipole moment and prevent any previously favorable interaction. The oxygen might also be moved out of range of any likely hydrogen bond donor in the binding site. On the other hand, it might not! Therefore, further conversion to an ether or ester as described previously would be worth carrying out.

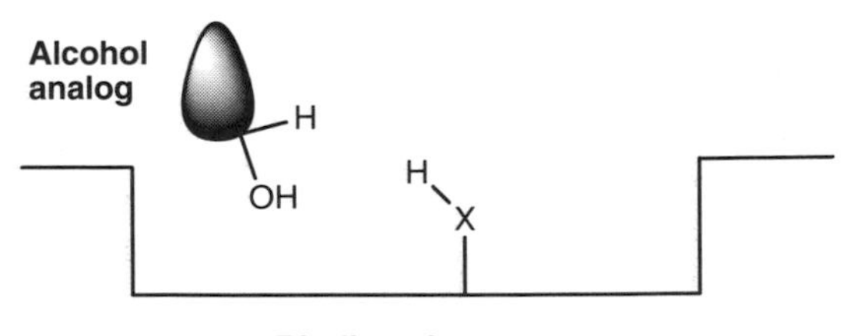

Fig. 8. Disruption of a ketone's binding interactions.

Esters

Several drugs contain **ester** functional groups, which could interact with target binding sites as hydrogen bond acceptors. However, esters are not particularly stable to metabolism. The body contains enzymes called **esterases**, which efficiently catalyze the hydrolysis of esters. The reason many drugs contain an ester group is that the group has been introduced to 'mask' a polar functional group such as a carboxylic acid, phenol or an alcohol. An ester group is less polar and this allows the drug to cross fatty barriers such as the gut wall more easily (*Fig. 9*). Once the drug is in the bloodstream, the ester is hydrolyzed to release the active structure. Compounds designed to act in this way are called **prodrugs** (see Topic I1).

Amides

Primary and secondary **amides** have the potential to act as hydrogen bond donors, whilst all amides have the potential to act as hydrogen bond acceptors (*Fig. 10*). It

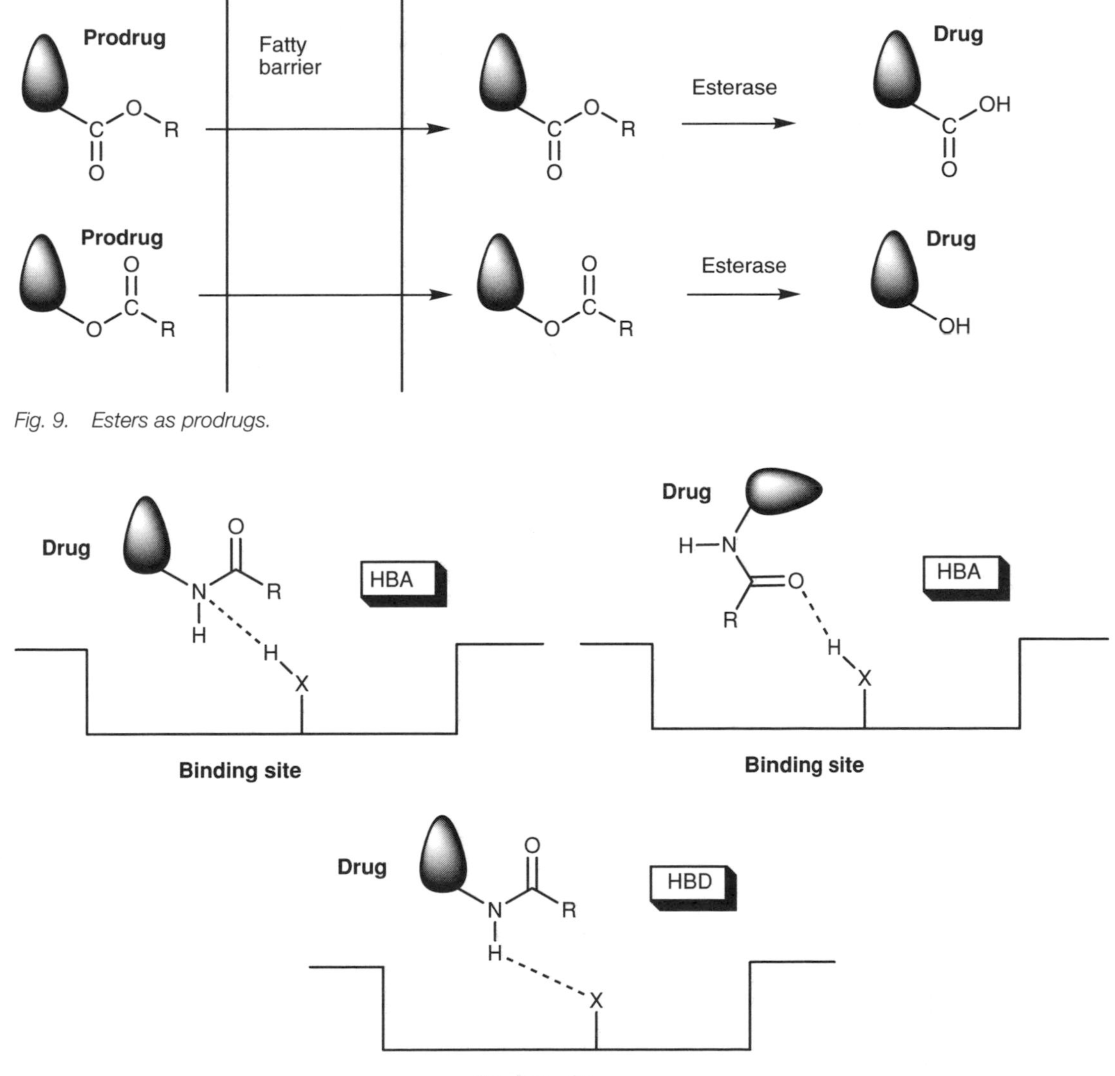

Fig. 9. Esters as prodrugs.

Fig. 10. Possible hydrogen bonding interactions of amides.

may sometimes be difficult to establish whether amide groups are important to binding or not. The easiest reaction to carry out on an amide is to hydrolyze the amide to its constituent carboxylic acid and amine. However, this may well split the drug in two, resulting in the loss of other important functional groups. An alternative reaction, which could be tried on secondary amides, is ***N*-methylation**. This would prevent the amide acting as a hydrogen bond donor (*Fig. 11*). The methyl group may also prevent the nitrogen atom acting as a hydrogen bond acceptor by acting as a **steric shield**. However, the *N*-methylated amide could still act as a hydrogen bond acceptor through the carbonyl oxygen.

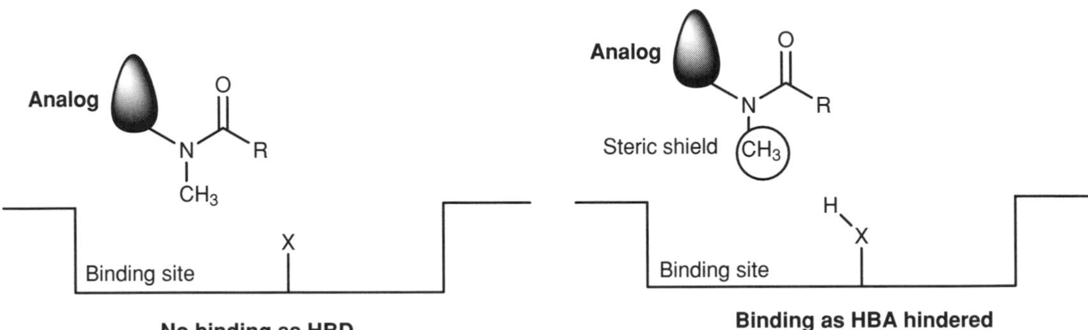

Fig. 11. Disruption of amide binding interactions.

Carboxylic acids

Some drugs have **carboxylic acid** groups, which could interact with their target sites through ionic bonding if the carboxylic acid group is ionized, or by hydrogen bonding if it is not (*Fig. 12*). In the latter case, the carboxylic acid could act as a hydrogen bond donor or hydrogen bond acceptor.

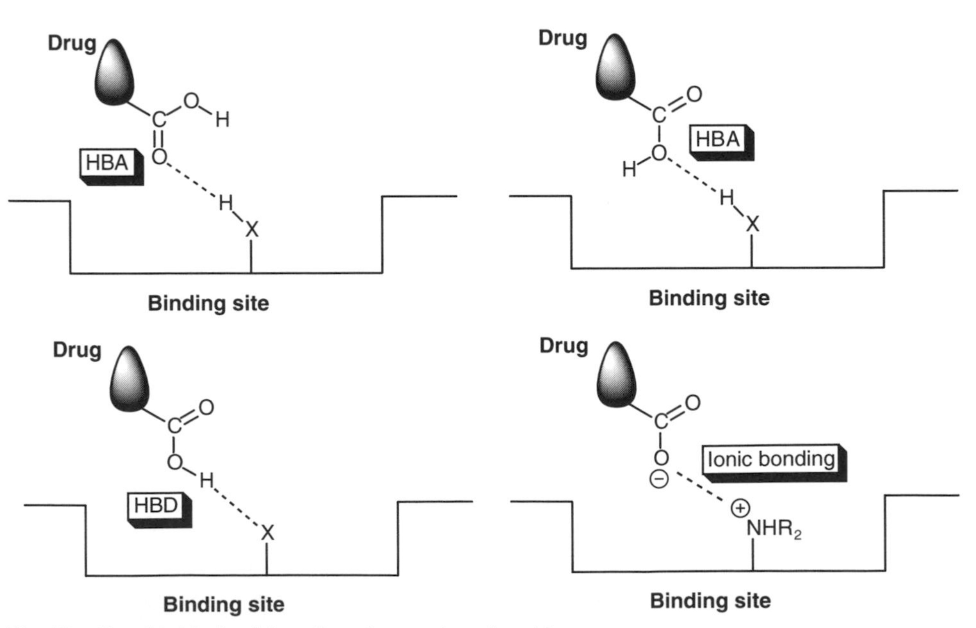

Fig. 12. Possible binding interactions for a carboxylic acid.

The importance of a carboxylic acid to binding can be determined by converting it to an ester. The ester is unable to take part in ionic bonding or as a hydrogen bond donor (*Fig. 13*). The steric bulk of the ester might also disrupt any hydrogen bonding where the carboxylic acid acts as a hydrogen bond acceptor.

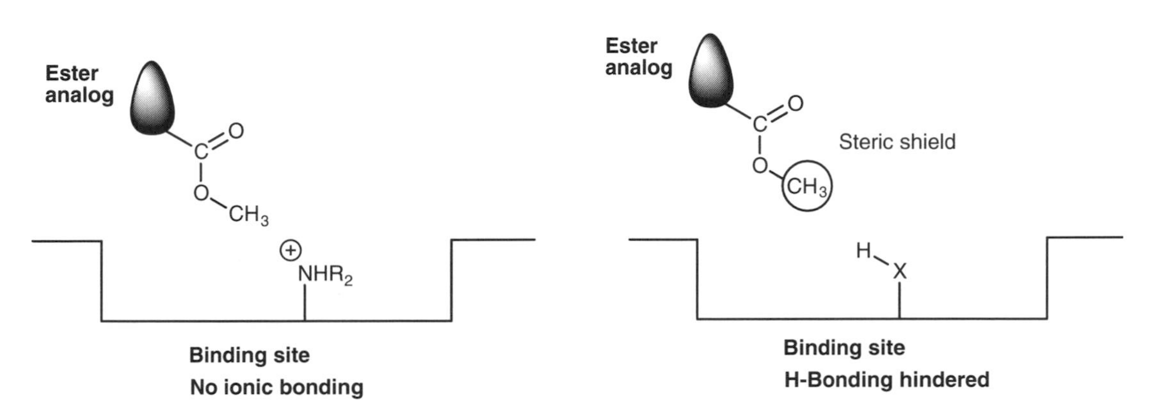

Fig. 13. Disruption of a carboxylic acid's binding interactions.

Alkenes and aromatic rings

Alkenes and **aromatic rings** are hydrophobic planar functional groups, which can interact with target binding sites by means of van der Waals interactions (*Fig. 14*).

The relevance of an alkene to binding can be determined by reducing the group to an alkane (*Fig. 15*). This alters the stereochemistry of the group making it bulkier. If the alkene is interacting with a planar surface, the resulting alkane would not be able to approach as closely and the interactions would be weaker. In theory, an aromatic ring could be reduced to a bulkier cyclohexane ring, which might be unable to enter a hydrophobic pocket. However, the reduction conditions involved are quite fierce and might not be feasible if they cause the drug to degrade. It is more likely that this type of analog would have to be prepared using a full synthesis.

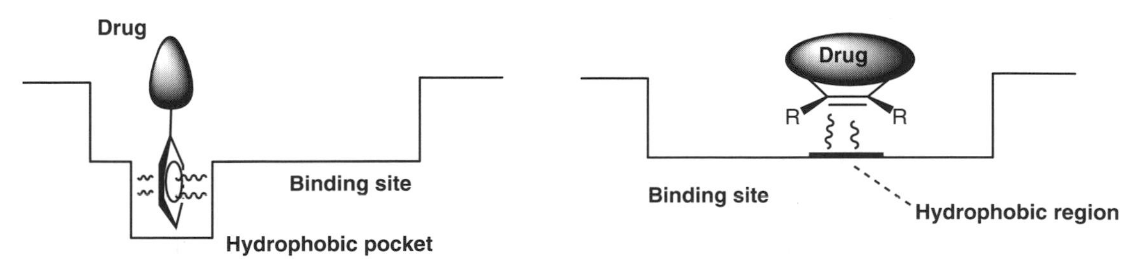

Fig. 14. van der Waals interactions for aromatic rings and alkenes.

Alkyl halides

Alkyl halides are present in a variety of important anticancer drugs. However, they do not form intermolecular bonds with the binding site. Since a halide ion is a good leaving group, the alkyl halides are strongly electrophilic and can react

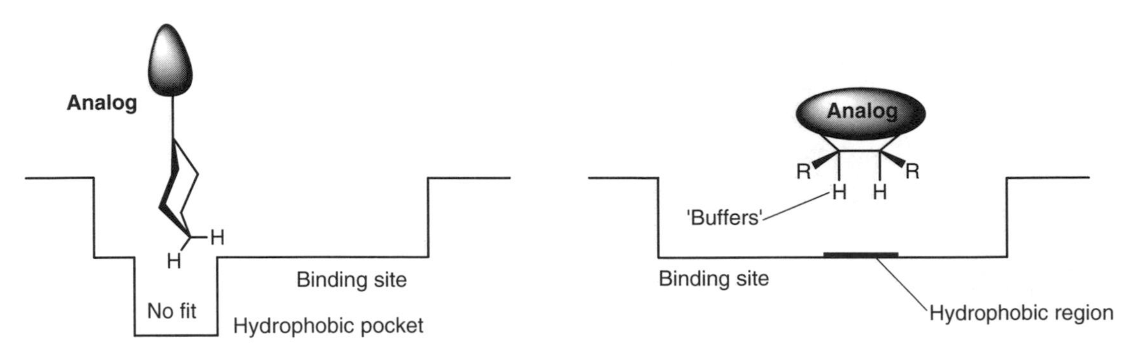

Fig. 15. Disruption of the binding interactions of an aromatic ring and alkene.

with nucleophilic groups on proteins and nucleic acids (see Section B). Nucleophilic substitution displaces the halide ion leading to the formation of a covalent bond between the macromolecule and the drug, resulting in irreversible alkylation of the macromolecule. Replacing the halide with a poorer leaving group such as a hydroxyl group would prevent this alkylation and demonstrate whether the alkyl halide is important or not.

Alkyl fluorides are the exception to the rule. The carbon–fluorine bond is strong and is not easily broken. Therefore, alkyl fluorides do not act as alkylating agents. Several drugs contain a fluorine atom, but it is not involved in any binding interactions. Instead, it has been introduced, either as an electronegative group to influence the electronic distribution of the molecule, or to block metabolism (see Topic L3).

Thiols

The **thiol** group (R–SH) is not that common in drugs, but it is useful for agents that are designed to act as inhibitors of **zinc metalloproteinases**. These are enzymes that contain a zinc ion as a cofactor. The sulfur atom can interact efficiently with the zinc ion and is present in important drugs, such as the antihypertensive agent **captopril**. Alkylation or oxidation of the thiol group is likely to be detrimental to binding and demonstrates the importance of this group.

Miscellaneous groups

There are a variety of other functional groups present in different drugs. Many of these have been introduced to influence the electronic properties of the molecule rather than as binding groups (e.g. aryl halides, nitro groups, nitriles). Other groups are present to restrict the shape or conformation of a molecule (e.g. alkynes) (see Topic H4). Functional groups have also been used as metabolic blockers (e.g. alkynes, aryl halides) (see Topics I2 and L3).

Alkyl groups

Alkyl groups are not functional groups. However, they can play an important role in the binding interactions of a drug with its target. Alkyl chains are hydrophobic and can interact with hydrophobic regions in the binding site through van der Waals interactions. It is not possible to modify alkyl chains directly. However, it may be possible to 'clip' an alkyl group off a lead compound and then replace it with a variety of other alkyl groups having different lengths and branching (*Fig. 16*). For example, an *N*-methyl group or a methyl ether can be removed and replaced with larger alkyl groups. An ester or an amide could be hydrolyzed and new alkyl groups introduced. The analogs prepared would demonstrate whether the alkyl groups are important to binding

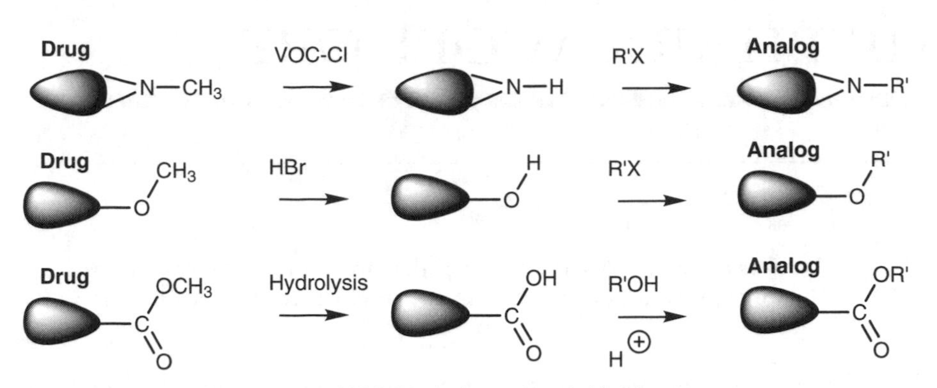

Fig. 16. Variation of alkyl groups. VOC-Cl, vinyloxycarbonyl chloride.

and would also allow an 'exploration' of any hydrophobic region in the binding site. For example, long alkyl chains would be beneficial if the binding region is a long cavity or slit, whereas short bulky alkyl groups would be beneficial for hydrophobic pockets (*Fig. 17*).

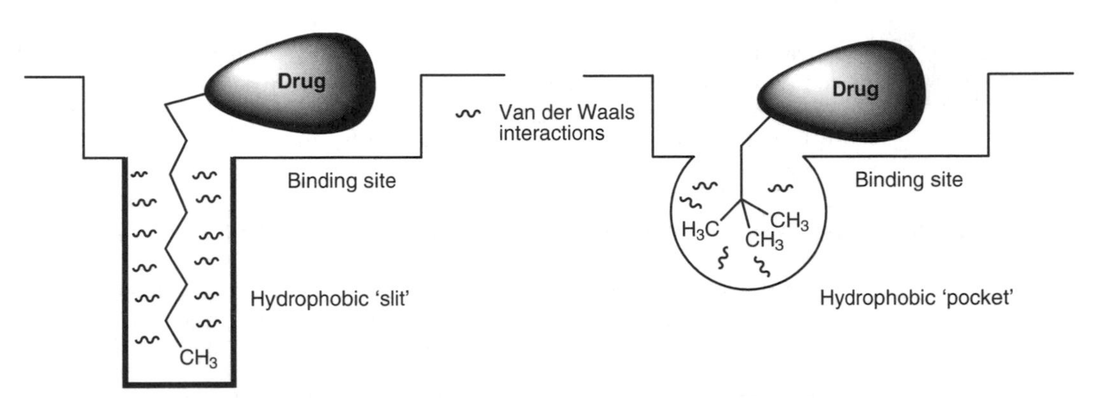

Fig. 17. Hydrophobic slits and pockets

G4 THE PHARMACOPHORE

Key Notes

Definition — The pharmacophore of a drug defines the important functional groups that are required for binding or activity, and their relative positions in space.

The active conformation — The active conformation is the conformation adopted by a drug when it binds to a binding site. It is necessary to know the active conformation in order to define the pharmacophore.

Target-based pharmacophores — If the structure of the binding site is known, it is possible to define a pharmacophore based on the amino acid residues present in the binding site. Molecular modeling can be used to define what kinds of groups are required to bind to the available amino acids and where they should be positioned.

Database searching of pharmacophores — Once a pharmacophore has been defined it can be used for the computerized search of compound databanks to see whether known structures contain the same pharmacophore. Such compounds may then be tested to see if they show activity and can be used as new lead compounds.

Related topics

Synthetic sources of lead compounds (E3)
Definition of SAR (G1)
Functional groups as binding groups (G4)
Quantitative structure–activity relationships (G5)
Computer aided drug design (H2)
Conformational restraint (H4)
Modeling studies (L4)

Definition

Once structure–activity relationship studies have been carried out, it should be possible to identify which functional groups are important for binding and/or activity in the lead compound and its analogs. The **pharmacophore** defines these groups and also defines their relative positions in space. Some pharmacophores are described by linking the important functional groups by a common skeleton. For example, the important functional groups in opiate analgesics are the aromatic ring, phenol and amine groups. A common pharmacophore linking these groups can be defined as shown in *Fig. 1*.

However, it is possible using **molecular modeling** software to define the relative positions of the important groups without specifying any connecting skeleton. Thus, the opiate pharmacophore can be defined by measuring the distance from the center of the aromatic ring to the nitrogen atom, along with the various angles (α and β) that define the relative positions of these two groups (*Fig. 2*). The advantage of defining a pharmacophore in this way is that one can rationalize why molecules having different structures or carbon skeletons can interact with the same binding site. As long as they have the important binding groups in the same positions, they share the same pharmacophore.

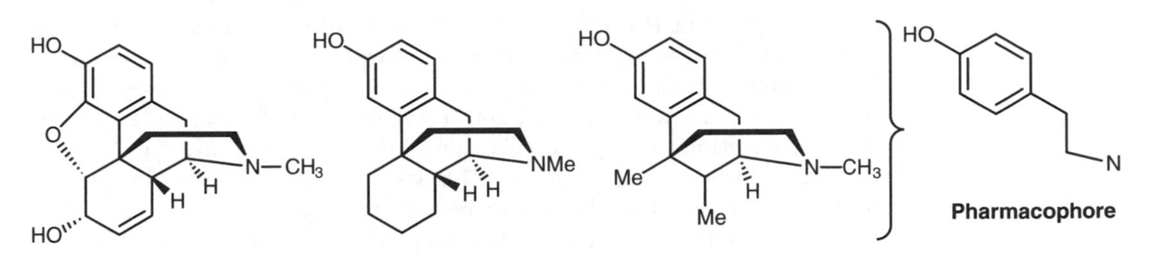

Fig. 1. Skeletal pharmacophore for opiate analgesics.

An even more general method of defining the pharmacophore would be to define the type of bonding interactions that are possible for the important functional groups (*Fig. 3*). Thus, the phenol group could act as a hydrogen bond donor (HBD) or a hydrogen bond acceptor (HBA). The aromatic ring can interact by van der Waals interactions and acts as a hydrophobic group (Ar), while the nitrogen could take part in ionic bonding if it is protonated (basic center) or as a hydrogen bond acceptor if it is not. The revised pharmacophore would then define these groups as points in space with particular bonding characteristics. This pharmacophore could then be used to explain a larger range of analgesic structures which do not necessarily contain a phenol, aromatic ring or nitrogen, but which contain functional groups that can bind by the same interactions. For example, molecules containing an amine instead of the phenol, or a heteroaromatic ring instead of the aromatic ring might show analgesic activity if these groups are in the correct positions defined by the pharmacophore.

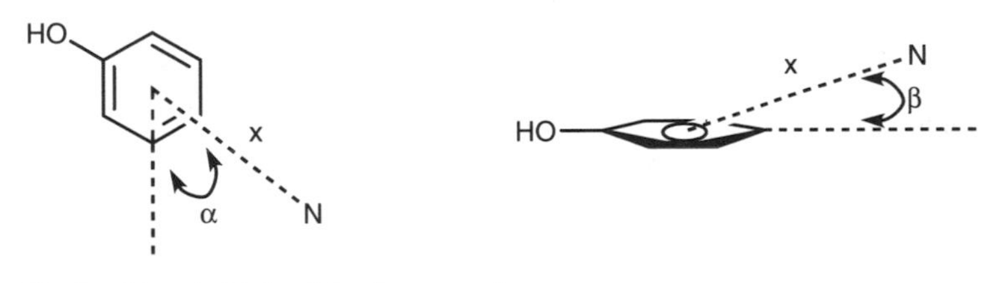

Fig. 2. Non skeletal opiate pharmacophore.

The active conformation

In order to identify the pharmacophore, it is important to identify the **active conformation** (i.e. the conformation which the drug adopts in order to bind to its binding site). With rigid molecules, such as the opiate analgesics above, this

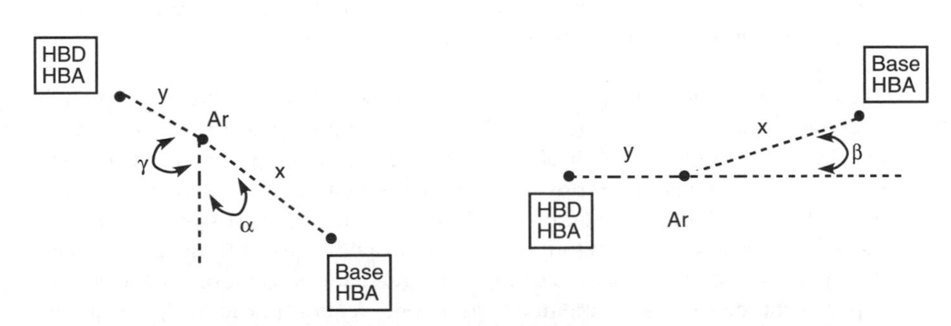

Fig. 3. Opiate pharmacophore defining bonding interactions.

poses little problem. However, more flexible molecules can adopt a large variety of different shapes. Take, for example, the neurotransmitter **dopamine** (*Fig. 4*). The important binding groups here are the two phenol groups, the aromatic ring and the amine. It is possible to construct a molecule of dopamine in its most stable conformation using molecular modeling and to measure the distances and angles between these groups. This certainly gives a possible pharmacophore, but it might not be correct, because molecules are not necessarily in their most stable conformation when they bind to receptors. Therefore, the active conformation has to be identified before the pharmacophore can be defined. A molecule such as dopamine can adopt a large variety of conformations due to single bond rotation in the alkylamine side chain, and there is no way of predicting which conformation is likely to be more active than another. Therefore, the best way of identifying the active conformation is to make a series of rigid analogs that contain the dopamine skeleton within their framework. This is usually done by introducing extra rings to prevent rotation of previously rotatable bonds. For example, the bicyclic compounds (I) and (II) are rigid analogs of dopamine where the highlighted bonds were previously rotatable in dopamine, but are now constrained within the ring system (*Fig. 4*). If structure II shows activity and structure I does not, this is good evidence that the active conformation of dopamine is 'trapped' within the skeleton of II but not I.

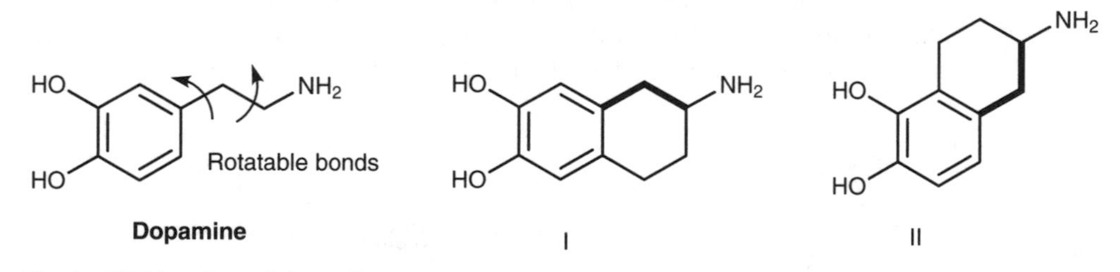

Fig. 4. Rigid analogs of dopamine.

Unfortunately, it is not always possible to obtain a suitably active rigid analog. However there are several molecular modeling software programs that can be used to search for potential active conformations from a range of active compounds. The software can be used to analyze each structure for a range of possible conformations. All the possible conformations for each active compound are then compared to identify conformations that are common to each active compound. In this way, it is possible to whittle down the number of possible active conformations to one or two likely candidates.

Target-based pharmacophores

The pharmacophores already mentioned are based on known lead compounds and their analogs. An alternative method of obtaining a pharmacophore is from the binding site of the target itself. The structures of many proteins can be determined by **X-ray crystallography**, and by including a bound ligand it is possible to identify the binding site and to study the structure of that site using molecular modeling software. By studying the amino acids present in the binding site, it is possible to generate a 'negative image' pharmacophore. For example, suppose the binding site contained the amino acids aspartic acid, serine and phenylalanine (*Fig. 5*). These groups would be capable of forming ionic bonds,

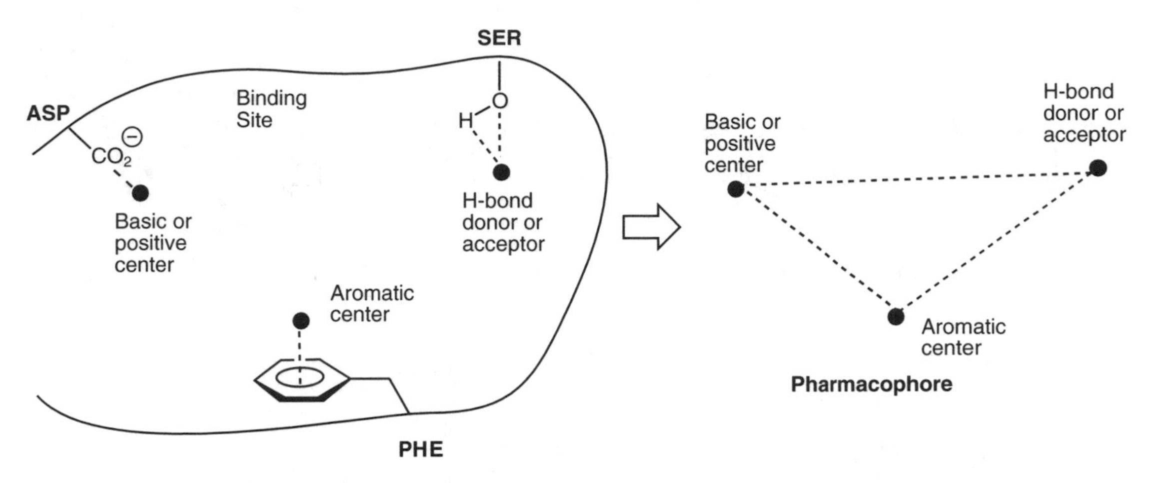

Fig. 5. Target-based pharmacophore.

hydrogen bonds and van der Waals interactions respectively. Therefore, the drugs that would interact with this binding site would have to contain an amine, a hydrogen bonding group, and a hydrophobic group. The required pharmacophore for such drugs could be defined by identifying the regions in space that are within bonding range of the said amino acids and defining what sort of groups would have to be present at those positions. Compounds known to bind to this binding site could then be studied to see whether they contain this pharmacophore. (It is also possible that potential lead compounds could fit the binding site and take part in hydrogen bonding to available peptide links. This too could be included in the pharmacophore analysis.)

This approach may also be useful in finding lead compounds for the various novel targets being discovered by genomic projects carried out on the DNA of humans and other organisms. Since these targets are totally novel, even the natural ligand or chemical messenger may not be known, and so the problem is to identify a lead compound as quickly as possible. If the binding site of the new target can be defined, a pharmacophore could be worked out as described above, then used to search databases of known compounds for any structures that contain the desired pharmacophore, and which might prove to be suitable lead compounds.

Database searching of pharmacophores

The definition of a pharmacophore allows the possibility of a computerized search of **compound databases** to see whether any compounds present contain the desired pharmacophore. If they do, they can then be tested to see whether they bind to the target of interest and show any activity. One method of carrying out this search is to define **pharmacophore triangles**. This involves defining all the possible triangles from a single pharmacophore, then searching the database to see whether the compounds contain some or all of these triangles. For example, the pharmacophore for **dopamine** contains four points and is made up of four pharmacophore triangles (*Fig. 6*).

The advantage of such a search is that it may identify compounds that have no apparent structural similarity to existing drugs. For example, the hypothetical structure I has no obvious relation to the opiate analgesics but contains the opiate pharmacophore mentioned earlier (*Fig. 7*). This might not seem obvious

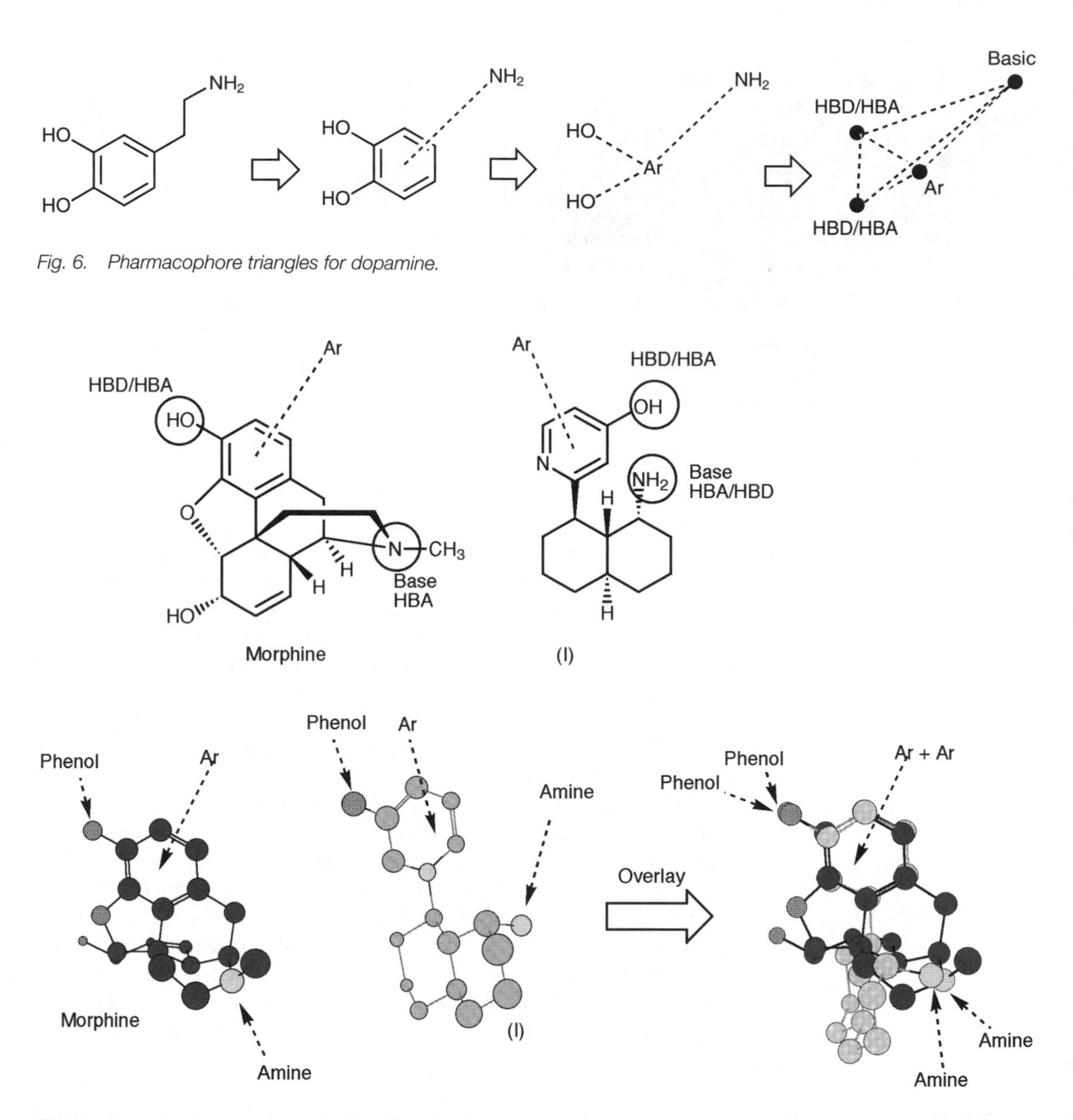

Fig. 6. Pharmacophore triangles for dopamine.

Fig. 7. Use of a pharmacophore to identify novel lead compounds

from the two-dimensional drawing, but an overlay of the 3D structure with morphine demonstrates that the groups required for the pharmacophore are in the same regions of space. (If anyone succeeds in synthesizing this molecule, the author will be interested to know whether it is active or not!)

G5 QUANTITATIVE STRUCTURE–ACTIVITY RELATIONSHIPS

Key Notes

Introduction

Quantitative structure–activity relationships (QSAR) involve the derivation of a mathematical formula which relates the biological activities of a group of compounds to their physicochemical properties. Traditional QSAR is carried out on a range of analogs sharing a common skeleton, but having different substituents.

Procedure

An initial QSAR equation is derived relating biological activity to one or two physical features. The equation can then evolve by introducing other physical features. Three physical features are of particular importance: hydrophobicity, electronic factors and steric factors.

Hydrophobicity of the molecule

The hydrophobicity of a molecule is measured by its log$P$ value, where P is the partition coefficient. The partition coefficient is the relative solubility of the compound in octanol and water.

Substituent hydrophobicity constant

The hydrophobic character of a substituent is given by the substituent hydrophobicity constant (π). π values can be used to calculate log$P$ values and can be introduced into QSAR equations to determine whether hydrophobic substituents at specific regions of a structure are important to activity.

Electronic properties

The electronic properties of aromatic substituents are measured as Hammett substituent constants (σ). Substituents with positive σ values are electron withdrawing, whereas substituents with negative σ values are electron donating.

The steric factor

Taft's steric factor (E_S) is measured experimentally. Molar refractivity (MR) is a measure of size calculated from a substituent's molecular weight, index of refraction and density. Verloop steric parameters are calculated by a computer software program.

Hansch equation

The Hansch equation is the name given to the QSAR equation and typically contains physical factors such as log$P$, π, σ, F, R, E_S, and MR. The substituents used to derive a Hansch equation must represent a good spread of values for each physical parameter and also distinguish between the parameters.

Craig plots

Craig plots compare two physical properties for different substituents. They are used to identify which substituents are valid for the derivation of a Hansch equation including these properties.

Other factors

Other physicochemical factors such as dipole moments, basicity, and the presence of specific substituents may be included in QSAR equations.

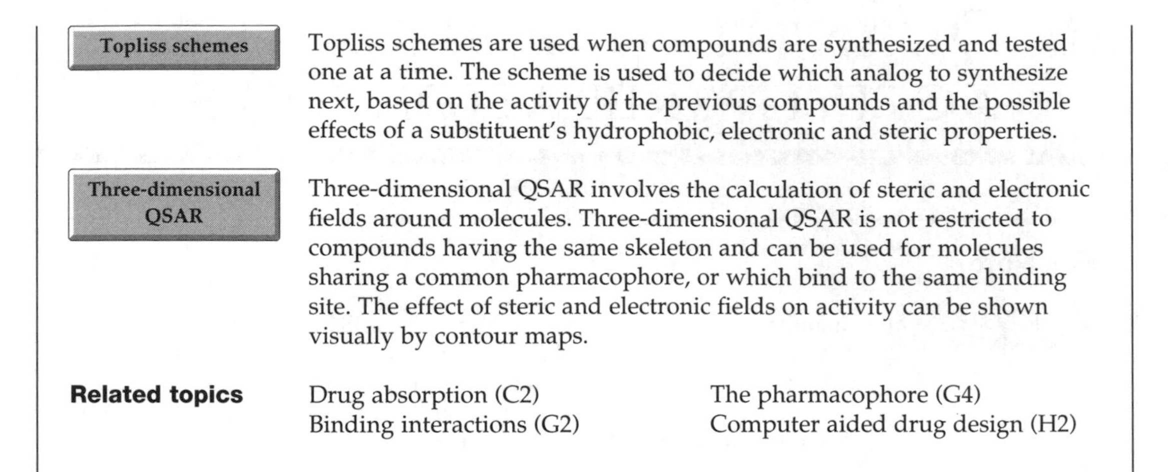

Topliss schemes

Topliss schemes are used when compounds are synthesized and tested one at a time. The scheme is used to decide which analog to synthesize next, based on the activity of the previous compounds and the possible effects of a substituent's hydrophobic, electronic and steric properties.

Three-dimensional QSAR

Three-dimensional QSAR involves the calculation of steric and electronic fields around molecules. Three-dimensional QSAR is not restricted to compounds having the same skeleton and can be used for molecules sharing a common pharmacophore, or which bind to the same binding site. The effect of steric and electronic fields on activity can be shown visually by contour maps.

Related topics

Drug absorption (C2)
Binding interactions (G2)
The pharmacophore (G4)
Computer aided drug design (H2)

Introduction

Quantitative structure–activity relationships (QSAR) is the study of how the physicochemical properties of a series of compounds affect their biological activity. Quantitative values are measured or calculated for the physical features and these are related to biological activities using a mathematical equation.

In order to carry out traditional QSAR, a range of analogs is synthesized which have a common skeleton, but which have different substituents. The activities of these analogs are measured and a formula is worked out relating these biological activities to the physical properties considered to be important (e.g. size, hydrophobicity, electronegativity, dipole moment, hydrogen bonding ability, etc.). Several of these physical features are likely to influence activity, and it would be ideal if one could synthesize analogs where one physical property was varied independently of any other. However, this is rarely possible. Changing one substituent for another usually results in several physical properties being altered at the same time. This also makes it difficult to identify whether one physical property is more important to activity than another. Therefore, in order to make sense of the data, it is necessary to make use of suitable computer software packages. Intuition alone is not enough.

There are several reasons why the physical properties of a compound should be important to biological activity. The overall hydrophobic character of a compound influences how efficiently it can cross cell membranes; the hydrophobic character and the size of individual substituents may influence how well the compound interacts and fits into its binding site, while the electronic character of substituents can influence the basicity of the compound, affecting both absorption and receptor binding. These are just a few of the factors that have to be taken into consideration.

Procedure

There are many software programs that help in deriving QSAR equations, but it is up to the medicinal chemist to decide what data to put in. Clearly, the biological activity of each compound has to be included, but the chemist has to decide which physical features might be most important to biological activity then derive an equation to test whether they really *are* important. QSAR is not a case of putting as much data as one can into a computer and hoping that the machine will make sense of it all. In fact, it is usually best to derive an initial equation based on only one or two physical features. Hopefully, the initial equation will give calculated activities close to the experimentally measured activities, but there will almost certainly be compounds which do not obey the mathematical equation – outriders. Such compounds should not be viewed as 'nuisances', quite the opposite in fact. The medicinal chemist can study these molecules and try to identify a physical feature which these molecules have which the others do not, then search for a more advanced formula that includes that property. QSAR equations constantly evolve and it is the medicinal chemist who directs that evolution.

Three physical properties are almost always considered in a QSAR equation. These are the hydrophobicity (or fatty character) of the molecule and/or its substituents, the electronic properties of its substituents and the size of its substituents.

Hydrophobicity of the molecule

The **hydrophobicity** of a molecule is normally measured by its **log*P*** value, where P is known as the **partition coefficient**. P can be measured experimentally by measuring the relative solubility of a compound in an octanol-water mixture where

$$P = \frac{\text{Conc. of compound in octanol}}{\text{Conc. of compound in water}}$$

The more hydrophobic the compound is, the greater proportion of it will dissolve in the organic layer, and the higher the value of P or $\log P$.

With most drugs, *in vivo* activity tends to increase as the $\log P$ value increases. In other words, activity increases with increasing hydrophobicity. This is usually an indication that increasing hydrophobicity allows easier passage of the drug through cell membranes in order to reach target sites. This might imply that one could keep increasing activity by continually increasing hydrophobicity. In fact, this is not the case. Most QSAR experiments are carried out on compounds that have a limited range of $\log P$ values. If compounds were synthesized with a much broader range of $\log P$ values, an optimum $\log P$ value would be found beyond which activity would fall. A parabolic curve would be the result (*Fig. 1*) having the formula:

$$\log(\text{activity}) = -k_1(\log P)^2 + k_2\log P + k_3$$

where k_1, k_2 and k_3 are constants.

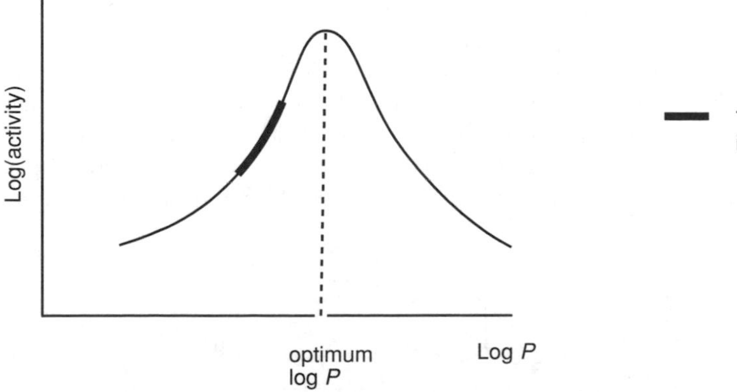

Fig. 1. Parabolic curve relating log(activity) vs. logP.

Looking at this equation, the $-(\log P)^2$ entry has a negative effect on activity, whereas the $\log P$ entry has a positive effect. When P is low, $\log P$ is more important than $-(\log P)^2$. Therefore, in the first part of the curve, $\log P$ is more significant and activity increases as $\log P$ increases. In the second part of the curve, the $-(\log P)^2$ factor becomes dominant and activity falls.

Log$P$ is present in most QSAR equations involving *in vivo* activity, since activity is dependent on drugs crossing cell membranes. If activity is measured by *in vitro* tests, the $\log P$ factor may be less significant and may even be absent.

Substituent hydrophobicity constant

The partition coefficient describes the overall hydrophobicity of a molecule, but it is also possible to quantify the hydrophobic character of individual substituents using sets of tables that give hydrophobicity constants (π) for each substituent. The **substituent hydrophobicity constant** is a measure of how hydrophobic a substituent is, relative to hydrogen. The value is measured experimentally by comparing the $\log P$ values of a standard compound with and without the substituent. The hydrophobicity constant (π_X) for the substituent (X) is then obtained using the following equation:

$$\pi_X = \log P_X - \log P_H$$

where P_H is the partition coefficient for the standard compound, and P_X is the partition coefficient for the analog containing the substituent. If π is positive, then the substituent is more hydrophobic than hydrogen. If π is negative, then the substituent is less hydrophobic than hydrogen.

It may not seem obvious why this would be useful if one has already measured the hydrophobicity of the molecule as a whole. However, there are two reasons why such constants can be useful. First of all, hydrophobicity constants can be used to calculate $\log P$ values for different compounds, avoiding the need to measure each $\log P$ value experimentally. For example, the $\log P$ value for *para*-bromoanisole can be calculated as 2.97, given the $\log P$ value for benzene (2.13) and the π constants for bromine and methoxy (0.86 and –0.02 respectively) (*Fig. 2*).

The π constants are only truly relevant for the structures that were used to determine them. Therefore, the aromatic π constants are relevant for substituted benzenes, but are less reliable for heteroaromatic ring systems. Similarly, the aromatic π constants are not relevant for aliphatic substituents and there is a separate table of constants for the latter. Ideally, accurate π values should be obtained experimentally for the molecular skeletons being studied. However, this may not always be possible and the values derived from different systems may need to be used as an approximation.

The second reason π values can be useful in QSAR is that they might be introduced into the QSAR equation itself to identify whether hydrophobic substituents at particular parts of the molecular skeleton have any localized influence on activity. In fact, it is perfectly possible for a QSAR equation to contain both $\log P$ and π. The former measures how the hydrophobic character

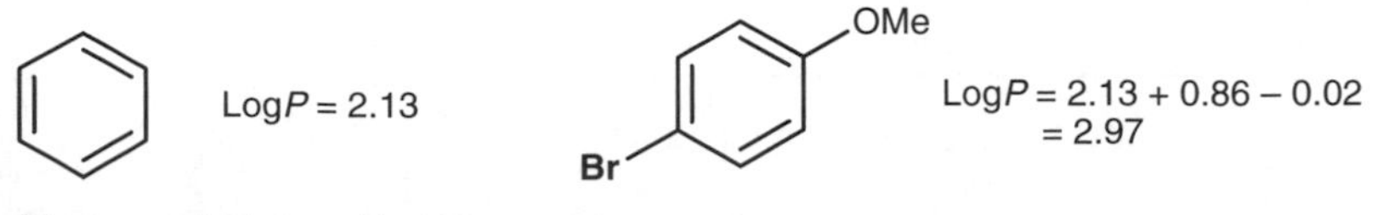

Fig. 2. Calculation of logP for para-bromoanisole.

of the molecule as a whole influences activity through such factors as the ability to cross cell membranes, while the latter demonstrates any localized hydrophobic influences that are at work. For example, the discovery that hydrophobic substituents at the *para* position of an aromatic ring are beneficial to activity might indicate that there is a hydrophobic binding pocket in the binding site that can accommodate such substituents.

Electronic properties

The **electronic properties** of different substituents can also play an important role in affecting biological activity. The electronic properties of aromatic substituents are described by the **Hammett substitution constant**, σ. These constants are available in tables and were measured experimentally by measuring what effect the substituents had on the dissociation of benzoic acids. Benzoic acid itself is a weak acid and only partially ionizes in water (*Fig. 3*). An equilibrium is set up between the ionized and nonionized forms, where the relative proportions of these species is known as the equilibrium or **dissociation constant** K_H (the subscript H signifies that there are no substituents on the aromatic ring):

$$K_H = \frac{[\mathrm{Ph\ CO_2^-}]}{[\mathrm{Ph\ CO_2H}]}$$

Substituents on the aromatic ring will affect this equilibrium. Electron withdrawing groups will stabilize the carboxylate anion and the equilibrium will shift to the ionized form and result in a larger equilibrium constant. An electron donating group will destabilize the carboxylate ion such that the equilibrium shifts to the left and results in a smaller equilibrium constant. The Hammett substituent constant (σ_X) for a particular substituent (X) is defined by the following equation:

$$\sigma_X = \log \frac{K_X}{K_H} = \log K_X - \log K_H$$

As with hydrophobic constants, these constants are only accurate for the molecular structures from which they were derived. Electron withdrawing substituents (e.g. Cl, CN, CF_3) have positive σ values while electron donating substituents (e.g. CH_3, CH_2CH_3) have negative σ values. The value of the Hammett substituent takes into account both the substituent's inductive and resonance effects, and so the value depends on whether the substituent is *meta* or *para to* the rest of the molecule (σ_m and σ_p). For example, σ_m for a phenol group is 0.12, reflecting the electron withdrawing influence felt at the *meta* position due to induction (*Fig. 4*). When the phenol group is at the *para* position, σ_p is –0.37 reflecting the fact that the group is now electron donating at that position due to resonance. It should be noted that aromatic substitution constants for *ortho* substitution are unreliable since *ortho* substituents can have a steric as well as an electronic effect.

As mentioned above, the Hammett substitution constants take into account both the inductive and resonance effects of the substituent. However, there are

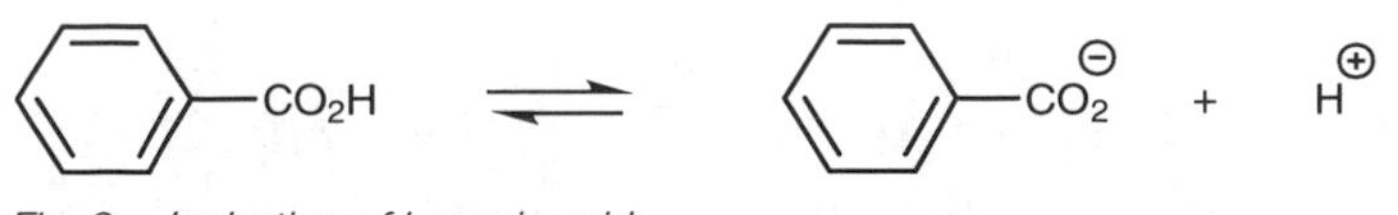

Fig. 3. Ionization of benzoic acid.

meta-substitution (inductive effect of phenol predominates at *meta* position)

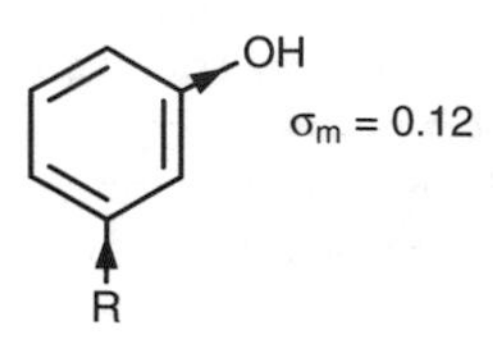

para-substitution (resonance effect of phenol predominates at *para* position)

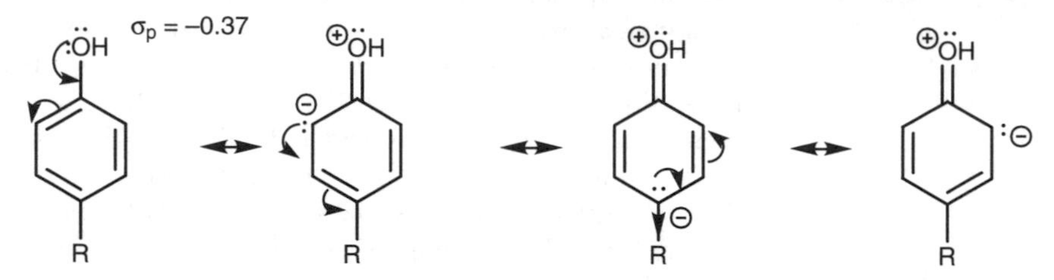

Fig. 4. Electronic influence of a phenol at the ortho and meta positions.

other constants which separately quantify the inductive effect (*F*) or the resonance effect (*R*) of aromatic substituents.

Aliphatic electronic substituent constants have been obtained by measuring the rates of hydrolysis for a series of aliphatic esters (*Fig. 5*), where methyl ethanoate is the parent ester. The extent to which the rate of hydrolysis is affected is a measure of the substituent's electronic effect, which arises purely from inductive effects. Electron donating groups reduce the rate of hydrolysis and therefore have negative values. Electron withdrawing groups increase the rate of hydrolysis and have positive values. Bulky substituents may also have a steric effect on the rate of hydrolysis by shielding the ester from attack. It is possible to separate out these two effects by measuring hydrolysis rates under basic and acidic conditions. Under basic conditions, steric and electronic factors are important, whereas under acidic conditions only steric factors are important. By comparing the rates, values for the electronic effect (σ_I), and for the steric effect (E_S) (see below) can be determined.

The steric factor

The size of different substituents can clearly be important to the activity of compounds. Bulky groups may lower activity by preventing drugs from fitting properly into the binding site. On the other hand, a bulky substituent may increase activity by forcing a compound to adopt the required active conformation for binding. Measuring the steric properties of substituents is not as straightforward as the measurement of a substituent's hydrophobic or electronic character. However, there are three methods that are generally used.

Taft's steric factor (E_s) is a measure of a substituent's size and is determined experimentally by measuring the effect different substituents have on the rate of a chemical reaction carried out on the parent structure (see above). Large

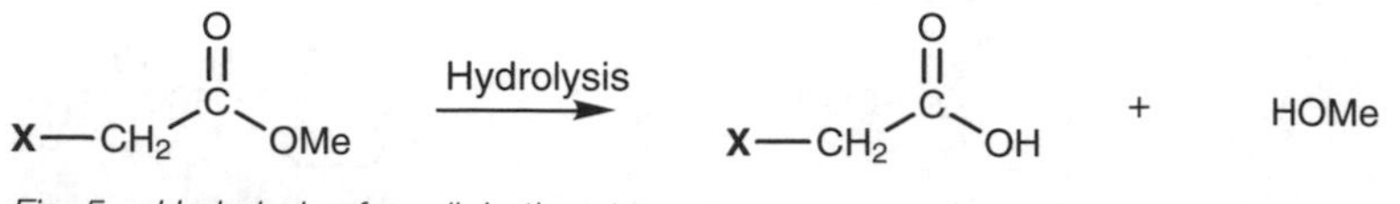

Fig. 5. Hydrolysis of an aliphatic ester.

substituents next to the reaction center hinder the reaction more than smaller substituents, so differences in reaction rate lead to a measure of a substituent's size.

It is also possible to calculate the **molar refractivity** (*MR*) of a substituent as a measure of its size, using the substituent's molecular weight (*MW*), density (*d*) and index of refraction (*n*):

$$MR = \frac{(n^2 - 1)}{(n^2 + 2)} \times \frac{MW}{d}$$

The term MW/d defines a volume, while the $(n^2 - 1)/(n^2 + 2)$ term provides a correction factor by defining how easily the substituent can be polarized. This is particularly significant if the substituent has π electrons or lone pairs of electrons.

A third method of determining size is to use a computer software program called **Sterimol**, which calculates steric factors known as V**erloop steric parameters**. In order to do this, the program measures the standard bond angles, van der Waals radii, bond lengths, and possible conformations for the substituent. The advantage of this approach is that a Verloop steric parameter can be calculated for any substituent without the need for any experimental measurements. For example, the Verloop steric parameters for a carboxylic acid group are shown in *Fig. 6*. L is the length of the substituent while B_1–B_4 are the radii of the group.

Hansch equation

The *Hansch equation* is the QSAR equation that relates physicochemical properties to activity. Typically, these equations will include a variety of parameters, the most common being log*P*, π, σ, *F*, *R* , *MR* and E_S. For example, a typical Hansch equation would have the following format.

$$\log\left(\frac{1}{C}\right) = -k_1(\log P)^2 + k_2\log P + k_3\pi + k_4\sigma + k_5\mathrm{ES} + k_6$$

where k_1–k_6 are constants. These constants would be determined by computer in order to get the best fitting line. Activity is often measured by $1/C$, where *C* is the concentration of a drug required to produce a specific effect (e.g. the concentration required to produce 50% inhibition of an enzyme). The more active the drug, the smaller the concentration required, and the larger the value of $1/C$.

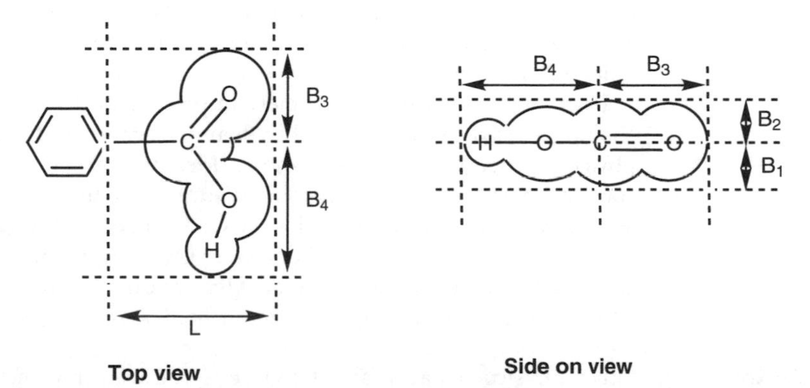

Fig. 6. Verloop steric parameters for a carboxylic acid.

An example of a Hansch equation is shown in *Fig. 7* for the inhibitory activity of a series of *N*-(phenyloxyethyl)cyclopropylamines against the enzyme **monoamine oxidase**. This shows that hydrophobic substituents that are electron withdrawing are good for activity since positive values of π and σ increase activity. The $E_{S(3,5)}$ factor represents the Taft steric parameters of any substituents which are at the *meta* position. This shows that bulky groups are bad for activity since such groups have a negative E_s value.

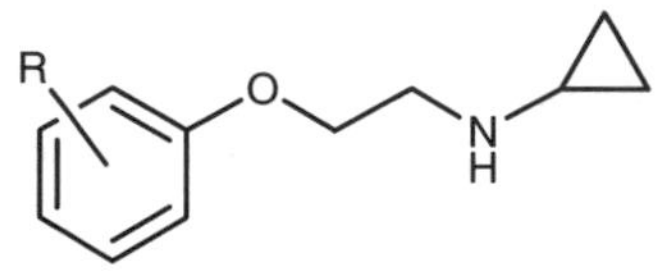

$$\log \frac{1}{C} = 0.398\pi + 1.089\sigma + 1.03E_{s(3,5)} + 4.541$$

Fig. 7. Hansch equation for monoamine oxidase inhibition.

It should be appreciated that a QSAR equation is only as good as the data that has been entered. Usually, a range of compounds are synthesized to quantify the effect that two or three physical parameters (e.g. π and σ) have on biological activity. However, it is crucial to choose a valid set of compounds in order to carry out the analysis. This means that the substituents present must give a good range of values for the physical parameters concerned, but they must also distinguish between the parameters such that they are not correlated (i.e. follow the same trend). For example, consider the substituents F, Cl, Br and I. The electron withdrawing effect of these substituents follows the trend, F, Cl, Br, I, where I is most electron withdrawing and F is least electron withdrawing. However, the hydrophobic character of these substituents also increases in the same sequence. Therefore, there is no way of knowing whether any similar trend in biological activity is due to the hydrophobic or the electronic properties of the substituents. In order to identify suitable substituents for QSAR studies, it is useful to consult Craig plots.

Craig plots

Craig plots compare two separate physical properties for different substituents. The example in *Fig. 8* compares the σ and π properties of different substituents by plotting the σ values on the y-axis and the π values on the x-axis. The plot reveals groups that are similar with respect to both properties and those that are not. For example, a cyano group and a methyl ketone group have similar σ and π values, whereas a cyano and trifluoromethyl group have similar σ values but quite different π values.

Plots such as these are extremely important in deciding which substituents should be used in a QSAR study when one wants to distinguish between two physical properties such as σ and π. Ideally, the compounds studied should contain substituents from all four quadrants of the plot. Therefore, choosing analogs with the substituents Cl, Br, I, CF_3 and NO_2 (all from the same quadrant) would not allow the derivation of a valid QSAR equation since it would be difficult to tell whether variations in activity are due to σ or π. Substituents such as Cl, CF_3, Et, OH and CN would be more valid.

Other factors

Other physicochemical factors can be introduced into the Hansch equation to try and improve the correlation between structure and biological activity. For

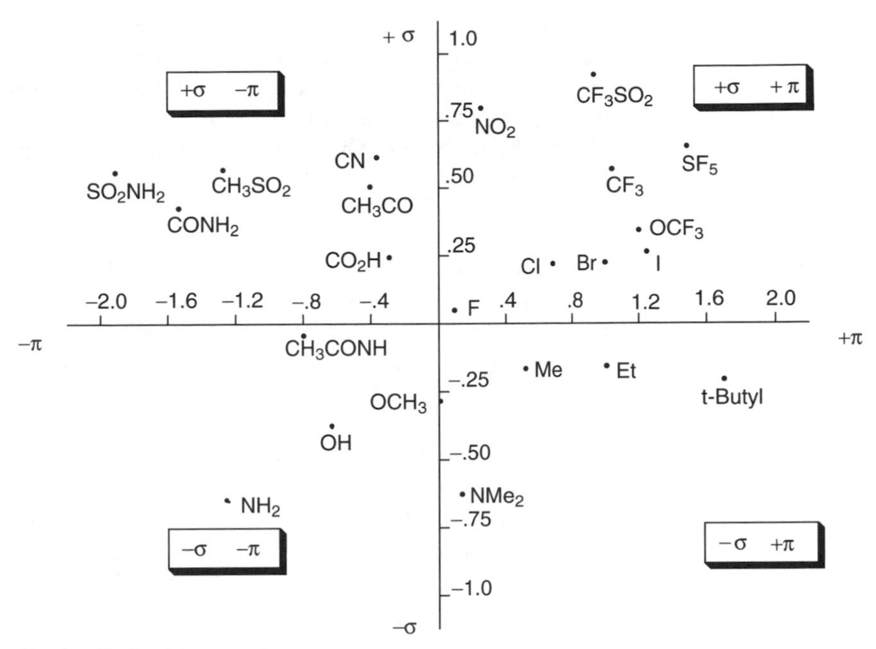

Fig. 8. Craig plot comparing σ vs π.

example, QSAR equations have been derived which measure how the orientation of a dipole moment affects biological activity. Other QSAR equations have been derived which relate molecular orbitals to activity.

Sometimes it is necessary to introduce a parameter that recognizes the importance of having a particular substituent at a specific position of the molecule. This involves putting an entry into the equation that contributes a specific value when the group is present, but contributes nothing when the group is absent – an approach known as the **Free-Wilson approach** to QSAR. For example, the QSAR equation for the antimalarial activity of 2-phenylquinolines was derived as shown in *Fig. 9*. This equation shows that activity increases slightly with increasing hydrophobicity up to an optimum value of log$P$. However, the variation is slight as shown by the small factors of 0.043 and 0.36. Activity also increases if the substituents are hydrophobic and electron withdrawing, where $\Sigma\pi$ and $\Sigma\sigma$ represent the combined π and σ factors for all the substituents present. If a substituent is present at position 4′, there is a significant steric effect

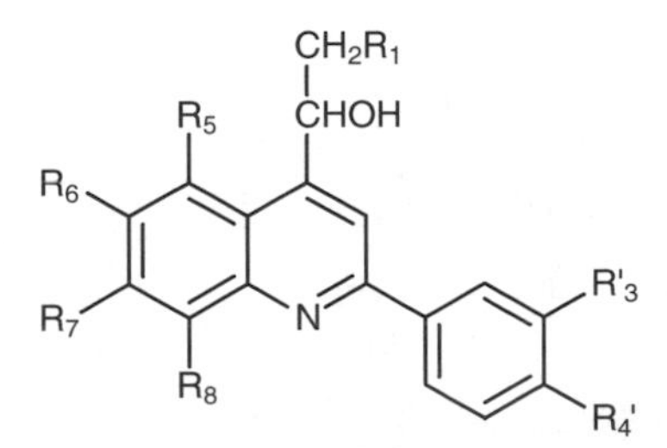

$$\log \frac{1}{C} = -0.043(\log P)^2 + 0.36\log P + 0.36\Sigma\pi + 0.62\Sigma\sigma - 0.48MR\text{-}4' + 0.40\ [\text{c-side}] + 0.34\ [\text{CH}_3\text{-}6,8] + 0.25\ [\text{2-Pip}] + 2.06.$$

Fig. 9. QSAR equation for the antimalarial activity of 2-phenylquinolines.

(represented by *MR*-4′). The remaining factors are examples of the Free-Wilson approach and are given a value of 1 if a particular substituent is present, and 0 if it is not. The factor [c-side] represents a cycloalkyl group at R_1. The presence of such a group increases activity by 0.40. The factor [CH_3-6,8] is given the value 1 if R_6 and R_8 are both methyl groups, thus increasing activity by 0.34. The factor [2-Pip] is given the value 1 if R_2 is a 2-piperidinyl group, thus increasing activity by 0.25.

Topliss schemes

Topliss schemes are useful in planning which analogs to synthesize if compounds are being synthesized and tested one at a time. There is a Topliss scheme for aromatic substituents (*Fig. 10*) and one for aliphatic substituents. The scheme is designed to rationalize differences in activity based on the various factors that we have already discussed: the hydrophobic, electronic and steric factors.

In order to use the Topliss scheme for aromatic substituents, the lead compound has to have an aromatic ring. The first analog to be synthesized would then be the 4-chloro analog. The Cl group is hydrophobic and electron withdrawing. If the activity of the chloro analog proves to be greater than the lead compound, it is assumed that both of these properties are important, and so a second chloro group is introduced at the *meta* position to enhance both effects.

If the activity of the 4-chloro analog drops, it suggests that both properties are bad for activity and so a more polar, electron donating group (OMe) is placed at that position instead. If activity drops again, it suggests that *para* substitution may be bad for activity for steric reasons, and so a *meta* chloro group is tried. On the other hand, if activity improves, further substituents are tried to determine the relative importance of hydrophobic and electronic properties.

The third possibility is that the activity of the 4-chloro-derivative is similar to the lead compound in which case hydrophobicity may be good for activity, but an electron withdrawing group is bad for activity. An electron donating, hydrophobic group (CH_3) would then be tried to test this theory out. Progress through the 'tree' is then continued based on similar arguments regarding the hydrophobic, electronic and steric properties of the substituents involved.

Three-dimensional QSAR

Three-dimensional QSAR is different from conventional QSAR in several respects and offers several advantages over it. In 3-D QSAR, the physical properties of each molecule are calculated by computer and no experimental

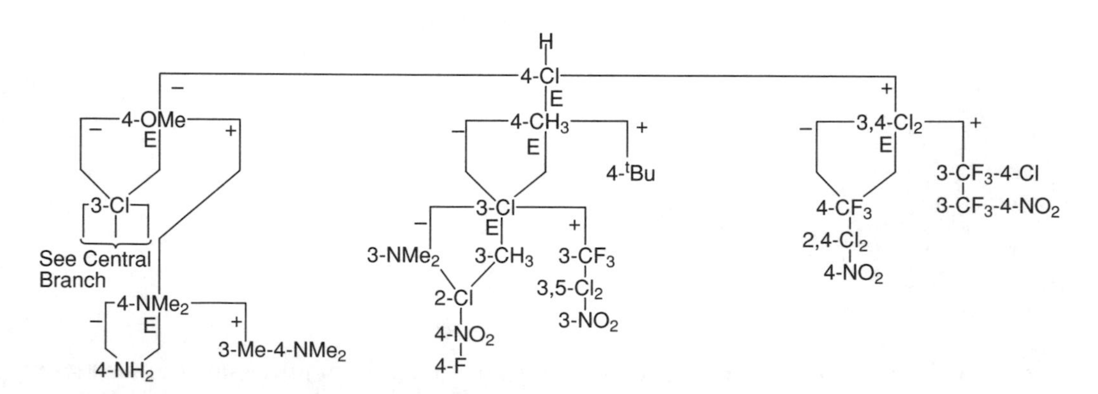

Fig. 10. Topliss scheme for aromatic substituents.

constants are involved. Secondly, the properties calculated are for the overall molecule and not for individual substituents. These properties are known as **fields** and most 3-D QSAR studies involve the calculation of a **steric field** and an **electrostatic field**. The former defines the shape of the molecule, and the latter defines the electronic character of the molecule. Interestingly, hydrophobic properties are much less important in 3-D QSAR than they are in normal QSAR.

Since the above fields are calculated, it is possible to study a larger variety of molecules than by QSAR. For example, traditional QSAR would not be able to cope with molecules having unusual substituents, since there would be no experimental constants recorded for those. Secondly, QSAR is restricted to molecules having the same basic skeleton, whereas this is not the case for 3-D QSAR. Thirdly, unlike normal QSAR, 3-D QSAR can be used to propose new molecules for synthesis.

The most popular method for carrying out 3-D QSAR is to use a software program called CoMFA (**comparative molecular field analysis**), which was developed by Tripos Software Ltd. Each molecule in the study is built on the computer using molecular modeling software. If the molecule can exist in several conformations, then the active conformation must be identified, otherwise comparison with other molecules would be meaningless. The pharmacophore for each molecule is then defined (*Fig. 11*).

The next stage involves fitting each molecule in turn into a lattice of grid points (*Fig. 12*). Here, it is important to ensure that each molecule is fitted into the lattice in the same relative position and orientation. This is helped if a

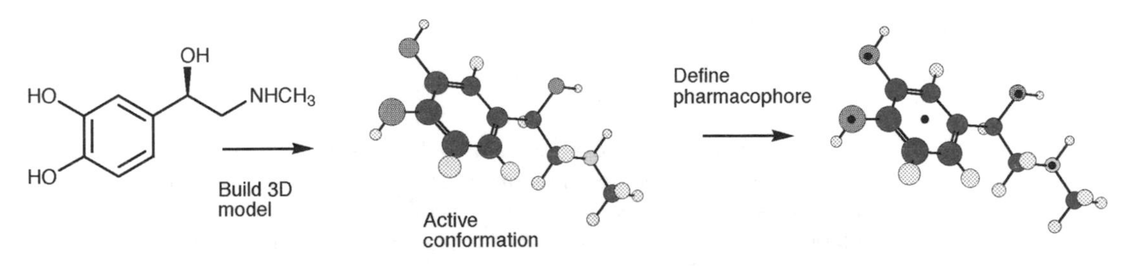

Fig. 11. Definition of active conformation and pharmacophore.

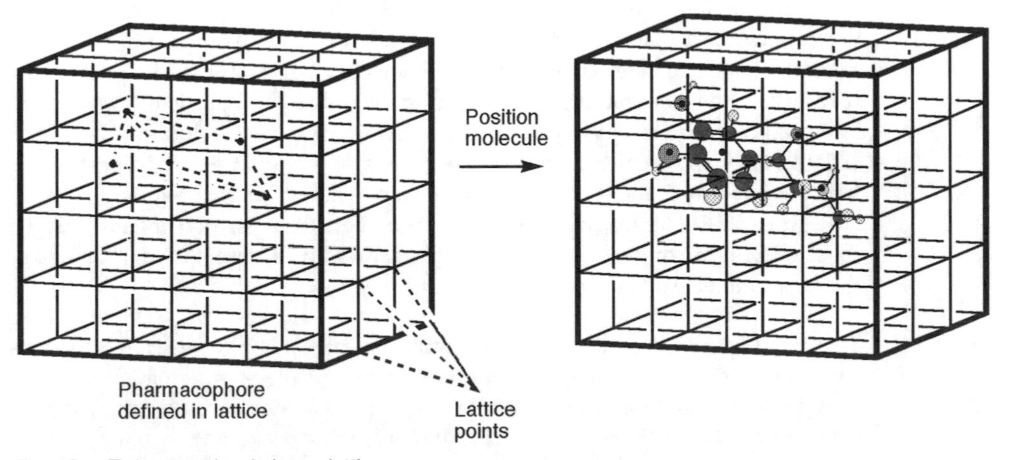

Fig. 12. Fitting a molecule into a lattice.

common pharmacophore for each molecule is known, in which case each molecule is positioned to ensure that the pharmacophore is positioned identically in the lattice.

Once a molecule has been fitted correctly into the lattice, the steric and electrostatic fields around it can be measured and defined. This is done by placing a **probe atom** (such as a proton or sp^3 hybridized carbocation) at each of the grid points in turn and calculating the steric and electrostatic interactions between it and the molecule (*Fig. 13*). The closer the probe atom is to the molecule, the higher the steric interaction energy will be. The grid points that have the same steric interaction energy can then be connected by contour lines to define the steric field. By choosing a suitable contour line, the shape of the molecule can be defined.

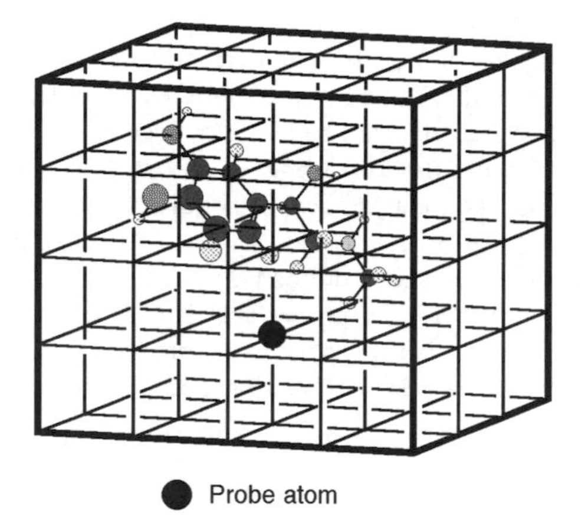

Fig. 13. Use of a probe atom to measure fields.

The electrostatic field can be calculated in a similar fashion. Since the probe atom is positively charged, favorable electrostatic interactions indicate regions of the molecule that are negative in character, while unfavorable electrostatic interactions will indicate regions of the molecule that are positive in character. Once again, grid points of the same energy can be joined by contour lines to indicate the positive and negative regions of the molecule.

Having defined the steric and electrostatic fields, the biological activities of the molecules can be compared with their calculated fields to identify whether there is any steric or electronic effect on activity. For example, the study may reveal that molecules having a bulky group positioned in a particular part of the lattice all show good activity, indicating that a group in this location is beneficial. Similarly, molecules having an electron-rich region at a certain part of the lattice may show that electron-rich groups are beneficial in that region.

Regions in the lattice that are shown to be beneficial (or detrimental) can be revealed with a contour map, constructed around a representative structure of the compounds tested (*Fig. 14*).

Case study

Tacrine (*Fig. 15*) is a reversible inhibitor of an enzyme called **acetylcholinesterase** and was the first drug to be used for the symptomatic treatment of Alzheimer's disease. Twelve tacrine analogs were synthesized with different substituents at positions 6 and 7 in order to study how the hydrophobic, steric and electronic properties of the substituents affected activity. The substituents

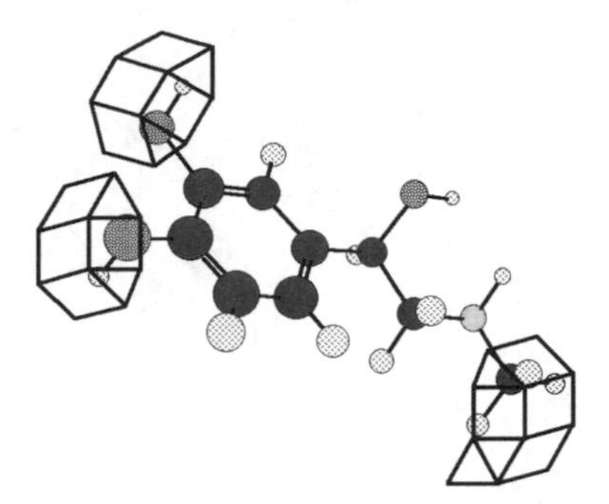

Fig. 14. Contour map revealing beneficial or detrimental regions for electrostatic or steric fields.

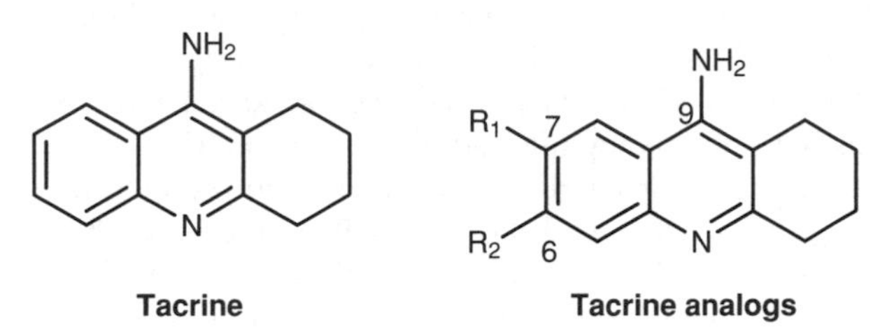

Fig. 15. Tacrine and tacrine analogs.

chosen were CH_3, Cl, NO_2, OCH_3, NH_2, and F since they provided a spread of values for each physical property and none of the properties were correlated across the range of substituents.

The physical parameters π, *MR*, *F* and *R* were tested and the best equation obtained was as follows:

$$\log(1/IC_{50}) = pIC_{50} = -3.09\, MR(R^1) + 1.43\, F(R^1, R^2) + 7.00$$

This demonstrated that the size of the substituent at position 7 was detrimental to activity since the value was negative. The combined electronic inductive effects of substituents 6 and 7 was beneficial when they were electron withdrawing, since positive values are better for activity. No hydrophobic influence was found to be significant. Three-dimensional QSAR was now tried using CoMFA, including another structural class of tetracyclic acetylcholinesterase inhibitors (II) (*Fig. 16*). Analysis of these compounds in the above QSAR study was not possible since QSAR is restricted to the same structural class of molecules. Alignment of the tacrine analogs and the tetracyclic compounds is apparently straightforward since they both share a quinoline ring system (*Fig. 17*).

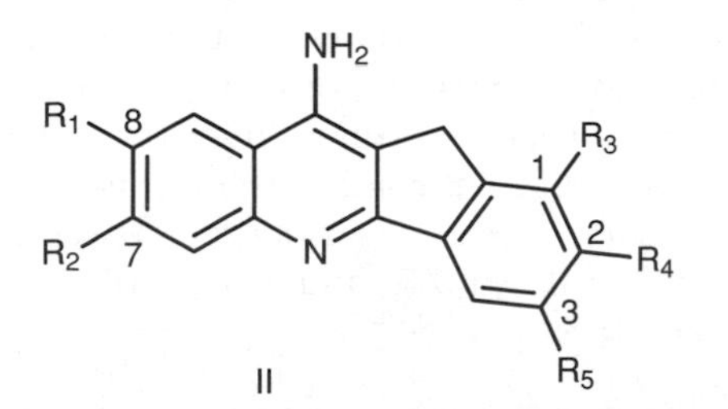

Fig. 16. Tetracyclic acetylcholinesterase inhibitors.

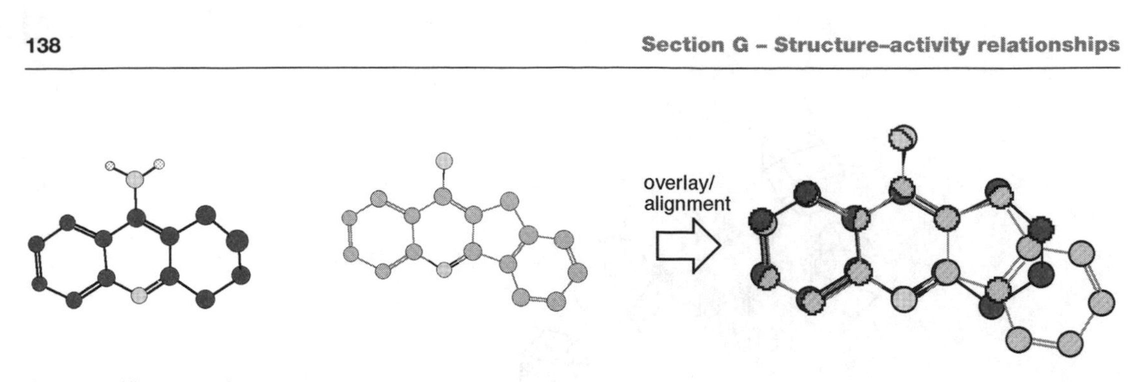

Fig. 17. Alignment of tacrine analogs and tetracyclic inhibitors.

It is therefore tempting to carry out the 3-D QSAR using this alignment. However, it is never safe to make assumptions and this is a case in point. It was possible to obtain an X-ray crystal structure of the tacrine–enzyme complex, which revealed how tacrine was bound to the active site of the enzyme. This was fed into a computer and molecular modeling was used to convert tacrine into the tetracyclic structure whilst still in the binding site. The complex was now minimized to find the most stable binding mode for the tetracyclic structure with the discovery that the binding mode for the tetracyclic structure was different from tacrine analogs. *Figure 18* shows the relative positions of the two structures when bound.

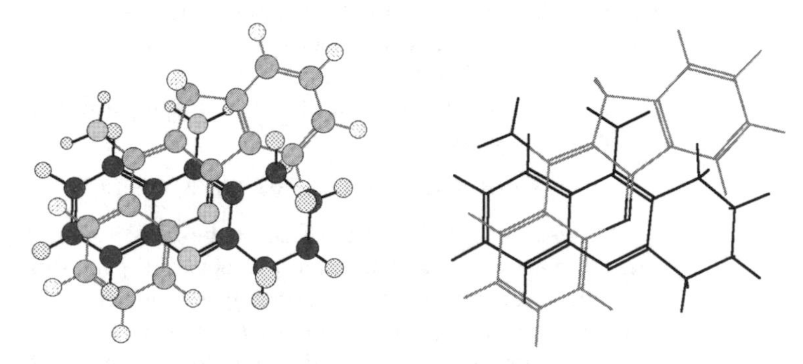

Fig. 18. Relative binding modes for tacrine analogs and tetracyclic inhibitors.

Therefore, the two sets of compounds were fitted into the active site and aligned with their parent structure before the 3-D QSAR was carried out, revealing that the steric field alone accounted for activity. The solid maps in *Fig. 19* show areas where the addition of steric bulk increases activity, whereas the dotted lines show areas where addition of steric bulk lowers activity. These results showed that substituents at position 6 of the tacrine analogs were favored, but that substituents at position 7 were not. The results also showed that adding steric bulk to the right-hand ring of tacrine would be favorable.

Based on these results, the tacrine analog (III) was synthesized (*Fig. 20*). QSAR predicted a pIC_{50} value of 7.31 and 3-D QSAR predicted a value of 7.40. The actual value was 7.18.

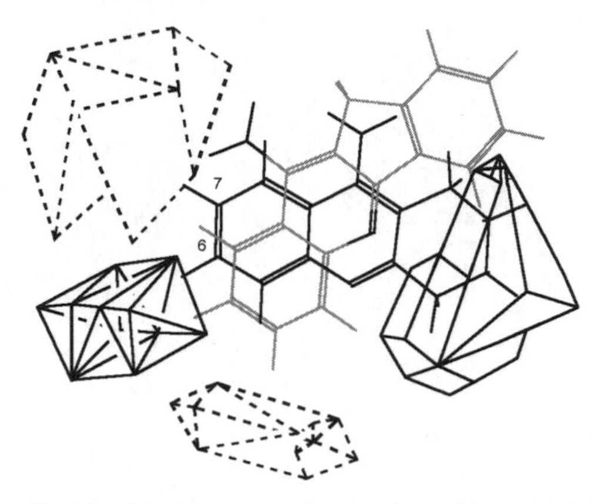

Fig. 19. Contour maps showing favorable and detrimental regions for steric bulk.

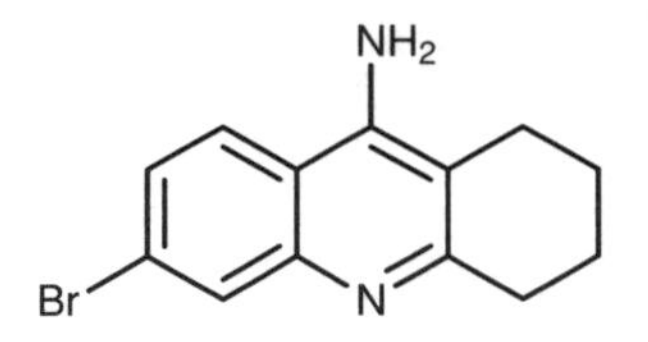

Fig. 20. Tacrine analog (III).

H1 AIMS OF DRUG DESIGN

Key Notes

Aims

The aims of target orientated drug design are to design easily synthesized drugs with optimal target binding and selectivity. New drugs must be the correct size and contain the required pharmacophore. It is also important to consider pharmacokinetic issues.

Drug design based on the lead compound

The traditional approach to drug design is to synthesize analogs of a lead compound, using classical design strategies. The results obtained can give clues as to the nature of the binding site.

Drug design based on the target structure

The structures of many target-binding sites have been worked out. *De novo* drug design involves the design of drugs based on the structure of the binding site. This has been useful in designing lead compounds.

Drug design based on the lead compound and target structure

Currently, a combination of traditional drug design and molecular modeling is used. Analogs are synthesized of a lead compound to determine the pharmacophore, which aids molecular modeling studies. Molecular modeling is used for the rational design of further analogs.

Related topics

From concept to market (A2)
Computer aided drug design (H2)
Simplification of complex molecules (H3)
Conformational restraint (H4)
Extra binding interactions (H5)
Enhancing existing interactions (H6)
The antibiotic age (1945–1970s) (M5)
The age of reason (1970s to present) (M6)

Aims

The aims of **target orientated drug design** are to improve the binding interactions between the drug and its target, and to increase the drug's selectivity for that target. Achieving the former should improve activity, while achieving the latter should reduce side effects. In order to achieve stronger and more selective binding, the drug must be the correct size to fit the binding site, have the required functional groups for binding, and have those functional groups in the correct relative position so that they can all bind at the same time. In other words, the correct **pharmacophore** must be present. There are another two considerations that should be kept in mind during such studies. It is no use designing drugs that are impossible to synthesize, so another aim of target orientated design is to design drugs that are as easy to synthesize as possible. Secondly, the drugs synthesized must be capable of reaching their targets, so it is important to keep in mind the limitations imposed by **pharmacokinetics** (see Section C).

Drug design based on the lead compound

The traditional approach to drug design has been to study the **structure activity relationships** of the lead compound and its analogs, and then to design new structures based on the information obtained. In the past, this was the only method of making any progress since little was known about the structures of drug targets. Indeed, the information obtained from SAR studies was often used to propose hypothetical structures for target binding sites. Pharmaceutical companies would employ teams of chemists to synthesize as many possible analogs of the lead compound as possible, using a standard synthetic route. The more compounds that were synthesized, the greater the chances of 'striking it lucky'. However, some design strategies emerged that could increase the chances of success, and which are still relevant today. Many of these are described in the following sections.

Drug design based on the target structure

Scientific advances have made the study of **drug targets** themselves more feasible. The gene for a particular receptor or enzyme can be cloned and inserted into the DNA of a fast growing tumor, yeast or bacterial cell. The protein is then produced in much greater quantity than in the natural cell and can either be isolated from the cell, or studied in the cell that produced it. The structures of drug targets and their binding sites have been studied using X-ray crystallography, NMR spectroscopy and molecular modeling. In theory, it is possible to design a drug based on the structure of the target binding site using molecular modeling, a process known as ***de novo* drug design**. However, this is not as straightforward as it may seem and, although lead compounds have been designed this way, no clinically useful drugs have been designed purely by *de novo* design.

Drug design based on the lead compound and target structure

Nowadays, most drug design projects involve a combination of the traditional and the modern approaches described above. The discovery of a lead compound is still crucial to the process, but once the lead compound has been discovered, the synthesis of analogs and molecular modeling studies are complimentary to each other. Analogs of the lead compound are synthesized for SAR studies and drug optimization, but molecular modeling studies are also carried out to discover how the lead compound interacts with the target binding site. The SAR studies can identify the pharmacophore and help in deciding how the lead compound binds to the binding site, and once the binding mode has been established, molecular modeling can be used to decide which analogs are worth making next. Therefore, instead of synthesizing every analog that could possibly be synthesized, molecular modeling helps to identify which analogs are most likely to bind to the binding site and to demonstrate how the traditional strategies of drug design can be used more effectively.

H2 Computer aided drug design

Key Notes

Introduction

Computers play an important role in many aspects of drug design and development.

3-D structures

The three-dimensional structures of compounds can be generated and studied using molecular modeling software.

Structure comparisons

Two structures can be compared to see how similar they are, or whether they share a common pharmacophore.

Protein structure

The structures of proteins can be determined by X-ray crystallography if suitable crystals can be grown. The structures of proteins belonging to the same family can be modeled using a known protein as a template.

Binding site studies

Molecular modeling can be used to study how a ligand binds to its binding site. Further analogs can be designed by identifying vacant regions in the binding site and adding substituents to the original ligand which will fit and bind to those regions.

***De novo* drug design**

De novo drug design involves the design of drugs based purely on the structure of the binding site. *De novo* drug design is useful in generating lead compounds. The lead compounds should be flexible and loose fitting to allow for any differences between the model binding site and the actual binding site.

Related topics

The pharmacophore (G4)
Quantitative structure–activity relationships (G5)
Extra binding interactions (H5)
Enhancing existing binding interactions (H6)
From lead compound to dianilino-phthalimides (L3)
Modeling studies (L4)
4-(Phenylamino)pyrrolopyrimidines (L5)
Pyrazolopyrimidines (L6)

Introduction

Computers are an integral part of the drug design process and have a large number of applications, which include structure analysis, structure comparisons, lead compound design, identification of active conformations and pharmacophores, combinatorial library design, protein and binding site structure, ligand binding, QSAR and 3-D QSAR. Some of these applications are described below while others (such as the design of lead compounds, pharmacophores, QSAR and 3-D QSAR) have been described in previous sections.

3-D structures

Molecular modeling software allows the medicinal chemist to create a model of a compound in 3D. The structure can first be drawn in two dimensions using a chemical drawing software package such as ChemDraw, then imported into a molecular modeling software package such as Chem3D. The 2-D structure is converted into a 3-D structure, which is quite clever, but is not error free, since the structure created is usually distorted (i.e. the bonds lengths and the bond angles are not ideal). Therefore, once a 3-D structure has been built or generated, it is important to carry out an operation called **energy minimization** (*Fig. 1*).

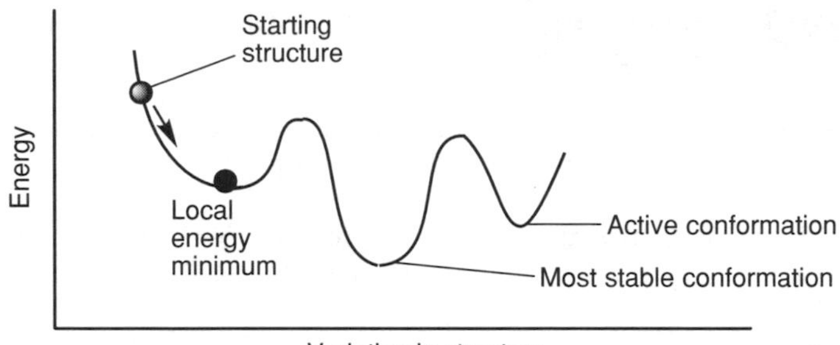

Fig. 1. Energy minimization.

This involves running a program that modifies the bond lengths and angles in the structure, then calculates the **steric energy** of the new model compared to the previous one. If the energy changes significantly, it means that neither structure is particularly stable and the process is repeated. Changes that decrease the total energy of the structure are retained, while those that increase the energy are not, and this continues until any modifications carried out have little effect on the total energy of the molecule. This corresponds to a stable structure or an energy minimum. Once this has been obtained, the molecule can be studied, and its dimensions measured (e.g. bond lengths, bond angles, torsion angles, atom–atom distances). However, it is important to realize that the energy minimized structure obtained may not be the *most* stable conformation possible since the energy minimization program will stop at the first stable conformation it meets (a **local energy minimum**). It is also wrong to assume that an energy-minimized conformation corresponds to an active conformation.

Structure comparisons

Once two compounds have been converted to their 3-D structures, molecular modeling can be used to compare how similar they are in structure. The molecules can be **overlaid** by identifying pairs of atoms where one atom of each atom pair belongs to one structure and the other atom is considered to be its equivalent in the second structure. An overlay operation is carried out where the program attempts to match up each defined pair of atoms. Once the overlay has been carried out, it is possible to measure how closely the corresponding atoms in each structure overlap with each other. This sort of operation is crucial when aligning molecules for 3D QSAR studies, and for comparing pharmacophores in different molecules. An example of overlaid molecules is given in Topic G4, where morphine is compared with a hypothetical structure containing the same pharmacophore. This overlay was carried out by pairing the aromatic atoms, the phenolic oxygens and the amino nitrogens in each structure (see also the overlay of acetylcholinesterase inhibitors in Topic G5).

Protein structure

Many important drug targets are **proteins**, hence identifying a protein's structure is extremely important to drug design. This requires the protein to be isolated and purified in the first place. However, in the past this was no easy task. A protein had to be extracted from the cells that normally produced it, and this could be a long, tedious process. Moreover, the concentration of a protein in a cell is often very low, making it impossible to obtain the protein in any great quantity. With the aid of genetic engineering, it is now possible to identify the gene responsible for a specific protein, introduce it into the DNA of a fast-growing cell, such as a microorganism or a tumor cell, and allow that cell to produce the protein in much greater quantity. Subsequent isolation and purification is more straightforward and higher yielding.

Once a protein has been purified, its structure can be identified. The **primary structure** (i.e. the amino acids present and their sequence) can be ascertained automatically using protein sequencing. The **secondary** and **tertiary structures** (i.e. the way in which the protein chain folds up into a three-dimensional structure) are best identified by crystallizing the protein and carrying out an X-ray crystallographic analysis. This gives the most accurate analysis of a protein's structure, but it would be wrong to assume that it is an accurate 'photograph' of the structure. All X-ray crystallographic analyses have an element of experimental error, so it is important to know the resolution of the analysis. In general, the position of atoms in the crystal structure is accurate to 0.2–0.4 Å.

Unfortunately, not all proteins can be crystallized – membrane-bound receptors are a case in point – so it is not possible to obtain the structures of these proteins from X-ray crystallography. However, it is still possible to create structural models of these proteins using molecular modeling if there is a closely related protein that *has* been crystallized and studied by X-ray crystallography. Fortunately, many proteins belong to families of proteins that have an overall structural similarity. This is because many proteins have evolved from a single protein ancestor. The specific amino acids may have altered in the primary sequence, but the overall secondary and tertiary structures have remained similar. For example, models of membrane-bound human **G-protein-coupled receptors** (Topic B2) have been constructed using a bacterial G-protein-coupled receptor called **bacteriorhodopsin** as a template (*Fig. 2*). The structure of bacteriorhodopsin consists of seven helices that traverse the cell membrane and are called the **transmembrane regions**. Intracellular and extracellular loops connect the transmembrane helices, and there is an *N*-terminal extracellular chain, and a *C*-terminal intracellular chain.

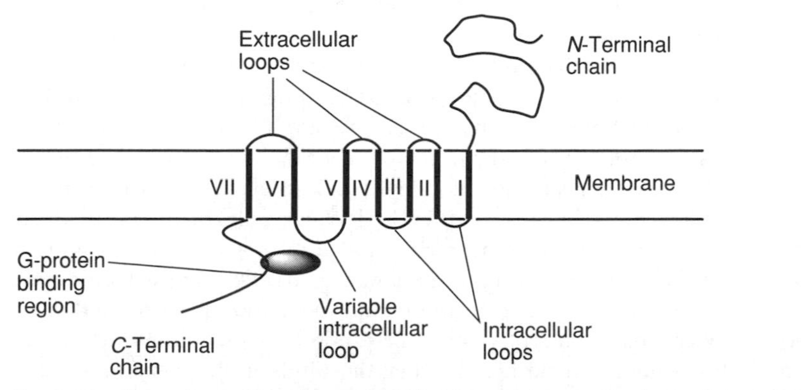

Fig. 2. Structure of bacteriorhodopsin. Modified from An introduction to medicinal chemistry.

Since human G-protein-coupled receptors belong to the same protein family as bacteriorhodopsin, it can be assumed that they too consist of seven transmembrane helices with an extracellular *N*-terminal chain and an intracellular *C*-terminal chain. In order to design a model of such a receptor, the primary sequence of the protein is first identified, then compared with the primary sequence of bacteriorhodopsin. Regions of the primary sequence that are dominated by hydrophobic amino acids are assumed to be the seven transmembrane helices embedded in the hydrophobic cell membrane. These are modeled first, using bacteriorhodopsin as the template, then the terminal regions and the regions linking the seven helices are identified and modeled by comparison with bacteriorhodopsin and other proteins which have similar sequences of amino acids.

Binding site studies

Several proteins have been crystallized allowing their structure to be identified by X-ray crystallography. However, this is of limited use in drug design for the following reasons: the structure of an enzyme or receptor gives no indication of where the binding site is; it gives no indication of the induced fit that arises following the binding of a ligand to its binding site; it also gives no indication of how the ligand binds to the binding site. For these reasons, it is better to bind a ligand (usually an antagonist or inhibitor, as they bind more strongly) to the protein's binding site, and then crystallize the protein–ligand complex. The X-ray structure of the complex then reveals where the ligand is bound to the protein, thus identifying the binding site. The structure of the complex can be downloaded into a computer and molecular modeling software used to study how the ligand binds to the protein. First of all, the program allows the operator to 'hide' most of the protein so only the binding site and the bound ligand are visible on the screen. The binding interactions can then be identified by measuring the distances between the atoms of the ligand and the closest atoms in the binding site. If atoms are within binding range and are capable of interacting with each other by some form of bonding, it can be assumed that this is an important binding interaction.

Having identified how the ligand binds to the binding site, the latter can be studied to see if there are any empty spaces not filled by the ligand. Once these have been identified, it is possible to design analogs of the ligand with extra substituents or groups attached that would fit into these vacant regions and allow the ligand to fit the binding site more snugly. Moreover, identifying the amino acid residues and peptide links that line these vacant regions would indicate what types of substituents could be added to the ligand in order to get further binding interactions. For example, if an empty hydrophobic pocket was identified, an extra alkyl substituent could be added to the ligand.

Once the interactions of the ligand with the binding site have been studied, the ligand can be 'removed' to leave the empty site. Potential new drugs can then be 'placed' into the binding site to see whether they fit and can form useful interactions. A **docking procedure** is carried out in a similar fashion to the overlay of two structures described earlier. In docking, pairs of atoms are identified that can interact with each other, with one atom of each pair belonging to the ligand and the other belonging to the binding site. Docking is then carried out where the program strives to fit the ligand in the binding site so that the distances between each pair of atoms correspond to the optimum binding distance. The operator can decide whether the binding site remains fixed in shape during the docking procedure or whether a certain amount of flexibility is allowed such that amino acid residues shift position. Docking procedures can be complicated by the fact that the positions of important atoms are not easily defined. For example, protons involved as H-bond donors are not fixed in one position in space due to single bond rotation.

Not all proteins can be crystallized and studied by X-ray crystallography. However, it is possible to construct a **model binding site** for such a protein by using molecular modeling if the binding site of a related protein has been identified. The known binding site is used as a template for the model site and the amino acids are altered to take account of the different amino acid sequences of each protein. Ligands can be docked by comparison with known binding interactions in comparable binding sites. Alternatively, docking experiments can be carried out to explore the various ways in which a ligand could bind to the model site and calculate which is the most stable docking interaction. In these experiments, the possible binding groups in the binding site and in the ligand are identified, but there is no constraint as to which binding groups are used in the ligand or the binding site.

De novo drug design

In theory, it should be possible to design a drug based on the structure of a binding site – a process known as ***de novo* drug design**. By knowing which amino acids are present in the binding site and where they are positioned, it should be possible to identify the binding interactions that could take place and then design a molecule that will fit and have the necessary functional groups to interact with the amino acid residues. In practice this is more difficult than it sounds and it has not been possible to design a clinically useful drug purely by *de novo* drug design. However, there has been success in designing lead compounds by *de novo* design, especially in the area of **thymidylate kinase inhibitors**. The procedure requires identification of a binding site with a bound ligand, as described above. The ligand molecule is then 'removed' to leave a vacant binding site. Novel drugs can be designed in a virtual fashion by studying how different molecules fit into the binding site. There is a temptation to design a ligand that would be a perfect fit for the binding site, but this is likely to result in failure for the following reasons.

- The position of the atoms in the binding site cannot be assumed to be precise as there is experimental error in the crystallographic structure. If the structure of the actual binding site is slightly different from the model and there is no room for flexibility, then a molecule that has been designed to be a perfect fit for the model may not bind at all to the actual binding site.
- It is possible that the designed molecule may not bind to the binding site exactly as predicted. If there is no room for flexibility, then the designed molecule may not bind at all.

Therefore, it is better to design a loose-fitting structure in the first instance and to check whether it binds as predicted. This is as far as *de novo* design can go since it is now necessary to synthesize the target structure, bind it to the protein and obtain an X-ray crystal structure of the ligand–protein complex to see how the structure binds in reality. Once this has been established, molecular modeling can be used as previously described to improve binding interactions.

In general, a lead compound designed by *de novo* techniques must bind in a stable conformation, be synthetically feasible, and be flexible rather than rigid. A flexible molecule is more likely to find an alternative mode of binding if the predicted mode does not occur.

H3 SIMPLIFICATION OF COMPLEX MOLECULES

Key Notes

Aims

Simplification is often used when the lead compound is a complex natural product. The aim is to find a simpler structure that is easier to synthesize, but retains the required pharmacophore.

Removing functional groups

Removing functional groups that are not part of the pharmacophore makes synthesis easier and may reduce side effects.

Removing rings

Removing excess rings makes synthesis easier and reduces the bulk of the molecule enabling it to fit the binding site more easily.

Removing asymmetric centers

The removal of asymmetric centers makes the synthesis of compounds easier and avoids the problem of different enantiomers having different activities.

Related topics

Stereochemistry (F2)
The pharmacophore (G4)
From lead compound to dianilino-phthalimides (L3)

Aims

Many lead compounds obtained from natural sources are **secondary metabolites** of quite complex nature that are extremely difficult to synthesize. In most cases, the full synthesis of a complex natural product is not viable, either in terms of the time taken or the cost involved. Therefore, the only way to obtain the compound is to extract it from its natural source. Unfortunately, the amount of secondary metabolite present may be very small, requiring the harvesting of large quantities of the natural source. The yield can be low and the extraction procedures time consuming, costly and environmentally damaging. Therefore, there is an advantage in designing simpler analogs that can be synthesized in the laboratory. Removing excess functional groups can also lower the chances of side effects, while reducing the size of the molecule may allow easier access to binding sites and greater activity.

Before simplifying the molecule, it is important to know the required **pharmacophore** so that it can be retained. This requires the identification of the important functional groups and the minimum carbon skeleton required to link those groups. The **simplification** process should then be carried out in gradual stages since oversimplification can have the opposite effect to what is desired, leading to a drop in activity and an increase in side effects (see Topic H4). (The beauty of complex molecules is that the carbon skeleton often holds the pharmacophore in a precise conformation that matches the active conformation.) Finally, it is important to realize that it is rarely possible to simplify molecules by merely 'pruning' excess groups off the lead compound. Normally, each simplified analog is synthesized from scratch, starting from simple starting materials.

Removing functional groups

Complex natural products tend to have several **functional groups**. Some of these functional groups will be important for activity, but others may cause side effects by allowing the molecule to bind to other targets. Removing such groups could therefore make the compound more selective in its action (*Fig. 1*). Moreover, the more functional groups that are present in a molecule, the more involved the synthesis will be. Not only do the functional groups have to be introduced at some point in the synthesis, but they also have to be protected so that they do not undergo unwanted reactions during the remainder of the synthesis.

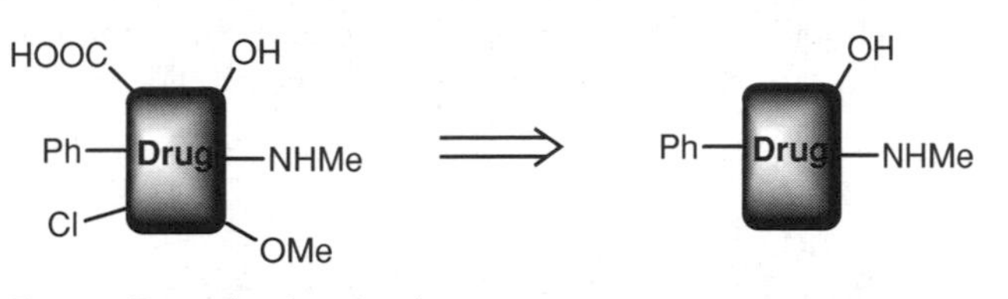

Fig. 1. Removing functional groups.

Removing rings

Many complex molecules contain several **rings**. However, some of these rings may be surplus to requirements and could be removed without affecting the pharmacophore. There is an advantage in removing rings from the carbon skeleton because it makes the synthesis much easier and may also remove several **asymmetric centers**. Moreover, the simplified compounds may show an increase in activity. Large complex molecules may struggle to fit the desired target binding site, and decreasing the size and bulk of the molecule may allow the analog to fit more easily. The strategy of removing rings has been beneficial in a wide range of drugs, including the **opiate analgesics** (*Fig. 2*). The removal of rings from morphine has resulted in analogs in which analgesic activity has been retained and which are much easier to synthesize. Note that the analgesic pharmacophore of phenol, aromatic ring and amine are retained in all three structures.

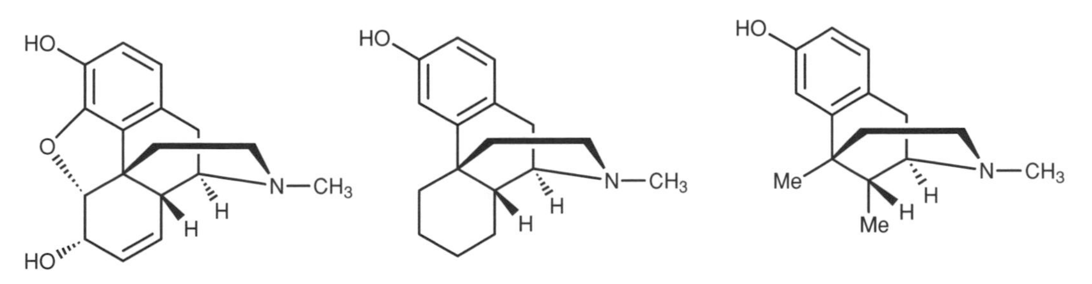

Fig. 2. Removing rings in opiates.

Removing asymmetric centers

Many complex natural compounds have **asymmetric centers** that are of a defined configuration. The more asymmetric centers present in a molecule the more involved the synthesis will be, since it is important to ensure the correct configuration at each center. The synthesis would be much easier if some or all of the asymmetric centers could be removed. Moreover, chiral drugs can exist as two enantiomers that may have different activities. Ideally, it would be better to design achiral drugs that do not have that complication.

Removing asymmetric centers is straightforward if the asymmetric centers are in nonessential regions of the skeleton, since that part of the skeleton can be removed entirely. However, other asymmetric centers may be present in portions of the skeleton that are responsible for holding the pharmacophore in place. In such cases, there are two strategies that can be used to remove the centers.

It is sometimes possible to replace an asymmetric carbon with a nitrogen atom (*Fig. 3*). The nitrogen atom is pyramidal (or tetrahedral if one includes the lone pair of electrons). However, it can also invert, which means that it cannot be an asymmetric center. This approach has been successful in several cases, but is not foolproof. Introducing a nitrogen atom means introducing a basic amine, which will alter the polarity, basicity and hydrophobicity of the molecule. This in turn can affect how easily the molecule reaches and binds to its target.

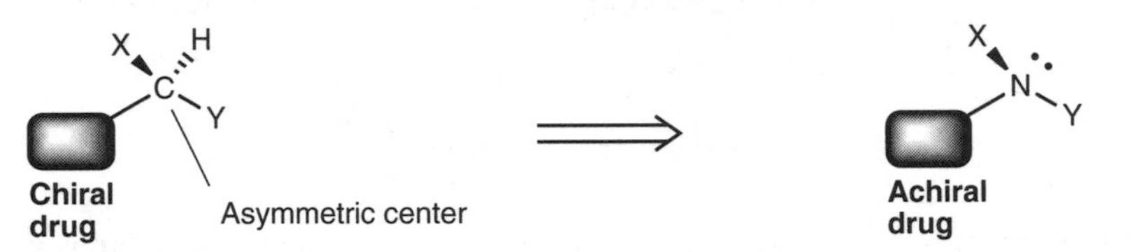

Fig. 3. Replacing an asymmetric center with nitrogen.

Another tactic which avoids introducing a basic center is to introduce symmetry to the molecule. For example, two different substituents at an asymmetric center could be made identical (*Fig. 4*).

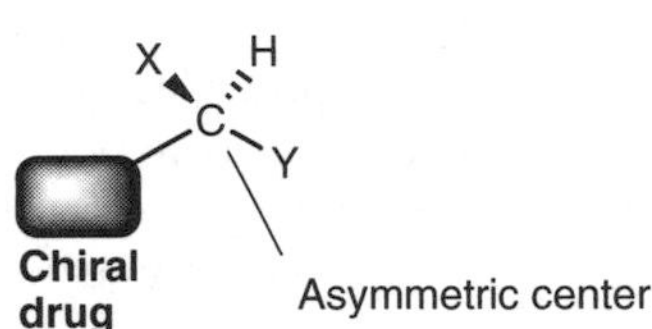

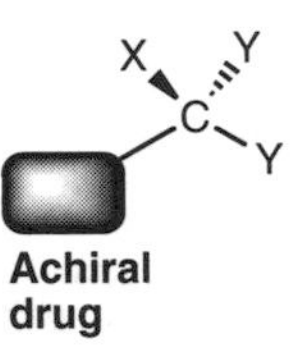

Fig. 4. Introducing symmetry.

H4 CONFORMATIONAL RESTRAINT

Key Notes

Aims	Conformational restraint is used on flexible molecules to design a more rigid molecule that retains the active conformation and loses inactive conformations. This should result in increased activity and reduced side effects.
Locking bonds within rings	The number of available conformations can be reduced by introducing a ring which includes previously rotatable single bonds.
Introducing rigid functional groups	Rigid functional groups such as an alkene, alkyne, amide or aromatic ring can be introduced into a flexible chain to reduce the number of possible conformations.
Steric blockers	Steric blockers are substituents that block particular conformations due to steric clashes. *ortho* Substitution on a ring will restrict the conformations of a neighboring side chain.

Related topics The pharmacophore (G4) Stereochemistry (F2)

Aims

Conformational restraint or rigidification is commonly used on a lead compound that is highly flexible and contains several rotatable single bonds (*Fig. 1*). The more flexible a molecule, the more conformations it can adopt and the less chance there is of it being in the active conformation when it meets its target. This results in low activity, and can also result in side effects if any of the alternative conformations allow the molecule to interact with a different type of target.

Therefore, the aim of conformational constraint is to design less flexible molecules which are either locked into their active conformation or which have fewer nonactive conformations available to them. Such molecules should show increased activity and greater selectivity for their target.

Conformational restraint can be a useful strategy to try where the lead compound is a naturally occurring neurotransmitter in the body. Many of these

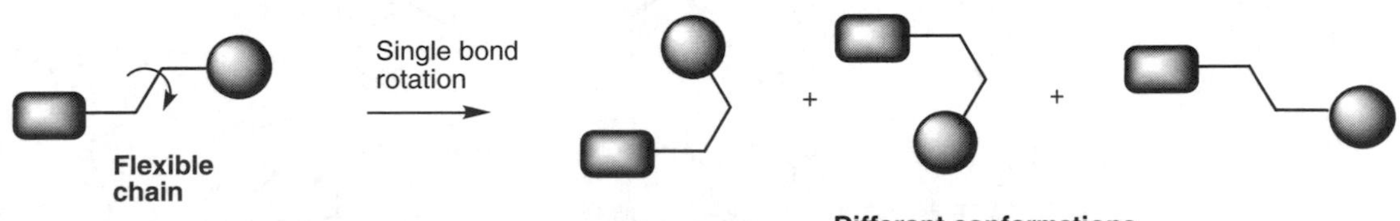

Fig. 1. Different conformations by single bond rotation.

are very simple molecules that are very flexible. For example, **norepinephrine** is a neurotransmitter that activates adrenergic receptors and has a simple, flexible structure that can adopt many different conformations (*Fig. 2*). It is not possible to predict which of these conformations is the active conformation. Moreover, the active conformation of norepinephrine may be different for the different types and subtypes of adrenergic receptors that are present in the body. Preparing more rigid analogs that retain the active conformation for one type of adrenergic receptor will increase both activity and selectivity for that receptor. Norepinephrine itself can be selective, since it is released by specific nerves to specific cells.

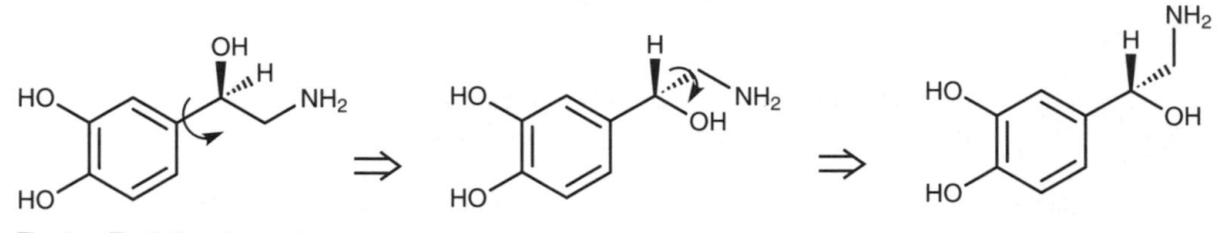

Fig. 2. Flexibility of norepinephrine.

There are potential problems with conformational restraint. Introducing the rigidity required usually involves making the structure more complex, which also makes the synthesis more difficult. Therefore, a compromise may have to be made between ease of synthesis and attaining rigidity. It is also possible that the rigidification process may result in the inactive conformations being retained, whilst abolishing the active conformation.

Locking bonds within rings

Introducing an extra ring into the lead compound is often used to prevent single bond rotation (*Fig. 3*). The extra ring is fused to an existing ring and includes bonds that were previously rotatable. Since these bonds are now part of a ring system, they are no longer free to rotate and fewer conformations are possible. The examples in *Fig. 3* demonstrate how the introduction of different rings can lock an ethylamine side chain into a variety of different conformations where the amine and the aromatic ring are in different relative positions. If both these groups are involved in binding the compound to its binding site, then the analog retaining the active conformation will be active.

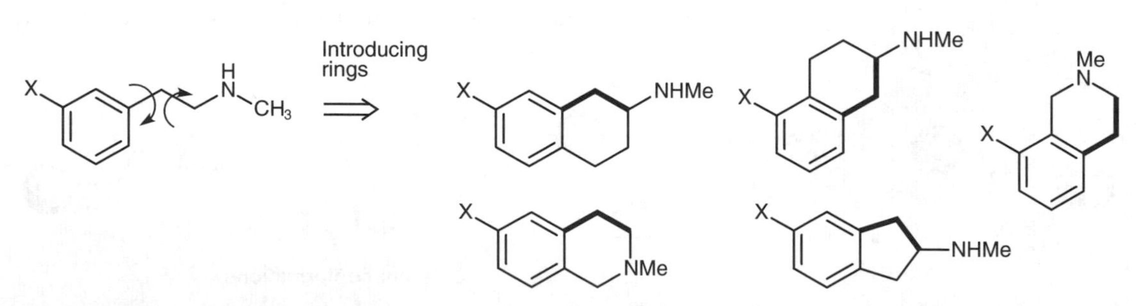

Fig. 3. Locking bonds within rings.

Introducing rigid functional groups

Another approach to conformational constraint is to introduce a **rigid functional group** into a flexible side chain where the chain connects two important binding groups (*Fig. 4*). The number of single bond rotations are thus reduced, making the active conformation more likely (assuming it is retained). Examples of rigid functional groups that have been used in this way include alkenes, alkynes, aromatic rings and amide groups.

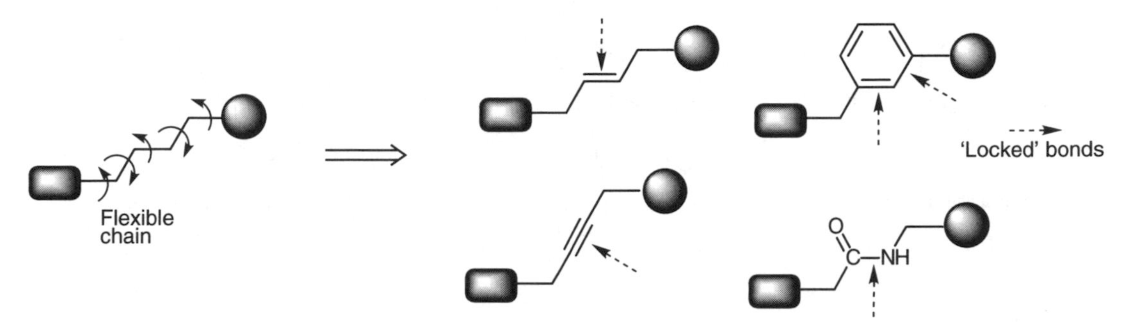

Fig. 4. Introducing rigid functional groups.

The anticancer agent, **combretastatin** (Topic E2), was made more rigid by the introduction of a double bond (*Fig. 5*). The *Z*-isomer was found to be more active than combretastatin, while the *E*-isomer was less active. Therefore, the *Z*-isomer retained the active conformation.

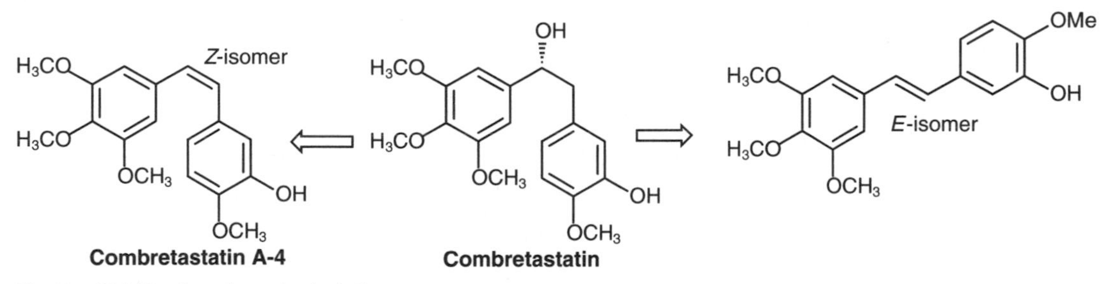

Fig. 5. Rigidification of combretastatin.

Steric blockers

An alternative approach to conformational restraint is to introduce substituents that make certain conformations unlikely due to unfavorable steric interactions. This can often be used to hinder the number of conformations available to a side chain attached to a ring system. For example, introducing a methyl group at the *ortho* position of an aromatic ring will prevent certain conformations (*Fig. 6*).

The tactic can also be used to force neighboring rings into particular confor-

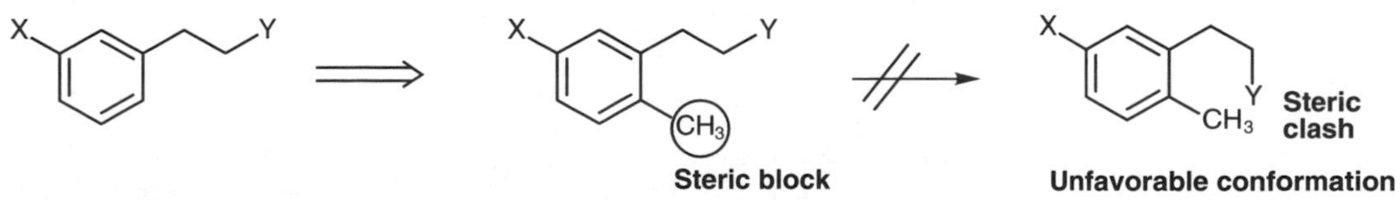

Fig. 6. Steric blocking.

mations. For example, introducing an *ortho* methyl group forces two neighboring rings out of plane with respect to each other (*Fig. 7*). This tactic was used to force two pyridine rings out of coplanarity in the serotonin antagonist shown in *Fig. 8*, resulting in increased target selectivity.

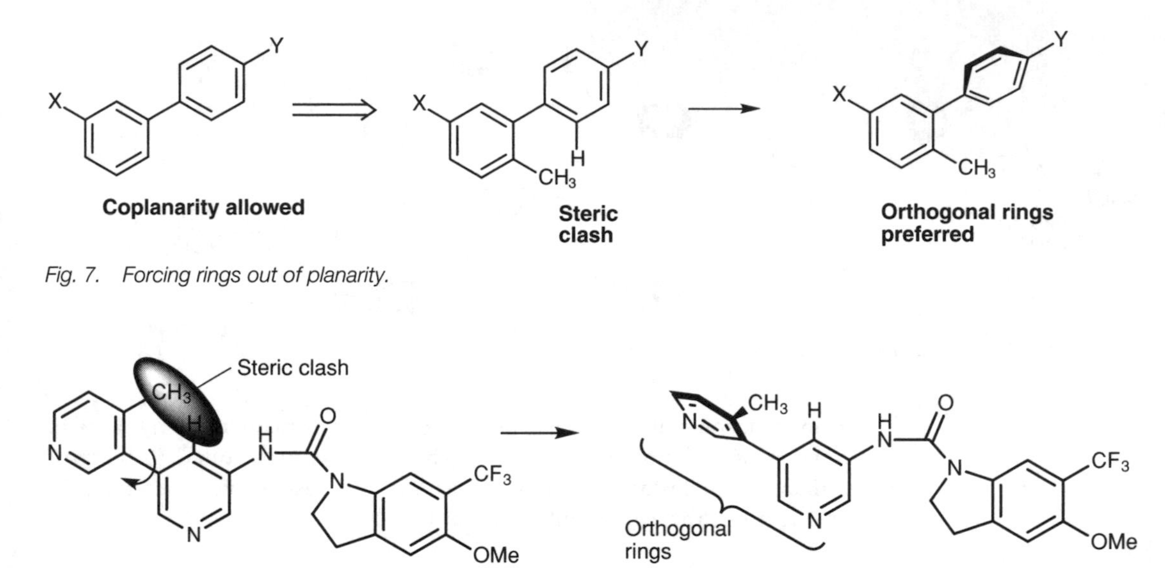

Fig. 7. Forcing rings out of planarity.

Fig. 8. Increasing target selectivity in a serotonin antagonist. From An Introduction to Medicinal Chemistry, *G.L. Patrick 2001, by permission of Oxford University Press.*

H5 EXTRA BINDING INTERACTIONS

Key Notes

Aims

A lead compound might bind to a binding site without using all the binding regions available in that site. Adding extra functional groups to interact with these regions should increase the affinity of the analogs to the binding site and may also increase target selectivity. Finding extra binding interactions can lead to stronger and more selective enzyme inhibitors and receptor agonists. The strategy can also be used to convert an agonist to an antagonist.

Synthetic strategy

Extra substituents can be added to functional groups already present on the lead compound, but care must be taken not to disrupt existing binding interactions. Substituents attached to the carbon skeleton usually require a full synthesis.

Modifying the lead compound

Extra substituents can be added to a lead compound by alkylating or acylating functional groups such as alcohols, phenols, aromatic rings and amines.

Skeletal modifications

A full synthesis is usually required if substituents are to be attached to the carbon skeleton of the lead compound, or if the carbon skeleton itself is to be modified. Aromatic rings can often be replaced with heteroaromatic rings, which allows the possibility of further hydrogen bonding interactions.

Related topics

Synthetic considerations (F1)
Binding interactions (G2)
Functional groups as binding groups (G3)
Computer aided drug design (H2)
Pyrazolopyrimidines (L6)

Aims

It is possible that the lead compound may not take advantage of all the possible binding regions available within a binding site (*Fig. 1*). This is particularly true if the lead compound is from a natural source or has been discovered from a random search of synthetic compounds. A secondary metabolite synthesized by a plant has not been produced to interact specifically with a receptor or enzyme in the human body. Therefore, any interaction that does take place is a matter of chance, and it is unlikely that the lead compound is the perfect ligand for the binding site. Clearly, it must have sufficient binding interactions to bind, but the chances that it interacts with all the available binding regions in the binding site are remote. Adding extra functional groups to the lead compound (**extension**) allows the possibility of finding extra binding regions, resulting in increased binding interactions and an increased affinity for the target. If the lead compound is an enzyme inhibitor, the extra binding interaction will make it a stronger inhibitor. If the lead

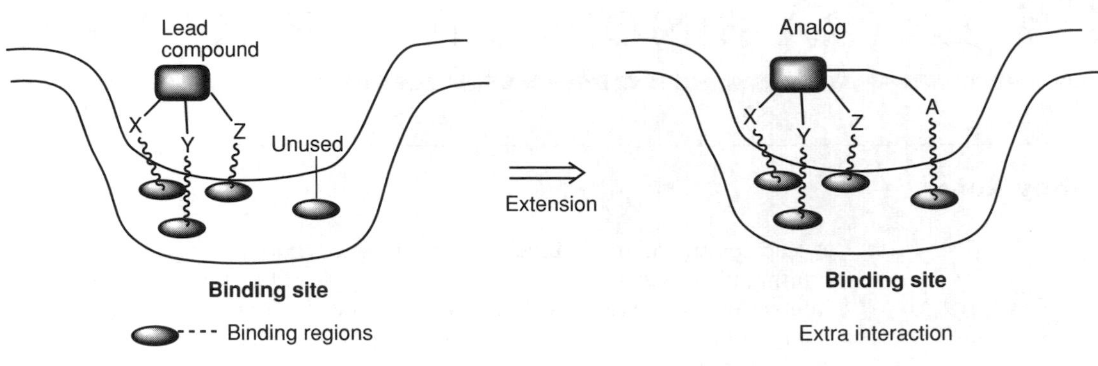

Fig. 1. Extension strategy.

compound is a ligand for a receptor, the extra binding interaction is likely to make it a stronger agonist if the extra binding region is used by the natural ligand. For example, adding a phenethyl group to **morphine** increases analgesic activity due to an increased hydrophobic interaction with the analgesic receptor. Similarly, the presence of a phenethyl group in **enalaprilate** (an antihypertensive agent that acts by inhibiting angiotensin converting enzyme) results in this molecule having greater binding interactions than **captopril** (*Fig.* 2).

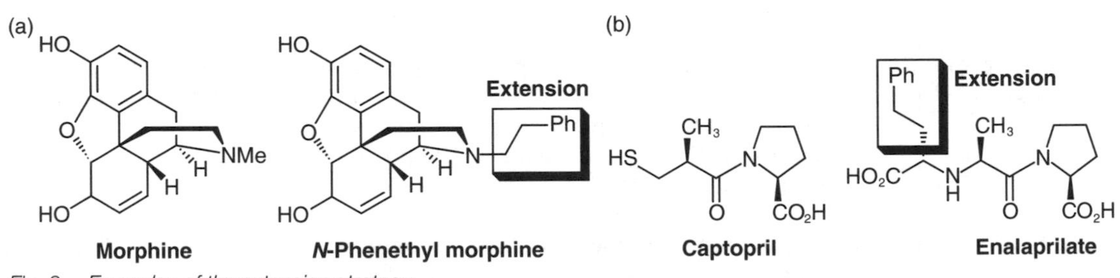

Fig. 2. Examples of the extension strategy.

The strategy of extension can also improve target selectivity. For example, there are subtle differences between the various types and subtypes of receptors as well as between various isozymes. For example, one receptor subtype may have an amino acid present in the binding site that is capable of hydrogen bonding, whereas the corresponding amino acid in another subtype is not. Adding a hydrogen-bonding group to the lead compound to take advantage of this could result in selectivity between the two subtypes.

Knowing the structure of the binding site and how the lead compound binds to that site is of enormous help in deciding which functional groups should be added and to which part of the skeleton they should be added. Such knowledge can be obtained from X-ray crystal structures of the protein/ligand complex or by making a model binding site. Molecular modeling allows a study of the unused space in the binding site and will indicate whether that space contains hydrophobic or hydrophilic amino acids. Substituents can then be chosen that are capable of fitting and interacting with these regions.

There may also be extra binding regions outside the normal binding site (*Fig.* 3). Finding such a region can be particularly useful when designing an enzyme inhibitor or receptor antagonist. Since the extra region is not used by the normal

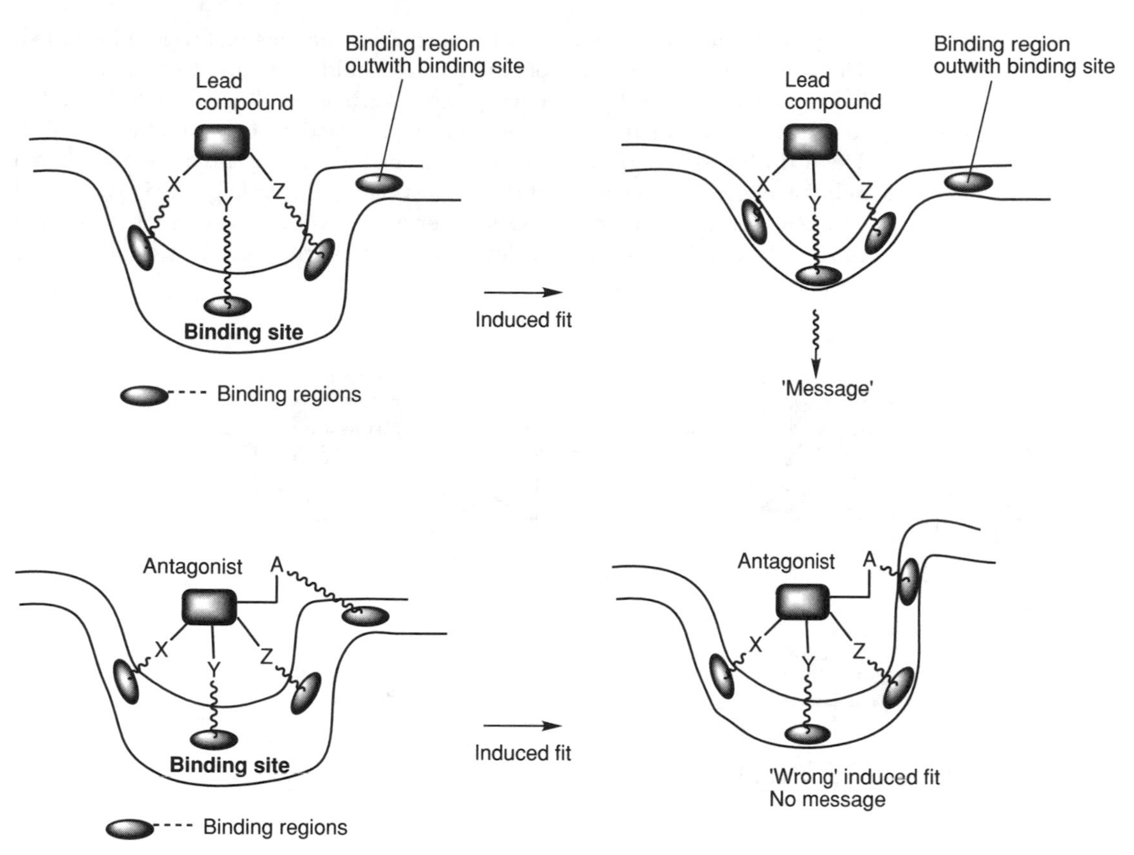

Fig. 3. Conversion of an agonist to an antagonist.

ligand, the extended analog will bind more strongly than the natural ligand. Moreover, taking advantage of an 'external' extra binding interaction is a good method of converting a receptor agonist into an antagonist. Since the normal ligand does not use the extra binding region, an analog interacting with this region is likely to produce a different induced fit. As a result, the analog will bind strongly to the receptor, but will fail to activate it, resulting in antagonism.

Synthetic strategy

There is no limit to the functional groups or substituents that could be added to a lead compound, but some consideration has to be given to where and how they are going to be added. The simplest method of addition is to add the new substituent to a convenient functional group already present in the lead compound. However, care has to be taken in case any of these functional groups are involved in binding the lead compound to the binding site, as adding a substituent to that group may prevent it binding because of steric clashes.

Attaching substituents to the carbon skeleton of the molecule is more difficult and usually requires a total synthesis. Moreover, the positions where the new groups can be placed will depend on the synthetic route used and the availability of the necessary reagents. However, this approach is less likely to interfere with other important binding interactions, since it leaves existing functional groups unaffected.

Molecular modeling studies of the binding site and the bound lead compound can be used to identify the sort of substituents that should be added. Therefore, alkyl groups or aromatic groups could be added to fit into empty hydrophobic pockets (*Fig. 4*). Polar groups, such as an alcohol, could be added to take advantage of any hydrogen-bonding interactions that might be possible. The extra functional group is usually attached as part of a hydrocarbon chain and the length of the chain is chosen to ensure that the extra group is positioned correctly (see also chain contraction/extension, Topic H6). If there is no knowledge of the binding site, then the length of the alkyl chain can be varied to probe for possible interactions.

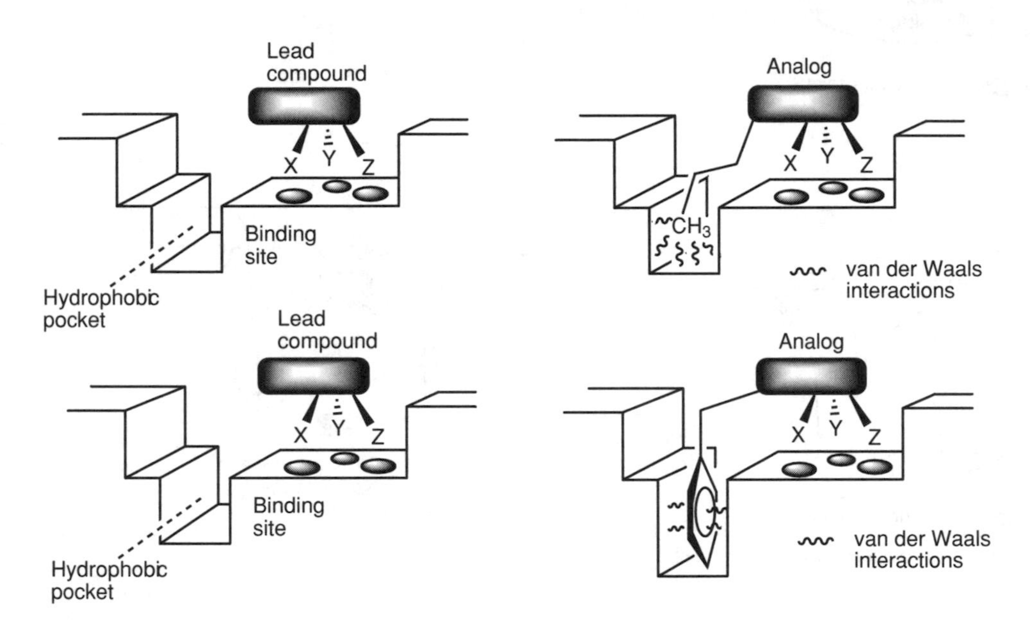

Fig. 4. Adding substituents to bind to hydrophobic pockets.

Modifying the lead compound

There are several methods by which extra functional groups can be added to a lead compound, depending on the functional groups already present (*Fig. 5*). Some of the most common functional groups found in lead compounds are alcohols, phenols, amines and aromatic rings. It is perfectly feasible to add extra substituents onto these functional groups by alkylation or acylation reactions. If the extra binding region is hydrophobic, an extra hydrocarbon chain may be sufficient to interact with it. Alternatively, hydrocarbon chains containing suitable functional groups can be added. The length of the chain can be varied (e.g. $n = 0$–4) to find which chain length is required to position the extra binding group in the correct location. A series of compounds such as this would be known as a **homologous series**, as they differ from one another by the number of methylene groups they contain.

Such modifications are suitable if the functional groups involved are not themselves involved in binding interactions. Otherwise, there is a risk that alkylation will disrupt that binding. For example, if an alcohol is acting as a hydrogen bond donor, alkylation will make such an interaction impossible. On the other hand, there are many examples where amines have been successfully alkylated despite being important binding groups. This is mainly due to the fact that amines are often protonated and interact with binding sites by ionic

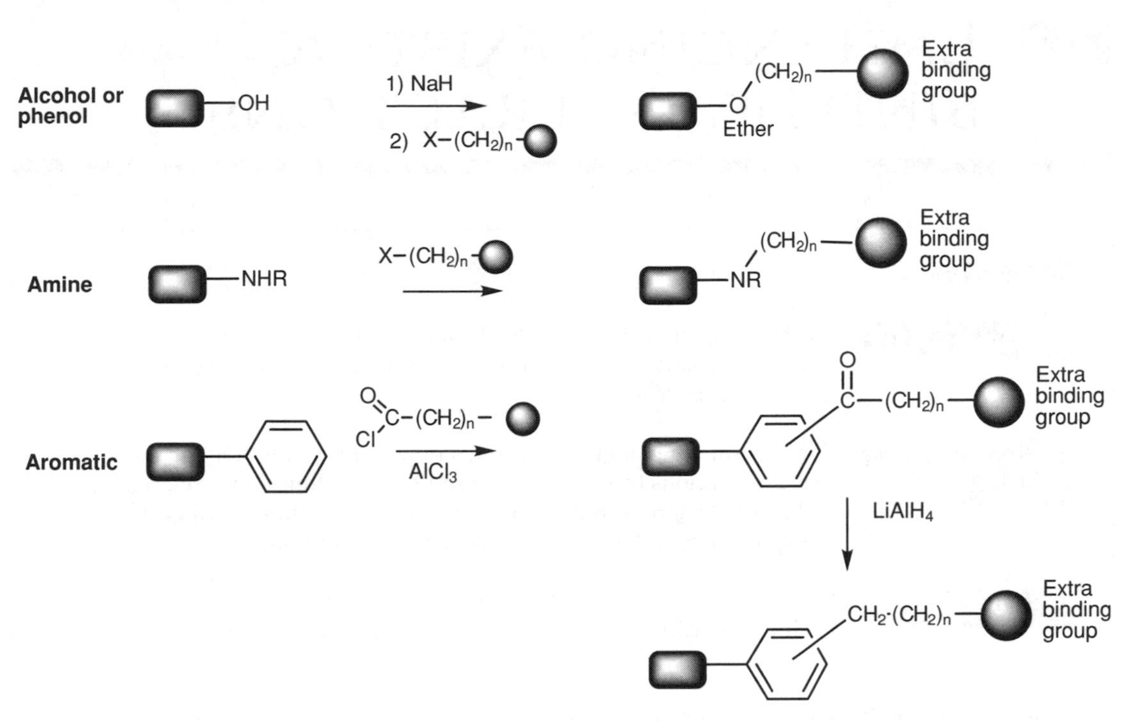

Fig. 5. Methods of adding extra alkyl groups or functional groups.

bonding. Alkylation of a primary or secondary amine gives a secondary or tertiary amine, which can still become protonated and interact with the binding site by ionic bonding, as long as the bulk of the added hydrocarbon chain does not interfere sterically.

Skeletal modifications

A full synthesis is usually required if the extra functional group or substituent is to be attached to the carbon skeleton of the lead compound. Such compounds are usually synthesized from simple starting materials by varying the reagents used in a particular synthesis. New functional groups could be introduced to different parts of the skeleton depending on the synthesis used and the availability of reagents. A full synthesis is also required if it is intended to replace a carbon atom in the original molecular skeleton with a heteroatom. Again, the ease with which this can be done depends on the synthetic route and the available reagents. For example, it is often possible to replace an aromatic ring with a heteroaromatic ring (*Fig. 6*). The heteroatom ring would still be able to interact with the same region of the binding site as the original aromatic ring, but the heteroatom may be able to provide an extra hydrogen-bonding interaction if there is a suitable binding region close by.

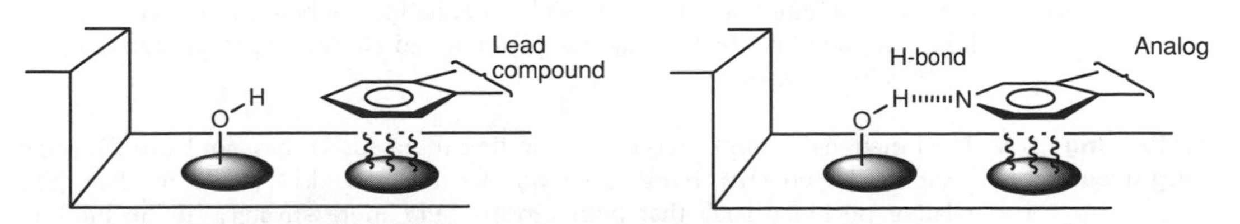

Fig. 6. Replacing an aromatic ring with a heteroaromatic ring.

H6 Enhancing existing binding interactions

Key Notes

Strategy

The binding interactions of a lead compound can be enhanced by fine-tuning the relative position of binding groups or by altering their electronic properties.

Repositioning binding groups

The repositioning of binding groups may improve binding interactions. There are various tactics that one can use, including chain contractions and extensions, ring contractions and expansions, functional group transpositions, and variation in aromatic substitution patterns.

Size matters

Increasing the size of an alkyl group or an aromatic ring system may lead to enhanced van der Waals interactions with a hydrophobic binding region.

Electronic factors

Ionic interactions can be enhanced by increasing the pK_a of important amines. Hydrogen-bonding interactions can be enhanced by replacing functional groups with others that will interact more strongly. Aromatic substituents can be varied to affect the binding strengths of aromatic functional groups.

Steric blocking

The introduction of a bulky alkyl group can prevent a ligand interacting with a receptor if it leads to a steric clash with part of the binding site. This can be used to achieve target selectivity for one type of receptor over another.

Related topics

Functional groups as binding groups (G3)
Computer aided drug design (H2)
From lead compound to dianilinophthalimides (L3)

Strategy

A lead compound is active because it can interact with some or all of the binding regions present in a binding site. However, that interaction might not be as strong as it could be. There are two possible reasons for this. First, the binding groups on the lead compound might not be in the ideal positions for effective bonding. Second, the interaction may be weak due to various electronic reasons. Therefore, it may be possible to enhance the binding interactions of a lead compound by 'fine-tuning' the position and electronic properties of important binding groups.

Repositioning binding groups

The functional groups involved in binding interactions may not be in the ideal positions for effective binding, so various tactics could be used to alter their relative positions such that both groups bind more strongly. If the binding groups are at either end of an alkyl chain then the chain could be extended or contracted to vary the separation and to improve the overlap of the binding groups with their relevant binding regions (*Fig. 1*). For example, a series of *N*-phenylalkyl analogs of **morphine** were synthesized before it was demonstrated that the ethyl link in ***N*-phenethylmorphine** (Topic H5, Fig. 2) was optimum for activity. Note that it is not possible to carry out an extension or contraction directly on a hydrocarbon chain and such analogs have to be synthesized from scratch.

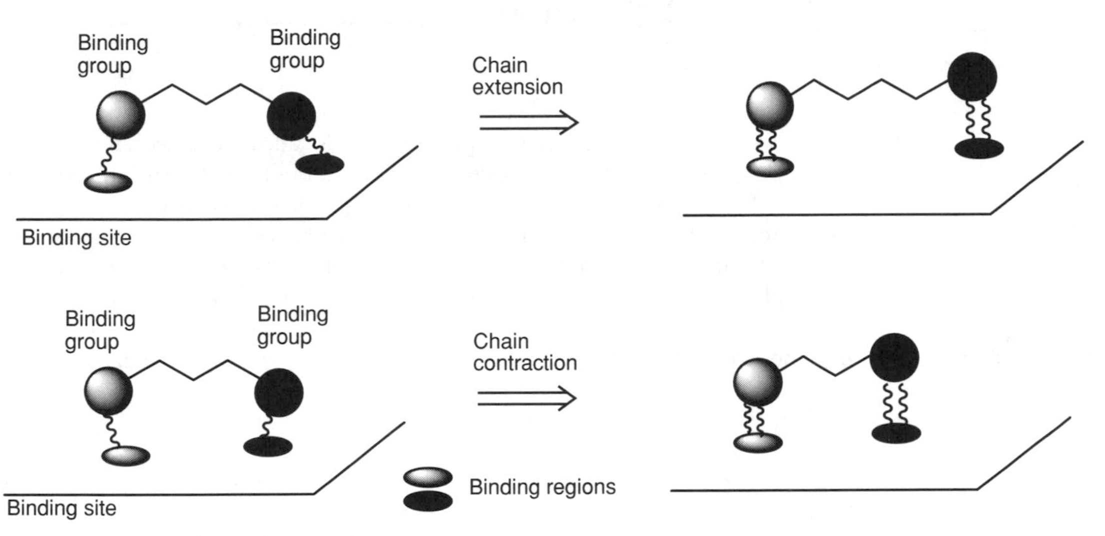

Fig. 1. Chain extension and contraction.

If the functional groups are part of the main skeleton or are attached to it, their relative positions can be altered by moving one of the groups to a neighboring position (**transpositions**) (*Fig. 2*). Once again, the required compound needs to be synthesized from scratch, since there is no reaction that will shift a functional group in this way.

If the skeleton is cyclic, the relative positions of important groups can be altered by using different ring sizes (**ring contractions** and **expansions**) (*Fig. 3*). Once again, these compounds have to be synthesized from scratch.

If the lead compound contains an aromatic ring, it is possible to carry out reactions on the lead compound such that substituents are attached to different positions of the aromatic ring. These can then be tested to determine the best position for the substituent.

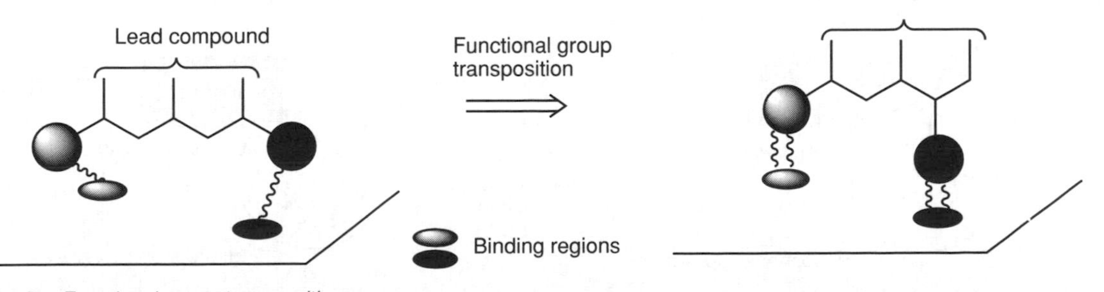

Fig. 2. Functional group transposition.

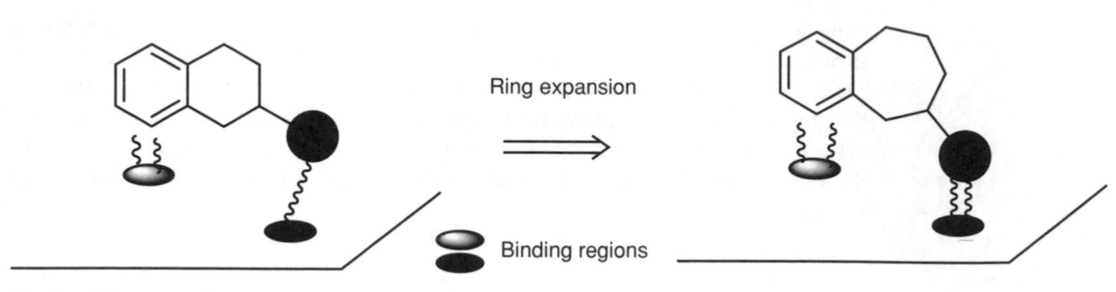

Fig. 3. Ring expansion.

Placing substituents at different positions of the aliphatic skeleton is trickier and usually requires a full synthesis for each compound made. The analogs that are possible depend on the synthetic route used and it may be necessary to completely change the synthetic route if a particular target molecule is desired. In that situation, one would have to weigh up the time and effort involved in developing such a synthesis versus the information that is likely to arise from it.

Size matters

In some cases, the interacting group may not be large enough to make the most of the binding region (*Fig. 4*). This is particularly true when considering the binding of an alkyl substituent to a hydrophobic region of the binding site. For example, a methyl group in the lead compound might fit into a large hydrophobic pocket. Increasing the size of the alkyl group would improve the interaction. If the pocket is long, then a longer chain could be used. If the pocket is short but wide, then a branched alkyl group could be used.

Another example might be a situation where an aromatic ring on the lead compound is interacting with part of a hydrophobic surface (*Fig. 5*). The interaction could be improved by synthesizing analogs where extra aromatic rings have been fused onto the original ring.

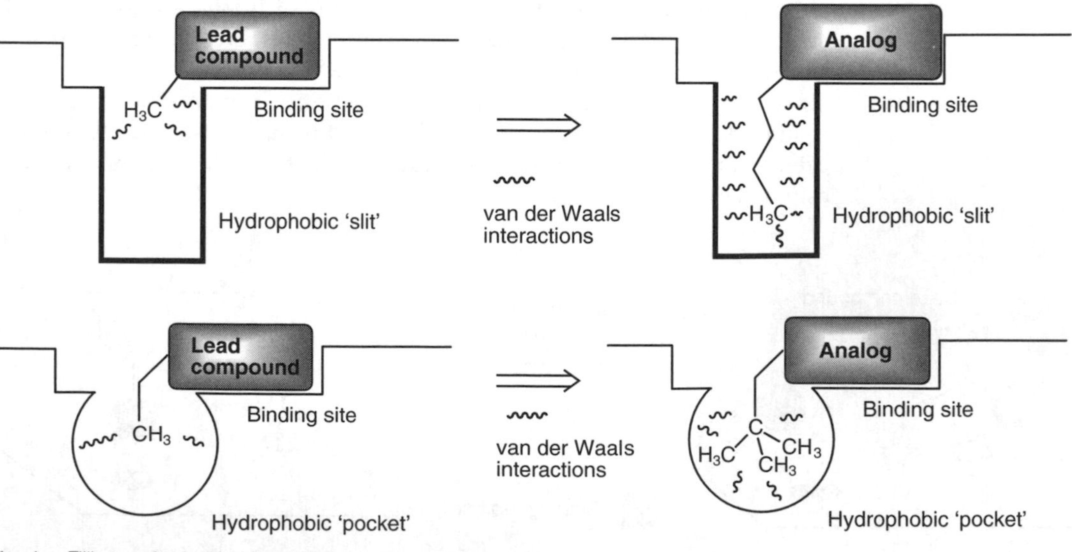

Fig. 4. Filling up hydrophobic regions.

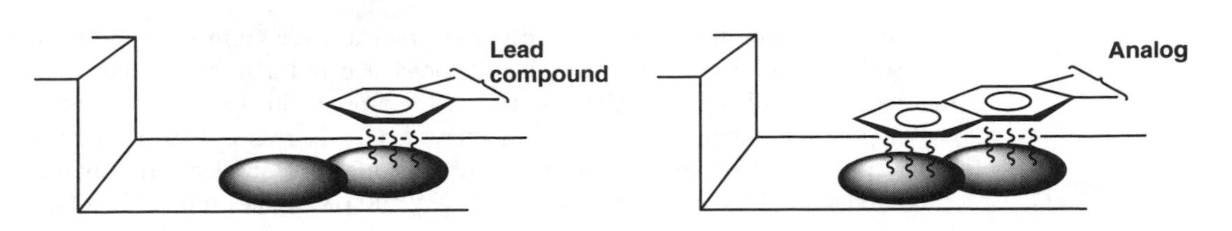

Fig. 5. Enhancing interactions with a hydrophobic surface.

Electronic factors

Electronic factors may play a role in how strongly a lead compound interacts with its target binding site. For example, the basicity of an amine in the lead compound may influence how strongly that group interacts by ionic bonding with a carboxylate ion in the binding site. Amines are weak bases, which can become protonated to form an ammonium ion. However, since they are weak bases, there is an equilibrium between the ionized form and the free base. The position of this equilibrium is defined by the pK_b of the amine, where K_b is the **equilibrium constant** for the following equation:

$$R—NH_2 + H^{\oplus} \rightleftharpoons R—\overset{\oplus}{N}H_3$$

$$K_b = \frac{[RNH_3^+]}{[RNH_2][H^+]} \qquad K_a = \frac{[RNH_2][H^+]}{[RNH_3^+]}$$

The above equilibrium can also be defined by pK_a where K_a is the equilibrium constant for the reverse reaction (i.e. where the free base is treated as the product). This equation can be rearranged to form the following equation:

$$\frac{1}{[H^+]} = \frac{1}{K_a} \times \frac{[RNH_2]}{[RNH_3^+]}$$

Rewriting this in pH notation gives the **Henderson–Hasselbalch** equation:

$$pH = pK_a + \log_{10}\frac{[RNH_2]}{[RNH_3^+]}$$

This equation tells us that pH = pK_a when the concentrations of free base and ionized base are equal (since $\log_{10} 1 = 0$). Therefore, a base having a pK_a of 5.4 is present as an equal concentration of ionized base and free base at pH 5.4. The equation can also be used to show that the ratio of free base to ionized base of this amine would be 100 : 1 at pH 7.4 (the pH of blood).

$$\log_{10}\frac{[RNH_2]}{[RNH_3^+]} = pH - pK_a = 2.0$$

$$\frac{[RNH_2]}{[RNH_3^+]} = 10^2 = \frac{100}{1}$$

The greater the percentage of amine that is ionized, the more likely it will be ionized when it enters the binding site, and the more likely it is that an ionic bonding interaction will take place. Therefore, it should be possible to enhance ionic binding by making the amine more basic (i.e. increase the pK_a). For example, increasing the pK_a from 5.4 to 6.4 would result in a ratio of free base to ionized base of 10 : 1 at blood pH, rather than 100 : 1.

The pK_a of an amine can be altered by varying its substituents. As far as an aliphatic amine is concerned, an extra alkyl group can be added to convert primary and secondary amines to secondary and tertiary amines respectively. The sort of alkyl groups that are already present could also be altered. Such modifications will certainly alter the pK_a of the amine, although it is not often easy to predict whether the pK_a will be increased or decreased. This is because of two conflicting effects (*Fig. 6*). Alkyl substituents have an electron donating effect and should therefore increase the basicity and pK_a of the amine. Therefore, tertiary amines should be more basic than secondary amines and secondary amines should be more basic than primary amines. However, there is also a **solvation effect** to consider. A primary amine is more solvated than a secondary amine when it is ionized, and a secondary amine is more solvated than a tertiary amine. This is because the primary ammonium ion has more hydrogen atoms that can participate in hydrogen bonding with water than secondary or tertiary ammonium ions. Solvation helps to stabilize the ionized form of the amine, so solvation factors alone would make primary amines stronger bases than secondary amines, and secondary amines stronger bases than tertiary bases. Since the two effects are conflicting, the effect of varying or adding extra substituents may have to be determined by experimentation.

Inductive effects of alkyl groups

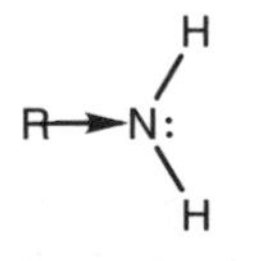

Primary amine

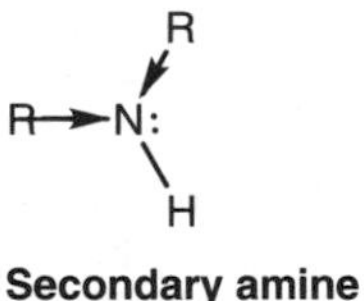

Secondary amine

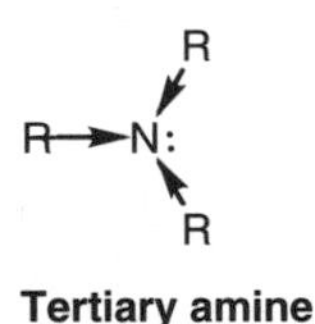

Tertiary amine

Solvation effects of alkyl groups

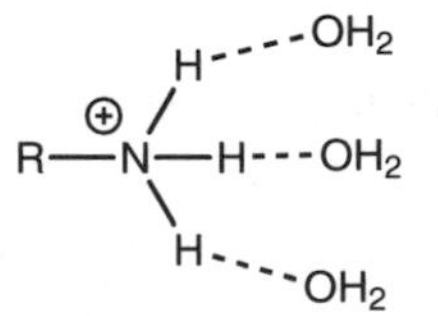

Primary ammonium ion

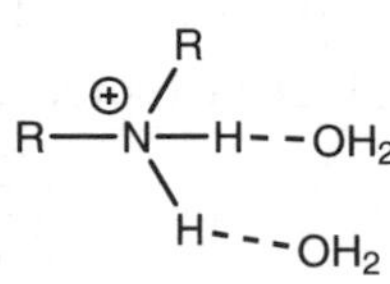

Secondary ammonium ion

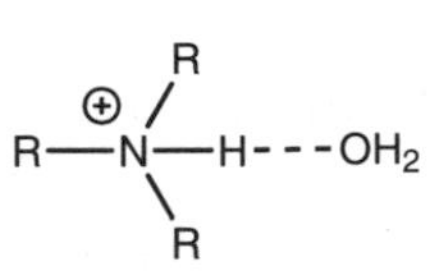

Tertiary ammonium ion

Fig. 6. Effects influencing the pK_a of amines.

The basicity of aromatic or heteroaromatic amines can be modified in a more predictable fashion by varying the substituents on the aromatic/heteroaromatic ring. Aromatic amines tend to be less basic than aliphatic amines since it is possible for the nitrogen's lone pair of electrons to interact with the ring through resonance (*Fig. 7*). Such an effect means that the nitrogen's lone pair is less likely to form a bond to a proton. In order to counteract this, an electron donating

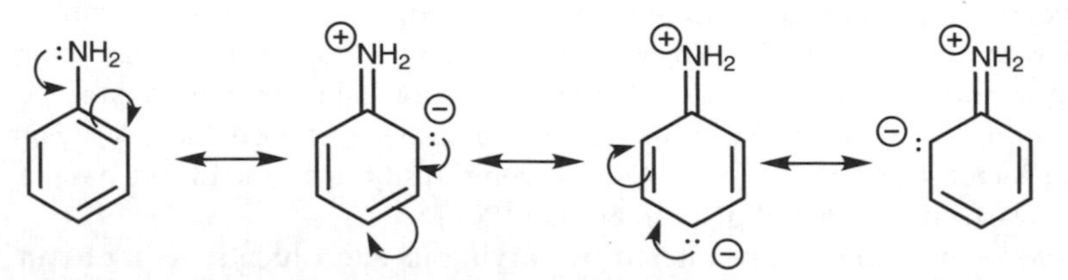

Fig. 7. Resonance in aromatic amines.

substituent could be introduced into the ring in order to increase its electron density (*Fig. 8*). This makes it less likely for the lone pair of electrons to be pulled into the ring, thus increasing the basicity of the amine. Conversely, an electron withdrawing group will make the ring more electron deficient and so decrease the basicity and pK_a of the amine.

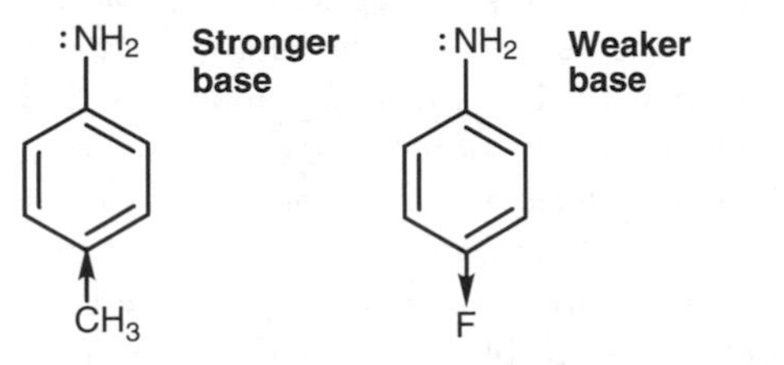

Fig. 8. Effect of substituents on the basicity of aromatic amines.

Increasing the basicity and ionization of an amine may be good for binding site interactions, but it is also important to consider the pharmacokinetic problems faced by drugs. If the amine is made too basic and is predominantly ionized, it will fail to cross fatty cell membranes in the body and could fail to reach its target. Therefore, a compromise has to be struck between maximizing the pK_a in order to maximize the ionic binding interactions with the target, and keeping the pK_a low so that the drug can reach the target. In general, it is found that the optimum pK_a for most amines is in the region 6–8. This means that there is a reasonable chance of the amine existing either as the free base or in the ionized form.

The strength of hydrogen bonding interactions can also be modified. The oxygen atoms of amide groups are strong hydrogen bond acceptors, compared to the carbonyl oxygens of ketones and esters. This is because the lone pair of electrons on an amide nitrogen is involved in a resonance interaction with the carbonyl group, which substantially increases the electron density of that oxygen (*Fig. 9*). Therefore, replacing an important ester group in the lead compound with an amide should increase the strength of any hydrogen bonding that is taking place and also increases the stability of the compound to hydrolysis (see Topic I2).

The oxygen of an aromatic ketone group could be made a stronger hydrogen bond acceptor by adding electron donating groups to the aromatic ring, thus making the ring more electron rich. Aliphatic ethers are better hydrogen bond acceptors than aromatic ethers, and both of these groups are better than a thioether. As far as hydrogen bond donors are concerned, ureas and phenols are stronger hydrogen bond donors than alcohols, which are in turn stronger donors than amides. (These trends are based on a hydrophobic environment, which is a reasonable assumption for most binding sites.)

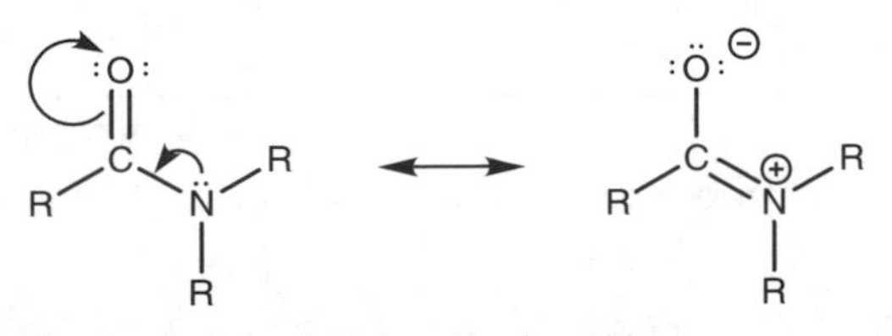

Fig. 9. Resonance interaction in amides.

Steric blocking

The above strategies can be used to enhance the existing binding interactions of a lead compound with its binding site. However, it is sometimes useful to modify a compound such that the existing binding interactions are hindered. This might seem illogical, but the strategy can be useful in attaining selectivity between two similar targets. For example, if the lead compound is capable of binding to two different types of receptor, it may be possible to add an alkyl group that prevents the analog binding to one type of receptor but not the other. In order to achieve such selectivity, the binding site for one receptor must be more spacious than the other. Molecular modeling can help in studying the binding sites of different types of receptor and identifying any such differences.

One example where **steric blocking** has led to improved target selectivity is in the field of adrenergic agonists (*Fig. 10*). There are two main types of adrenergic receptor – the α-adrenoreceptor and the β-adrenoreceptor. **Epinephrine** can bind to both types of receptor. **Isoprenaline**, on the other hand, is an agonist at the β-adrenoreceptor and is inactive at the α-adrenoreceptor. The only difference between epinephrine and isoprenaline is the size of the alkyl group attached to the amine. This is a methyl group in epinephrine, but a much bulkier isopropyl group in isoprenaline. These results indicate that the β-adrenoreceptor has more space available than the α-adrenoreceptor, and that a larger alkyl group prevents binding to the latter receptor (*Fig. 11*).

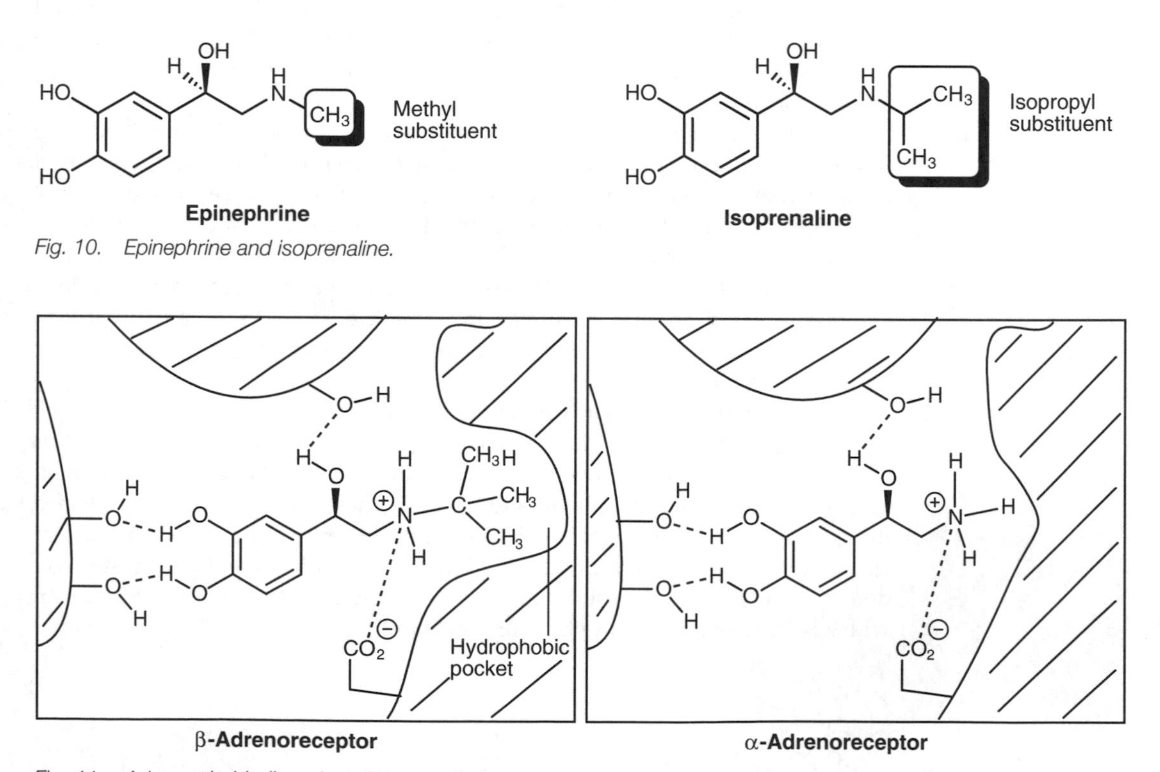

Fig. 10. Epinephrine and isoprenaline.

Fig. 11. Adrenergic binding sites. From An Introduction to Medicinal Chemistry, *G.L. Patrick 2001, by permission of Oxford University Press.*

Section I - Pharmacokinetic orientated drug design

I1 DRUG SOLUBILITY

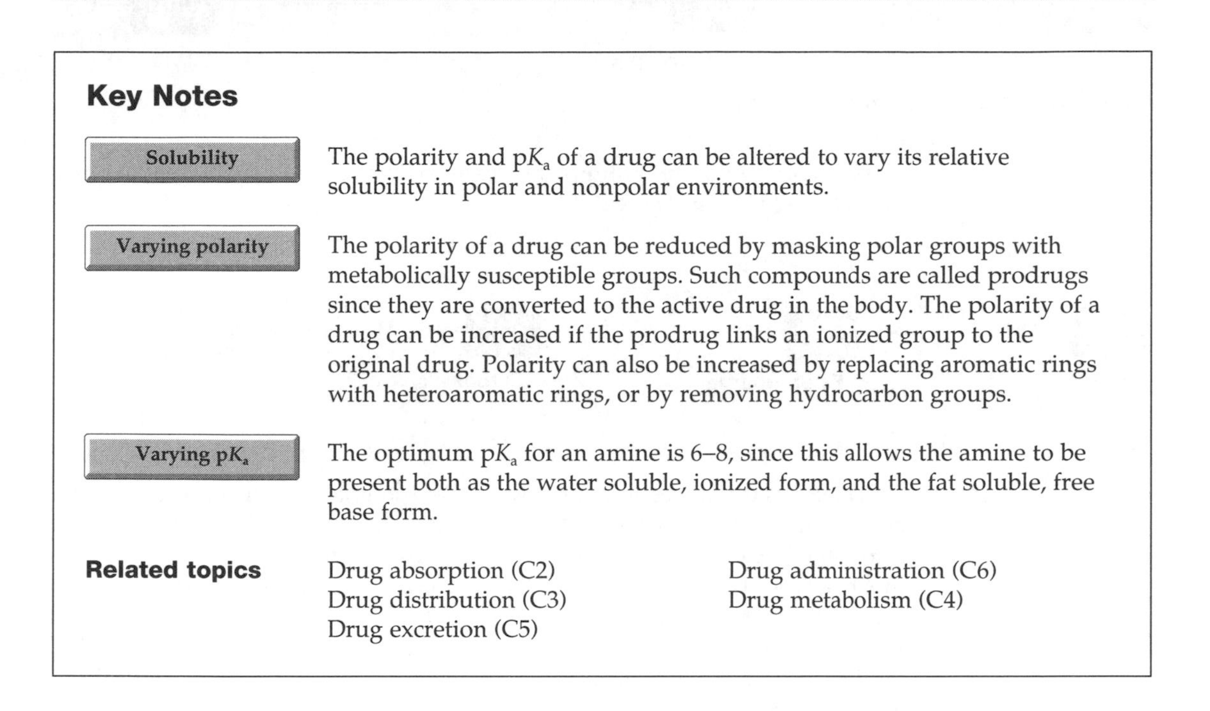

Key Notes

Solubility — The polarity and pK_a of a drug can be altered to vary its relative solubility in polar and nonpolar environments.

Varying polarity — The polarity of a drug can be reduced by masking polar groups with metabolically susceptible groups. Such compounds are called prodrugs since they are converted to the active drug in the body. The polarity of a drug can be increased if the prodrug links an ionized group to the original drug. Polarity can also be increased by replacing aromatic rings with heteroaromatic rings, or by removing hydrocarbon groups.

Varying pK_a — The optimum pK_a for an amine is 6–8, since this allows the amine to be present both as the water soluble, ionized form, and the fat soluble, free base form.

Related topics

Drug absorption (C2)
Drug distribution (C3)
Drug excretion (C5)
Drug administration (C6)
Drug metabolism (C4)

Solubility

Drugs need to be soluble in both polar and nonpolar environments. The relative solubility of a drug in these environments is important to its absorption, distribution and excretion. Modifying the polarity or pK_a of a drug will vary its relative solubility and may improve its pharmacokinetic properties.

Varying polarity

If a drug is too polar, its **polarity** can be reduced by removing or masking polar functional groups. If polar functional groups are removed, it is important to remove only those groups that are not part of the pharmacophore. However, the removal of excess polar groups is likely to have been carried out during the simplification of a lead compound (Section H), such that the polar groups remaining are indeed important to the pharmacophore. Therefore, the masking strategy tends to be more useful in reducing polarity.

Since the polar groups present are likely to be important for binding interactions, the masking must be temporary so that the masks aid drug absorption and distribution, but are lost by the time the drug reaches its target. The best way of achieving this is to take advantage of the body's own metabolic reactions. In other words, a masking group is added which is stable to normal conditions, but which is susceptible to metabolic enzymes (*Fig. 1*). A drug masked in this fashion is known as a **prodrug**. Since the prodrug has one or more of its important binding groups masked, it is normally inactive itself, but once it is metabolized in the body, the active drug is unveiled.

The polar functional groups usually masked in the prodrug approach are alcohols, phenols and carboxylic acids. Alcohols and phenols are usually

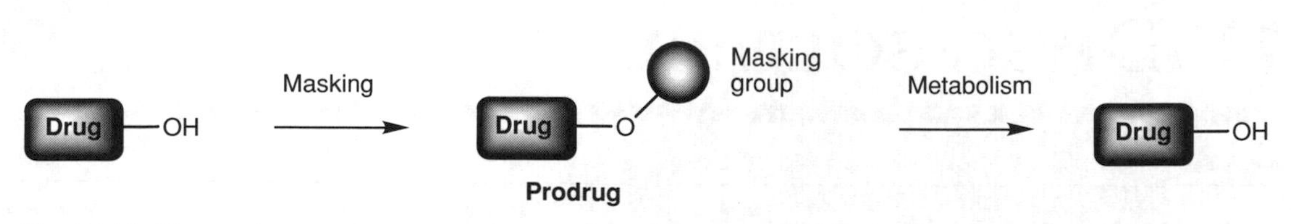

Fig. 1. Masking a polar group with a prodrug.

masked using an acetyl group, while carboxylic acids are usually masked as a methyl or ethyl ester (*Fig. 2*).

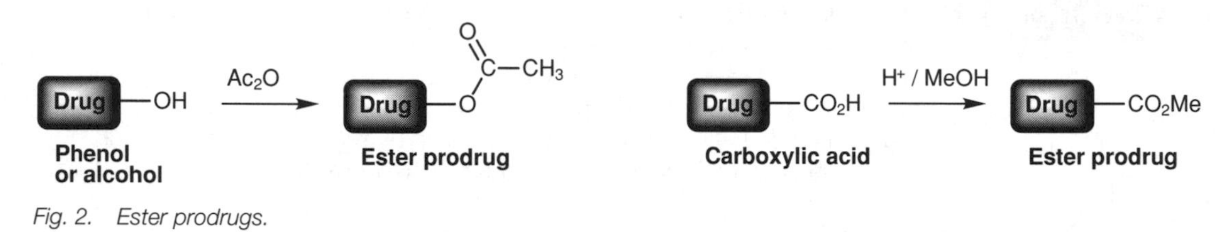

Fig. 2. Ester prodrugs.

Sometimes the strategy does not work. For example, **penicillin methyl esters** are not metabolized in the body because the penicillin ring system acts as a steric shield, blocking the esterase enzymes from hydrolyzing the ester (*Fig. 3*). In situations like this, an extended ester could be used (*Fig. 4*). This involves esterifying the carboxylic acid with a group that has a second ester attached. The second ester is more exposed and *can* be hydrolyzed by esterases. The resulting product is chemically unstable and spontaneously decomposes to reveal the original carboxylic acid.

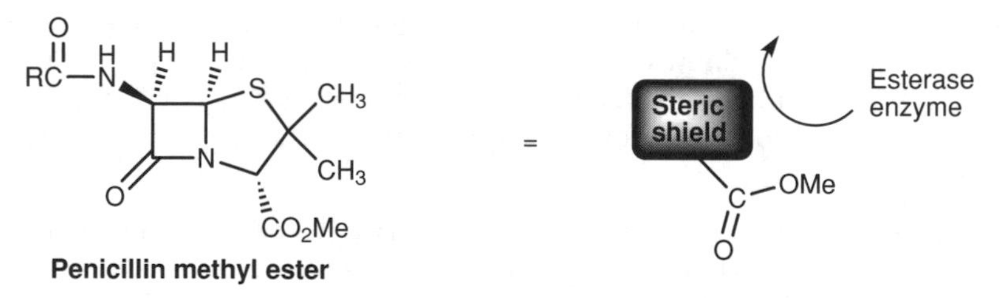

Fig. 3. Penicillin acting as a steric shield to esterases.

Occasionally, fine-tuning of the prodrug ester may be necessary if it is found that the ester is metabolized too slowly or too quickly. This can be done by varying the nature of the leaving group (*Fig. 5*). Ester prodrugs of carboxylic acids can be made less reactive by decreasing the stability of the alcohol leaving

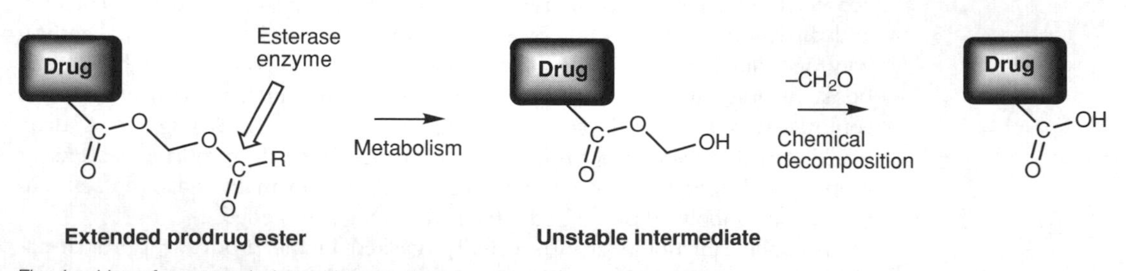

Fig. 4. Use of an extended ester in a prodrug.

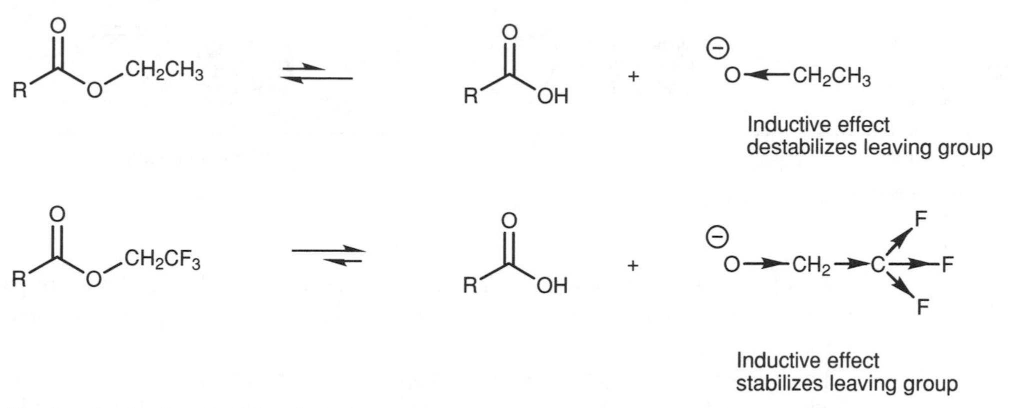

Fig. 5. Varying the reactivity of esters.

group. Electron donating groups will increase the electron density on the oxygen and destabilize it as a leaving group. The opposite tactic can be used to make the prodrug esters more reactive.

Prodrugs can also be used to increase the polarity of a drug that is too hydrophobic. This involves esterifying an available alcohol, phenol or carboxylic acid with a group containing an ionic functional group. For example, alcohol groups in various drugs have been converted to succinate, phosphate and amino acid esters (*Fig. 6*). All these groups are natural chemicals in the body, and produce no toxic effects once they are released from the prodrug. For example, the antitumor agent **combretastatin A-4** (Topic H4) showed poor water solubility and was therefore submitted for clinical trials as the more polar phosphate prodrug (*Fig. 7*).

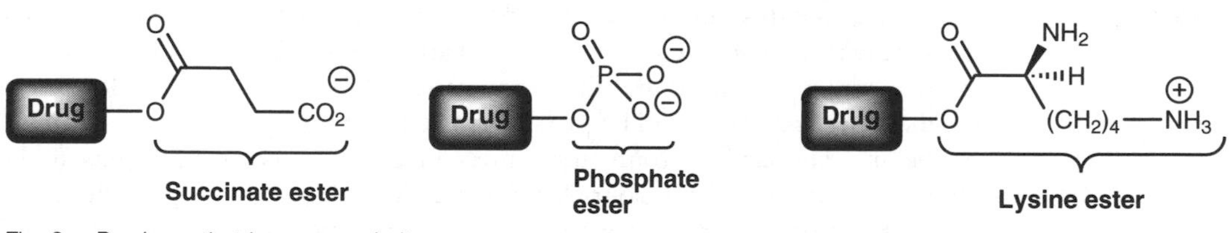

Fig. 6. Prodrugs that increase polarity.

Another popular method of increasing polarity is to replace an aromatic ring with a heteroaromatic ring (*Fig. 8*). If a heteroaromatic ring is already present, this could be replaced with a heteroaromatic ring containing more heteroatoms.

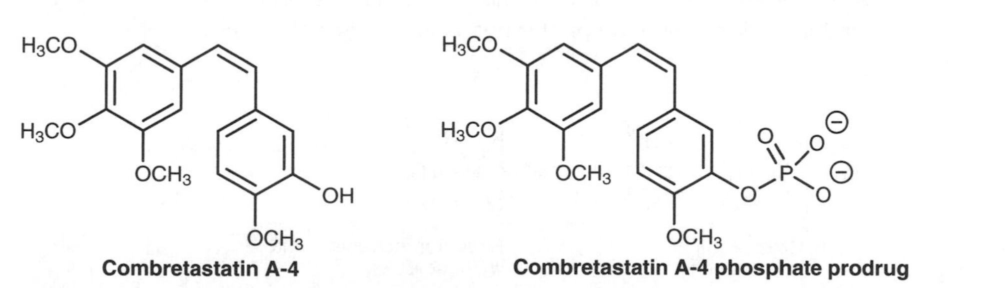

Fig. 7. Phosphate prodrug of combretastatin A-4.

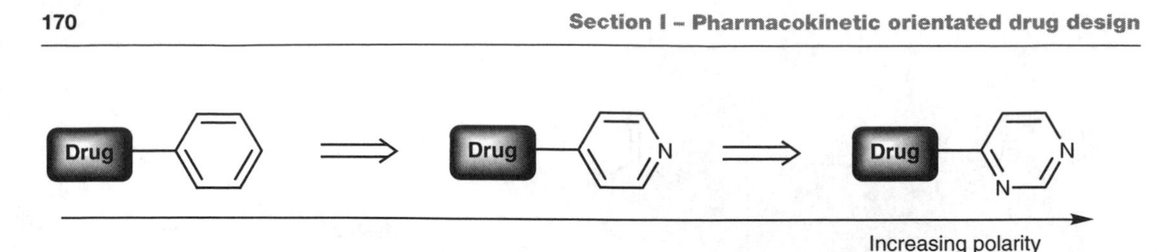

Fig. 8. Increasing polarity by introducing heteroatoms.

Finally, the polarity of a drug can be increased if hydrophobic alkyl groups are removed. This increases the relative importance of any polar groups present in the molecule.

Prodrugs with different solubility properties to the parent drug can sometimes be used to target a drug to particular parts of the body. For example, **sulfasalazine** (*Fig. 9*) is a prodrug used against gut infections. The prodrug is highly polar, due to the presence of a carboxylic acid and phenol. Therefore, it is poorly absorbed through the gut wall and remains in the gut where it is needed. Intestinal bacteria reduce the azo group to release **sulfapyridine** (a sulfonamide) and **aminosalicylate**.

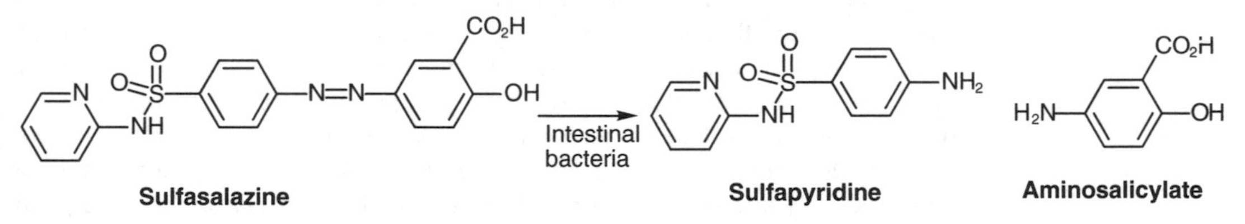

Fig. 9. Sulfasalazine.

Varying pK_a

If the drug contains an amine functional group, its relative solubility in polar and nonpolar solvents can be altered by varying the **pK_a**. The higher the pK_a, the more the amine is ionized and the more water soluble it will be. The lower the pK_a, the more the amine will be present as the free base and the more soluble it will be in nonpolar environments. Since an orally active drug needs to be soluble in both polar and nonpolar environments, the best pK_a is where the ionized and free base forms are equal (*Fig. 10*). This would correspond to a pK_a of 7.4, which is the pH of blood. In practice, a pK_a in the region of 6–8 would provide significant amounts of the ionized and free base forms for the drug to be soluble in both environments. Methods of varying pK_a have already been discussed in Topic H6, since this can also have an influence on receptor binding. A compromise may need to be made between the optimum pK_a for receptor binding and the optimum pK_a for pharmacokinetic properties.

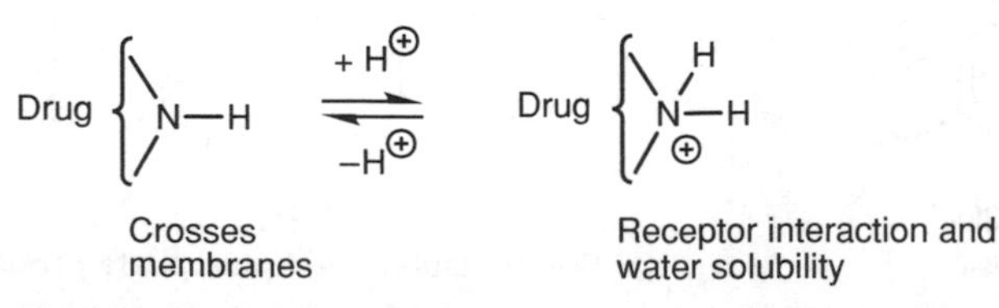

Fig. 10. Equilibrium between ionized and free base forms of an amine.

I2 DRUG STABILITY

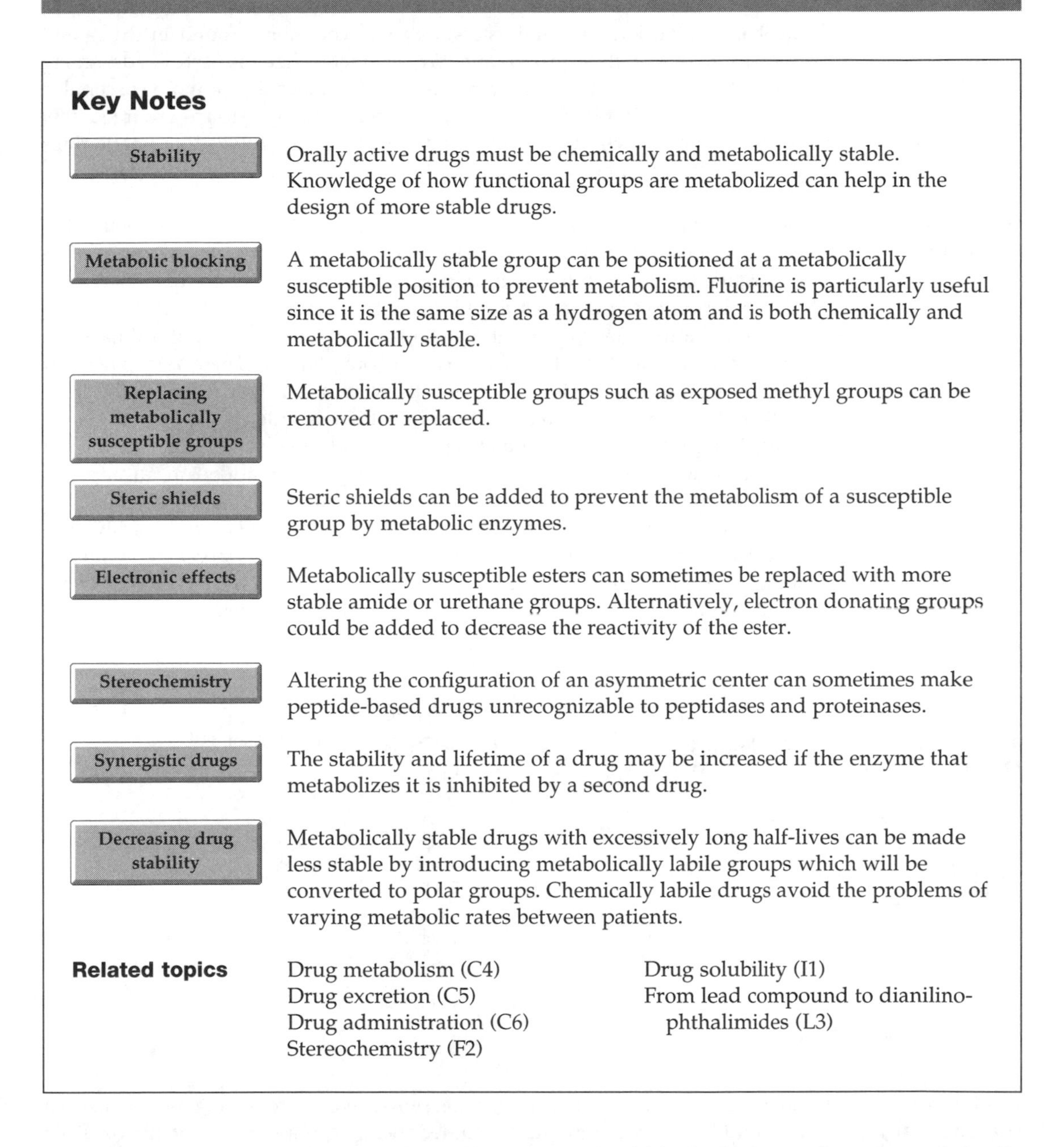

Key Notes

Stability	Orally active drugs must be chemically and metabolically stable. Knowledge of how functional groups are metabolized can help in the design of more stable drugs.
Metabolic blocking	A metabolically stable group can be positioned at a metabolically susceptible position to prevent metabolism. Fluorine is particularly useful since it is the same size as a hydrogen atom and is both chemically and metabolically stable.
Replacing metabolically susceptible groups	Metabolically susceptible groups such as exposed methyl groups can be removed or replaced.
Steric shields	Steric shields can be added to prevent the metabolism of a susceptible group by metabolic enzymes.
Electronic effects	Metabolically susceptible esters can sometimes be replaced with more stable amide or urethane groups. Alternatively, electron donating groups could be added to decrease the reactivity of the ester.
Stereochemistry	Altering the configuration of an asymmetric center can sometimes make peptide-based drugs unrecognizable to peptidases and proteinases.
Synergistic drugs	The stability and lifetime of a drug may be increased if the enzyme that metabolizes it is inhibited by a second drug.
Decreasing drug stability	Metabolically stable drugs with excessively long half-lives can be made less stable by introducing metabolically labile groups which will be converted to polar groups. Chemically labile drugs avoid the problems of varying metabolic rates between patients.

Related topics

Drug metabolism (C4)
Drug excretion (C5)
Drug administration (C6)
Stereochemistry (F2)
Drug solubility (I1)
From lead compound to dianilino-phthalimides (L3)

Stability

If a drug is to be orally active, it should be both chemically and metabolically stable. Certain functional groups are likely to be too unstable for use in drugs. Chemically reactive functional groups, such as acid chlorides or acid anhydrides, are quickly hydrolyzed in water, and have no role in drug design. Other chemically reactive functional groups, such as aldehydes and alkyl halides, are found in some drugs and are designed to interact with nucleophilic groups on a protein or nucleic acid target. However, if the group is too reactive, the drug will be toxic since it will react indiscriminately with any protein or nucleic acid.

Functional groups such as aromatic rings, heteroaromatic rings, phenols, alcohols, amines, amides and esters are more commonly found in drugs and knowledge of how these groups are normally metabolized is useful in designing drugs that are more stable. Certain features of a molecule are more susceptible to metabolism than others. For example, exposed methyl groups are frequently oxidized to carboxylic acids; esters and amides are hydrolyzed; aromatic rings can be oxidized to phenols. Phenols and alcohols easily form polar conjugates.

Metabolic blocking

Metabolism normally takes place at specific regions of a molecular skeleton. For example, the oxidation of an aromatic ring or an aliphatic cyclic system often occurs at a specific position to introduce a phenol or alcohol group respectively. **Metabolic blocking** involves the placement of a metabolically stable atom at that position in order to prevent metabolism. The introduction of a fluorine or chlorine atom has been frequently carried out (*Fig. 1*). There is a particular advantage in using fluorine. Fluorine is the same size as hydrogen, so the steric bulk of the drug is not increased. Therefore, there is little risk of preventing the drug fitting its binding site due to steric clashes. Moreover, the fluoride ion is a poor leaving group. Therefore, if the fluorine atom is added to an aliphatic skeleton, there is no chance of it being displaced in an alkylation reaction. The only problem with fluorine is that introducing it can often be synthetically difficult. Therefore chlorine may be used instead despite being larger. Bromine and iodine are not used as metabolic blockers because they are too large and are good leaving groups if they are attached to an aliphatic skeleton.

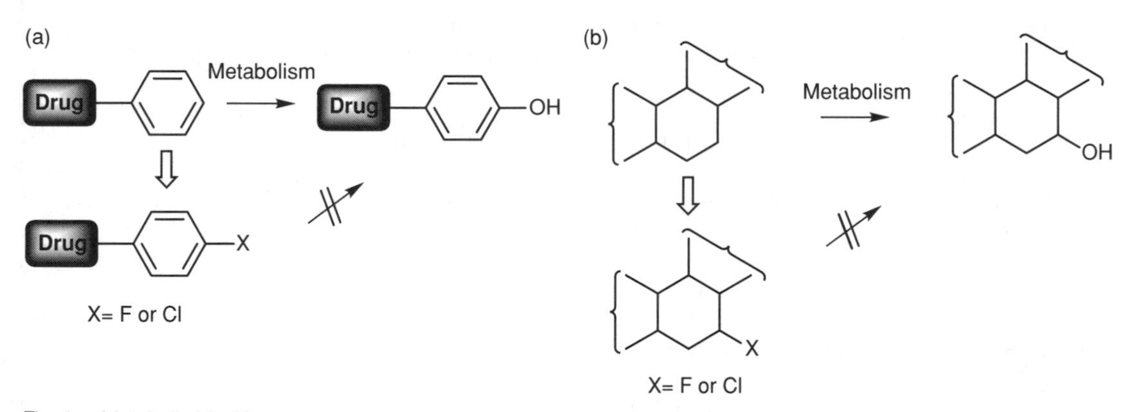

Fig. 1. Metabolic blocking.

Replacing metabolically susceptible groups

Sometimes, drug metabolism can be prevented by removing or replacing a susceptible group. For example, exposed methyl groups are frequently oxidized to carboxylic acids (*Fig. 2*). Removing the methyl group or replacing it with a metabolically stable group would prevent metabolism from taking place.

Steric shields

Steric shields are frequently used to protect susceptible groups such as esters and amides from hydrolysis. Usually, a bulky substituent, such as a methyl group, is added to the molecular skeleton so that it is close to the ester or amide and prevents enzymes attacking the carbonyl group (*Fig. 3*).

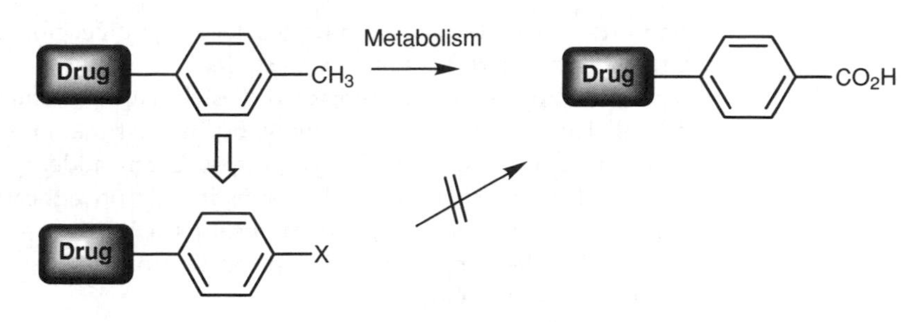

Fig. 2. Replacing a metabolically labile methyl group.

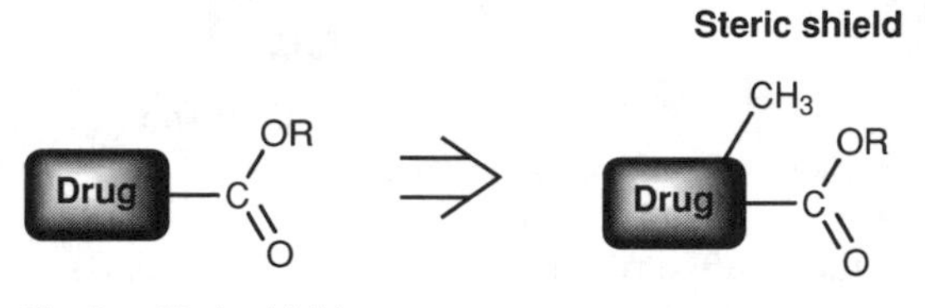

Fig. 3. Steric shields.

Peptides and proteins are also susceptible to hydrolysis by peptidases and proteinases. This often occurs at specific positions in the peptide chain. Introducing an *N*-methyl group can often block this hydrolysis (*Fig. 4*).

Electronic effects Esters are particularly prone to chemical and enzymatic hydrolysis, but there are several electronic tactics that can be used to counteract this. The ester could be replaced with an amide group (*Fig. 5*). Amides are less reactive to hydrolysis

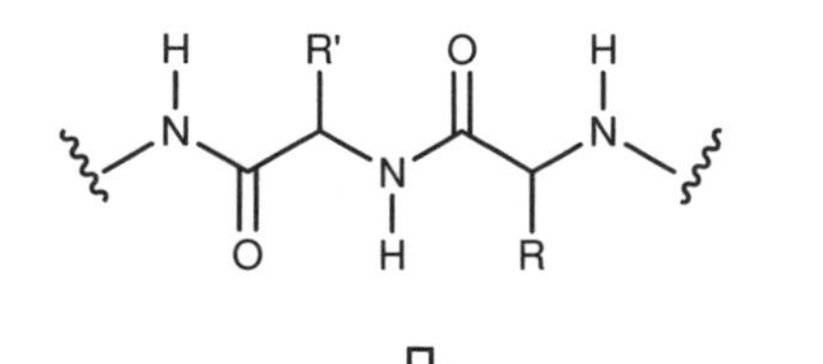

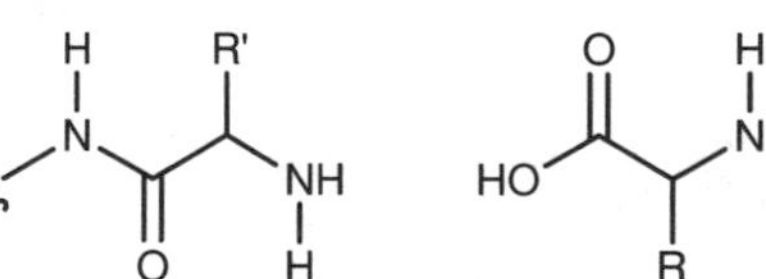

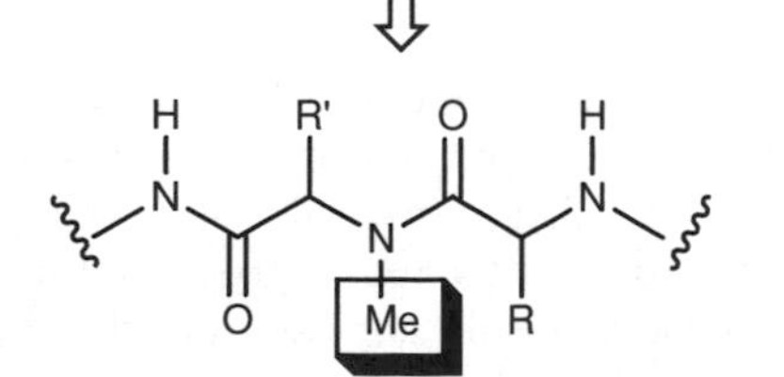

Fig. 4. Blocking the hydrolysis of peptides.

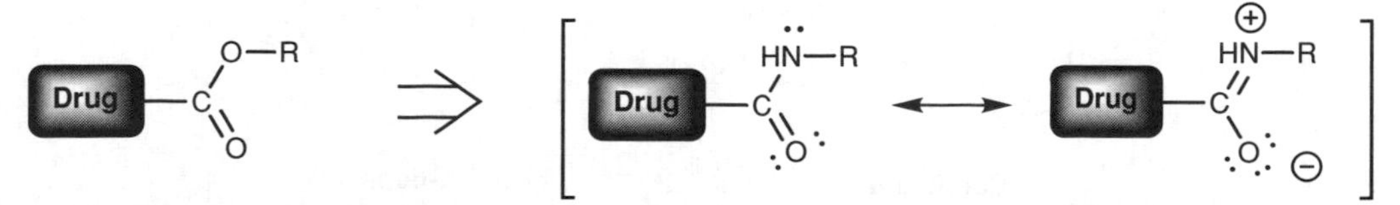

Fig. 5. Replacing an ester with an amide.

than esters since the amide nitrogen feeds its lone pair of electrons into the carbonyl group and decreases its reactivity.

Another approach is to replace the ester with a urethane functional group (*Fig. 6*). This retains the ester moiety, but the extra amino group has the same effect on the neighboring carbonyl group as in an amide.

A third approach is to modify substituents on either side of the ester to improve stability. Electron donating groups will decrease the reactivity of the ester, and if they are on the alcohol portion of the ester they will make that portion a poorer leaving group.

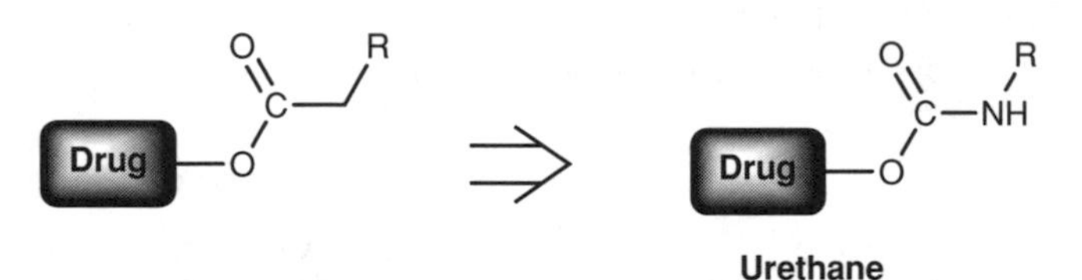

Fig. 6. Replacing an ester with a urethane.

Stereochemistry

Altering the configuration of an asymmetric center can sometimes prevent metabolism (*Fig. 7*). There is a danger that this will also make the drug unrecognizable to its target, but the strategy has worked in stabilizing several peptide-based drugs.

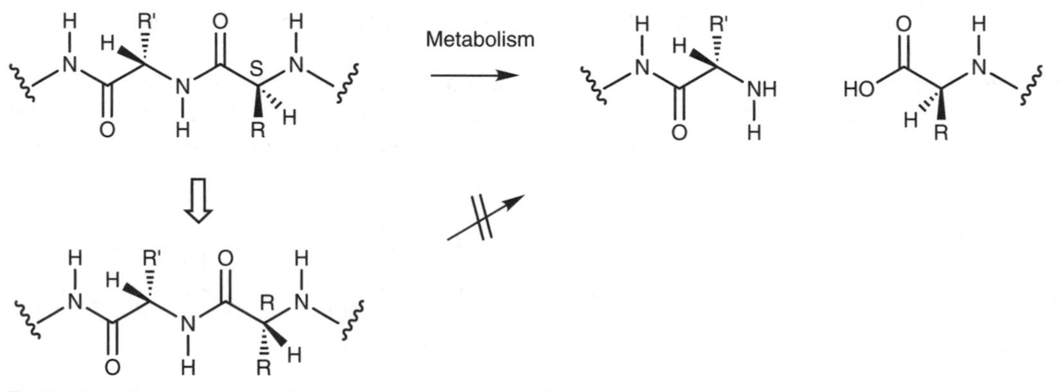

Fig. 7. Inverting an asymmetric center to prevent metabolism.

Synergistic drugs

Occasionally, it is possible to prevent a drug being metabolized by a specific enzyme if a second drug is administered to inhibit that enzyme. This is an example of **synergism**, where the activity of one drug affects the activity of another. For example, **carbidopa** is administered with **L-dopa**, a drug used in the treatment of Parkinson's disease (*Fig. 8*). L-Dopa is normally decarboxylated

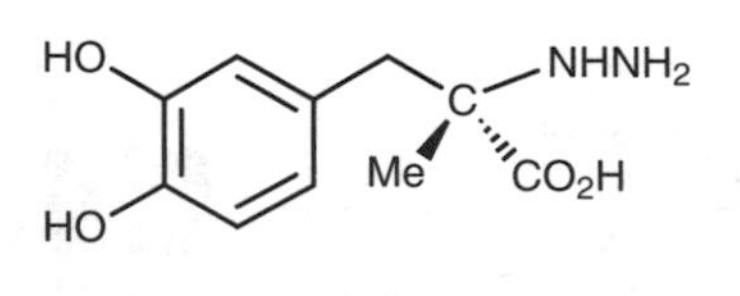

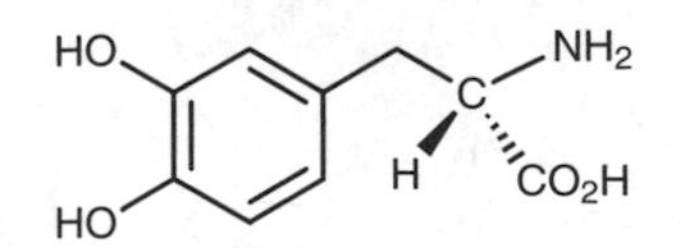

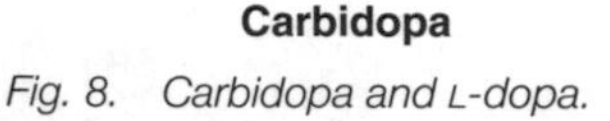

Fig. 8. Carbidopa and L-dopa.

by an enzyme called **dopa decarboxylase**, but the enzyme can be inhibited using carbidopa. This allows L-dopa to be administered in smaller and safer doses.

Decreasing drug stability

Occasionally, a drug is metabolically stable, has a long half-life in the body and is slow to clear from the system. This can cause as many problems as a drug that is metabolized too quickly, so it is sometimes necessary to make it more susceptible to metabolism. This can be done by introducing metabolically susceptible groups which will be converted to polar groups allowing the drug to be excreted more efficiently. For example, addition of an exposed methyl group may lead to that group being metabolized to a carboxylic acid. Introducing an ester group will mean that esterases will expose a polar alcohol or carboxylic acid, while introduction of a phenol or alcohol will allow the drug to be conjugated and excreted.

These examples rely on metabolic enzymes. However, the activity of metabolic enzymes varies between individuals. Chemically unstable drugs avoid this problem and the neuromuscular blocking agent **atracurium** (*Fig. 9*) is one such example. Atracurium is given as an intravenous drip during surgery and is designed to have a short half-life in order to speed up recovery times. Atracurium is stable under acid conditions, but is unstable in the slightly alkaline conditions of blood. It undergoes a **Hofmann elimination** resulting in degradation and inactivation of the drug (*Fig. 10*).

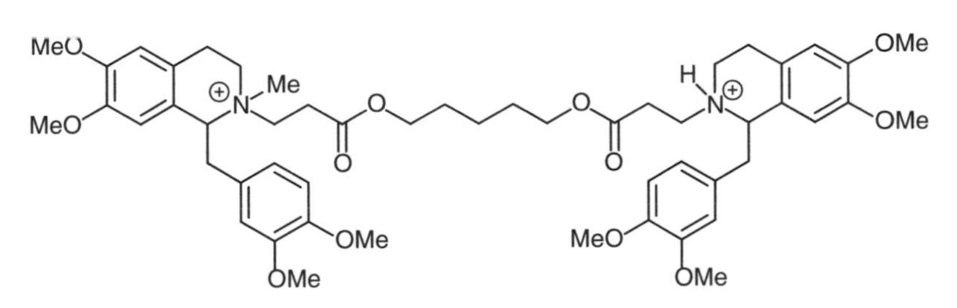

Fig. 9. Atracurium.

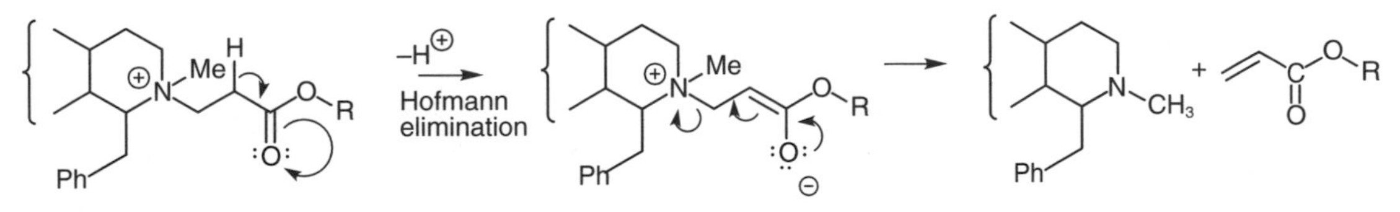

Fig. 10. Hofmann elimination.

J1 PATENTING AND CHEMICAL DEVELOPMENT

Key Notes

Patenting

Patenting is carried out as soon as a potentially useful drug is identified. Patents are taken out on groups of compounds rather than individual structures, since it cannot be predicted which compound will eventually reach the market place. Several years of patent protection are lost due to preclinical and clinical trials.

Chemical development

Chemical development involves the development of a synthesis that will be suitable for the large-scale production of a drug. The priorities are to produce a high quality product in high yield, using a synthesis that is cheap and efficient. The final synthesis must also be safe and environmentally friendly.

Development phases

The first phase involves the development of a synthesis capable of producing a kilogram of compound for initial tests. The second phase requires the synthesis of 10 kg for toxicological studies, formulation and initial clinical trials. The third phase requires the synthesis of 100 kg for clinical trials. Finally, a full production synthesis is required. The final stage of the synthetic route and purification of the product are optimized first, and specifications of the drug determined. These define the analytical tests that need to be carried out and the purity standards demanded.

Process development

Process development involves the development of the synthetic process so that it is suitable for use in a production plant. The number of operations involved should be cut to a minimum.

Potential problems of the initial synthesis

Typical problems include low yields, the number of synthetic steps involved, the cost of reagents and reactants, the potential hazards of individual chemicals and solvents, and the experimental procedures used.

Related topics

From concept to market (A2)
Synthetic considerations (F1)
Optimization of reactions (J2)
Scale-up issues (J3)
Process development (J4)
Specifications (J5)

Patenting

Once a potentially useful drug is discovered, it is patented as soon as possible. This is essential because competitors may be working in the same field of research, and may be studying the same kinds of compounds. **Patents** are taken out before toxicology tests, drug metabolism studies, clinical trials and chemical development are carried out. Therefore, it is not possible to say whether the most promising compound at that stage will be the one that eventually reaches the market. The results of preclinical and clinical trials may well see modifications to the structure in order to avoid undesirable side effects or other similar problems. Therefore, patents are normally taken out on a group of compounds belonging to the same structural class. The patent must be careful to cover all the possible analogs that could be synthesized by the synthetic route used, but there is no requirement to synthesize all of these for the patent. Several examples should be given to illustrate the synthetic procedure, and biological results should be supplied to illustrate the application to which these drugs are intended.

Since patenting is carried out before the drugs are rigorously tested, several years of the patent are lost while these tests are being carried out. Any unforeseen delay or difficulty during this time may see the project being dropped if it is felt that the remaining patent protection is too short to recover costs.

Chemical development

Once a novel compound has been identified as being potentially useful, it is necessary to develop a large-scale synthesis for it. This is the job of the chemical development laboratories. The starting point for **chemical development** is the synthesis used in the drug discovery laboratory and one might be tempted to think that a simple scale-up of the 'ingredients' would be sufficient. However, this is far from the case. During the drug discovery phase, the emphasis has been on producing as many different compounds as possible in as short a time period as possible. Considerations such as yield and cost are of secondary importance. For example, it does not matter if the yield is only 5% as long as there is sufficient compound to test. Cost is also not an issue since the reactions are carried out on small scale, allowing the use of expensive reagents or starting materials. However, once a useful compound has been discovered, increasing quantities of the compound will be required for the various preclinical tests and clinical trials, so the synthetic route has to be scaled-up and made more cost-effective. Eventually, the drug may reach the market and the synthesis will have to be adapted so that it is suitable for production scale.

The priorities in chemical development are to devise a synthetic route that will be cheap, efficient and high yielding, have the minimum number of synthetic steps, and consistently provide a high quality product that meets predetermined specifications of purity. Issues such as safety and the environment are also extremely important at the production scale. During chemical development, the reaction conditions for each reaction in the synthetic route will be closely studied and modified in order to get the best yields and/or purity. Different solvents, reagents and catalysts may be tried and the final reaction conditions may be radically different from the original conditions. It may even be necessary to abandon the original synthesis and devise a completely different route. For example, the synthesis of the antihistamine, **fexofenadine**, was originally carried out by the route used for the related antihistamine, **terfenadine** (*Fig. 1*).

Although this route was suitable for terfenadine, it proved unsatisfactory for fexofenadine since the Friedal-Crafts reaction gave a mixture of *meta* and *para* isomers, the major isomer being the unwanted *meta* isomer, which had to be removed by chromatography. Therefore, a more efficient and practical synthesis was devised for fexofenadine, starting from easily available materials, and which required no chromatographic separations (*Fig. 2*). Moreover, the route could be adapted to produce the pure active enantiomer in 96% ee. Thus, chemical development is more than just carrying out a simple scale-up exercise and such studies can take 5–10 years.

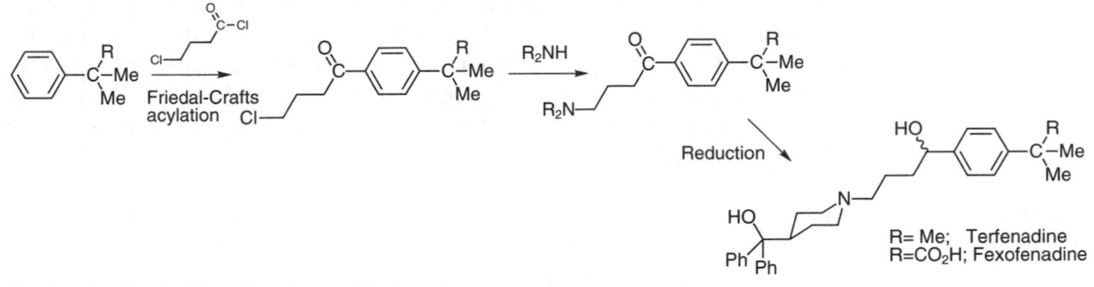

Fig. 1. Synthesis of fexofenadine and terfenadine.

Development phases

There are typically several phases in chemical development. In the first phase, about one kilogram of drug is required for short-term toxicology and stability tests, analytical research and pharmaceutical development. Often, the original synthetic route will be quickly developed and scaled-up in order to produce this quantity of material as time is critical. The next stage is to produce about 10 kg for long-term toxicology tests, as well as for formulation studies. The latter are carried out to find the best way of administering the drug (e.g. tablets) and requires large quantities of compound. Some of the material may also be used for phase I clinical trials. The third stage involves a further scale-up to the pilot plant, where about 100 kg is prepared for phase II and phase III clinical trials. When the drug reaches the market, several tons will need to be prepared.

Due to the time scales involved, the chemical process used to synthesize the drug during stage 1 may differ markedly from that used in stage 3. However, it is important that the quality and purity of the drug remain as constant as possible for all the studies carried out. Therefore, an early priority in chemical development is to optimize the final step of the synthesis and to develop a purification procedure that will consistently give a high quality product. The **specifications** of the final product are defined and determine the various analytical tests and purity standards required. These specifications must satisfy the various regulatory authorities whose permission is required before clinical trials are started. Any future batches of the drug must also meet these specifications. Once the final stages have been optimized, future development work can then look to optimize or alter the earlier stages of the synthesis.

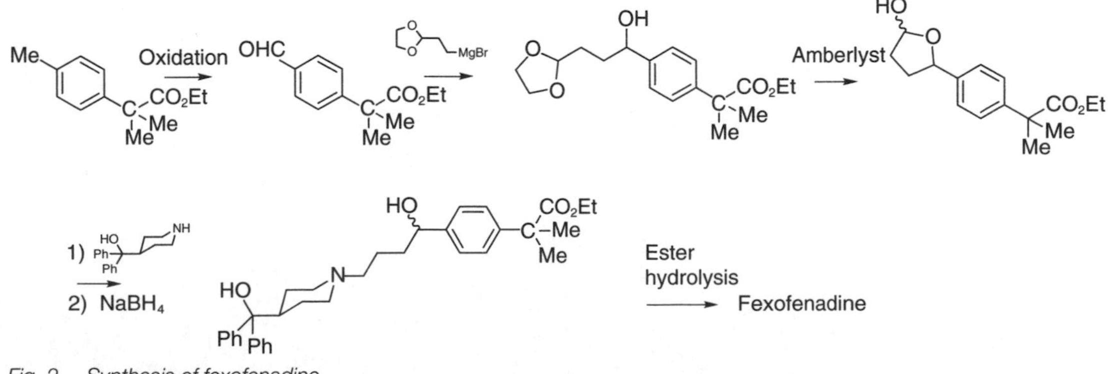

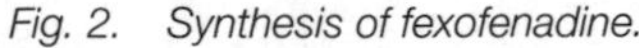
Fig. 2. Synthesis of fexofenadine.

In some development programs, the structure originally identified as the most promising clinical prospect may be supplanted by another structure that demonstrates better properties. The new structure may be a close analog of the original compound, but such a change can have radical effects on chemical development and require totally different conditions to maximize the yields for each synthetic step.

Process development

Process development is part of the chemical development program and aims, not only to develop the individual steps in the synthetic route, but to ensure that they are integrated with each other so that the full synthesis can be run smoothly and efficiently on a production scale. The aim is to reduce the number of operations to the minimum. For example, rather than isolating each intermediate in the synthetic sequence, it is better to move them directly from one reaction vessel to the next for the subsequent synthetic step. Process development is very much aimed at optimizing the process for a specific compound. If the original structure is abandoned in favor of a different analog, the process may have to be rethought completely.

Potential problems of the initial synthesis

The synthesis originally used in the drug discovery laboratory will be one that provides as wide a variety of different analogs as possible – it has not been designed for one specific compound. A development chemist has to develop a synthesis that is ideal for the large-scale production of a specific compound. In order to do this it is necessary to assess the initial synthesis for potential problems, such as low yields, the number of steps involved, the cost of reagents and reactants, the potential hazards of individual chemicals and solvents, and the experimental procedures used.

J2 OPTIMIZATION OF REACTIONS

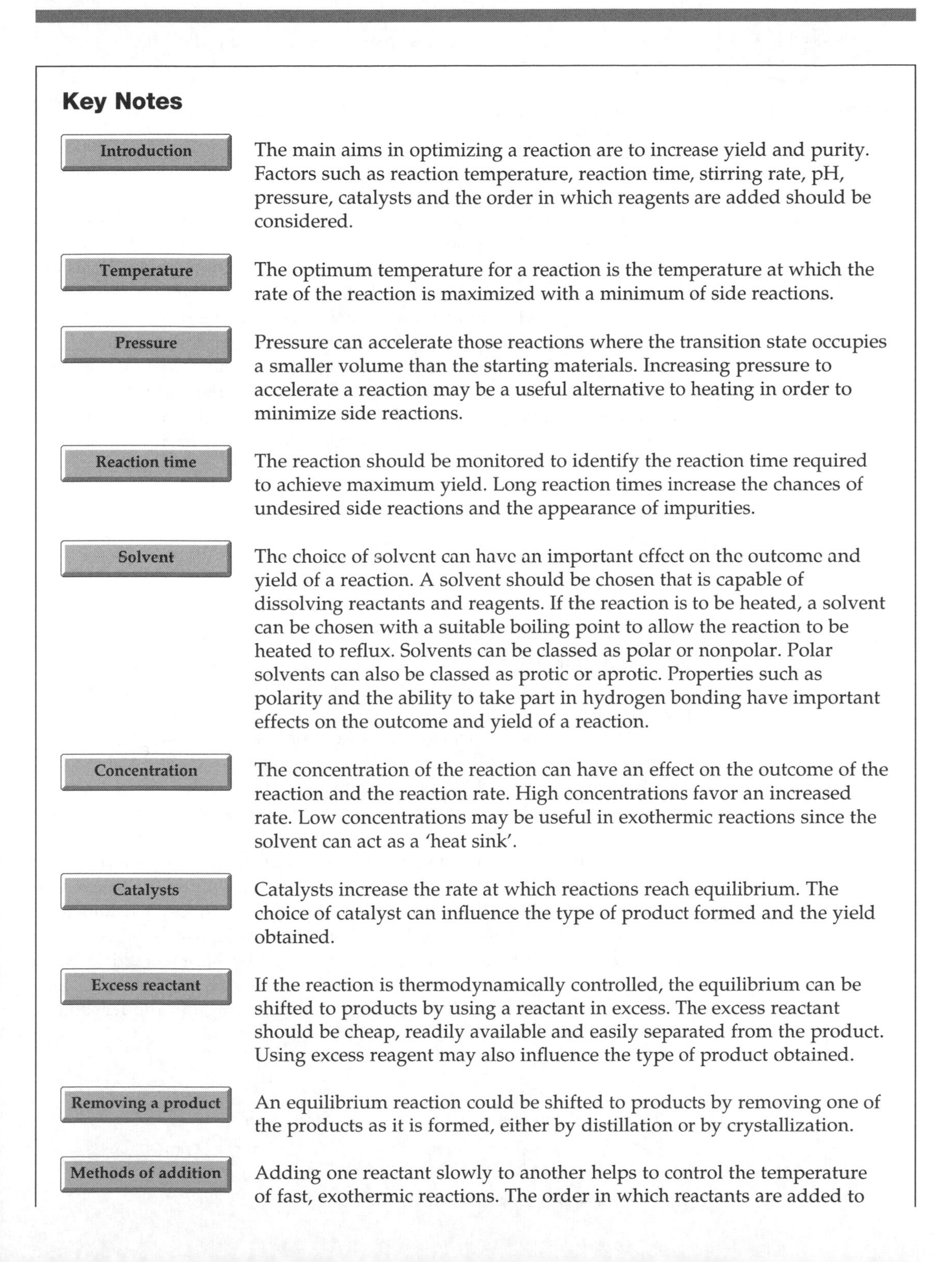

Key Notes

Introduction
The main aims in optimizing a reaction are to increase yield and purity. Factors such as reaction temperature, reaction time, stirring rate, pH, pressure, catalysts and the order in which reagents are added should be considered.

Temperature
The optimum temperature for a reaction is the temperature at which the rate of the reaction is maximized with a minimum of side reactions.

Pressure
Pressure can accelerate those reactions where the transition state occupies a smaller volume than the starting materials. Increasing pressure to accelerate a reaction may be a useful alternative to heating in order to minimize side reactions.

Reaction time
The reaction should be monitored to identify the reaction time required to achieve maximum yield. Long reaction times increase the chances of undesired side reactions and the appearance of impurities.

Solvent
The choice of solvent can have an important effect on the outcome and yield of a reaction. A solvent should be chosen that is capable of dissolving reactants and reagents. If the reaction is to be heated, a solvent can be chosen with a suitable boiling point to allow the reaction to be heated to reflux. Solvents can be classed as polar or nonpolar. Polar solvents can also be classed as protic or aprotic. Properties such as polarity and the ability to take part in hydrogen bonding have important effects on the outcome and yield of a reaction.

Concentration
The concentration of the reaction can have an effect on the outcome of the reaction and the reaction rate. High concentrations favor an increased rate. Low concentrations may be useful in exothermic reactions since the solvent can act as a 'heat sink'.

Catalysts
Catalysts increase the rate at which reactions reach equilibrium. The choice of catalyst can influence the type of product formed and the yield obtained.

Excess reactant
If the reaction is thermodynamically controlled, the equilibrium can be shifted to products by using a reactant in excess. The excess reactant should be cheap, readily available and easily separated from the product. Using excess reagent may also influence the type of product obtained.

Removing a product
An equilibrium reaction could be shifted to products by removing one of the products as it is formed, either by distillation or by crystallization.

Methods of addition
Adding one reactant slowly to another helps to control the temperature of fast, exothermic reactions. The order in which reactants are added to

a reaction vessel can influence the outcome and yield of the reaction. Stirring rates are also crucial.

Reactivity of reagents/reactants

Carrying out a reaction with a less reactive reagent may affect the outcome of the reaction.

Related topics

Patenting and chemical development (J1)

Scale up issues (J3)

Introduction

The major aims in developing a synthesis are to optimize the **yield** and **purity** of each step. These aims do not necessarily go hand in hand, and it may be necessary to accept a lower than optimum yield in order to obtain high purity. At the same time, such issues as **cost** and **safety** have to be considered. There is little point in spending a lot of time optimizing a reaction if the reagent used is so costly and hazardous that it is totally impractical for the production plant.

There are many factors that have to be taken into account when optimizing reaction conditions. These include temperature, reaction time, the order and rate of addition of reactants, pH control, stirring rate, pressure and the use of catalysts. Methods of purifying the product may also have to be developed.

Temperature

It is important to optimize the **temperature** for a particular reaction. If the temperature is too low, the reaction will be slow. If the temperature is too high, there is an increased chance of unwanted side reactions resulting in impurities and a lower yield. Some reagents or reactants may be unstable if the temperature is too high.

Pressure

Pressures above 5 kilobar (5000 atmospheres) can accelerate reactions that have negative **activation volumes** (i.e. where the transition state occupies a smaller volume than the starting materials). Common reactions that can be accelerated by pressure include esterifications, the quaternization of amines, hydrolyzes of esters, Claisen and Cope rearrangements, nucleophilic substitutions and Diels-Alder reactions. For example, the esterification of acetic acid with ethanol at 50°C is five times faster at 2 kbar than if the reaction is carried out at 1 atmosphere. If the pressure is increased to 4 kbar, the reaction proceeds 26 times faster.

Pressure can be used as an alternative to increased temperatures if the latter lead to undesired side effects. For example, the phosphonium salt below was prepared in good yield at 20°C and 15 kbar (*Fig. 1*). At 1 atmosphere and 20°C no reaction occurs, while at 80°C and 1 atmosphere, decomposition occurs.

The hydrolysis of esters is usually carried out under basic conditions with heating, but this can prove a problem if asymmetric centers are present, since heating in base may lead to their epimerization. Carrying out the reaction at room temperature under pressure may avoid such problems.

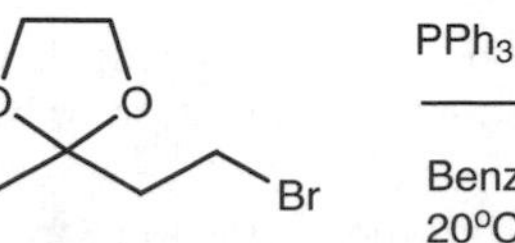

PPh_3

Benzene-toluene
20°C / 15 000atm

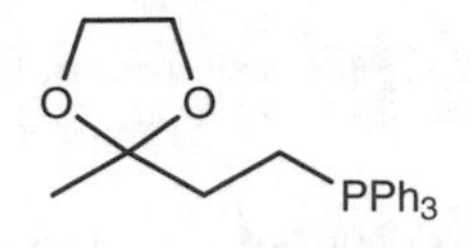

Fig. 1. Preparation of a phosphonium salt.

Reaction time

It is important to monitor a reaction to find the optimum **time** for the reaction. This can be done by analyzing aliquots of the reaction mixture using a variety of analytical methods, such as thin layer chromatography, gas chromatography, infra red spectroscopy, NMR spectroscopy or high performance liquid chromatography (HPLC). Ideally, the reaction should be allowed to run until all the starting material has been converted to product. However, it may not be as simple as this. If an equilibrium reaction is involved, there is no point continuing the reaction once equilibrium has been reached. Therefore, the reaction should be run for the minimum length of time required to give the highest yield.

Even if the reactant should eventually be used up in the reaction, it may not necessarily be good to run the reaction for this period of time. In some reactions, prolonged reaction times may result in the desired product undergoing further unwanted reactions, resulting in impurities and a decreased yield. Reaction times greater than 15 hours should also be discouraged since they prove costly in terms of overheads and labor at the production level.

Solvent

The choice of **solvent** can be crucial to the outcome and the yield of a reaction. Primarily, a solvent is required that will dissolve the reactants and reagents, and ensure that reaction will take place. For example, a polar solvent is needed to dissolve ionic reagents such as potassium cyanide. Sometimes, it is possible to use a solvent even though the reactant is only sparingly soluble. This is because the reactant that *has* dissolved will be converted to product and will disturb the equilibrium between the undissolved and dissolved states. As a result, more reactant is taken up into solution until it is all converted to product.

The solubility of products may also be an issue. If a product proves to be insoluble and precipitates from solution, it may help to push equilibrium to the right and increase yield. On the other hand, the precipitation of a product may be a problem. For example, the hydrogenolysis of the protected dipeptide (I) was carried out in ethanol over a palladium charcoal catalyst, but only went in 50% yield (*Fig. 2*). This was because the dipeptide product precipitated from solution and coated the catalyst, thus stopping the reaction. The desired dipeptide was soluble in water, but it was not possible to use water as the solvent since the protected dipeptide was insoluble. The answer to the problem was to use an ethanol/water mixture and to find the ideal ratio of ethanol to water to ensure that both the protected and the free dipeptide remained in solution. Using a 50:50 mixture of ethanol/water, the reaction went in quantitative yield.

Solubility is not the only issue to consider when choosing a solvent. For example, if the reaction is to be heated at a particular temperature, the easiest way to control the reaction temperature is to choose a solvent that boils at or

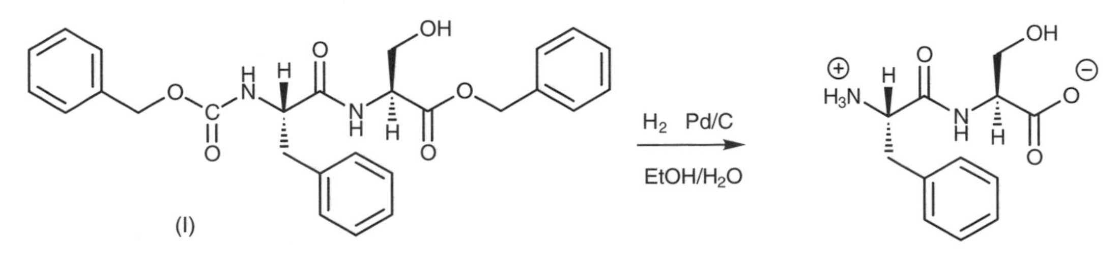

Fig. 2. Deprotection of a dipeptide.

near that temperature. This means that the reaction can be heated to reflux and avoids the problems of monitoring and controlling the temperature.

The solvent should also be compatible with the type of reaction being attempted. Solvents can be classed as **polar** (e.g. ethanol, water, acetone) or **nonpolar** (e.g. toluene, chloroform, petroleum ether). Polar solvents can be further classified as **protic** or **aprotic**. Protic solvents are capable of acting as hydrogen bond donors (e.g. ethanol and water), whereas aprotic solvents cannot (e.g. dimethylformamide, dimethylsulfoxide). These properties have important influences on reactions. For example, protic solvents give higher rates in S_N1 reactions compared to aprotic solvents since they aid the departure of the anionic leaving groups in the rate-determining step. S_N2 reactions are not aided by protic solvents since the rate-determining step involves the anion acting as a nucleophile. If the anion is solvated, it is less reactive. A dipolar, aprotic solvent such as dimethylsulfoxide (DMSO) is better for an S_N2 reaction since it strongly solvates cations, but leaves anions relatively unsolvated and so much more reactive. For example, NaCN in DMSO reacts with primary and secondary alkyl chlorides by an S_N2 reaction to give nitriles in 0.5–2 hours (*Fig. 3*). The same reaction carried out in aqueous alcohol takes 1–4 days.

Changing a solvent can often have an effect on the outcome and yield of a reaction and it is useful to try out the various solvents to see what effect they have. For example, toluene and nitrobenzene can be tried as solvents for the Friedal-Crafts acylation to see which solvent gives the best result.

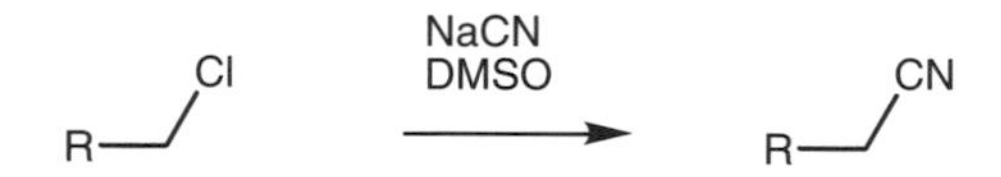

Fig. 3. Reaction of an alkyl chloride with NaCN.

Concentration

The effect of varying the **concentration** of the reaction is worth studying. At high concentrations (small volume of solvent), the rate of the reaction may be expected to increase since there is more chance of the reactants meeting and reacting with each other. However, care has to be taken in case higher concentrations also encourage unwanted reactions. Low concentrations (large volume of solvent) may be useful for very fast, exothermic reactions since the solvent can act as 'heat sink' and allows easier control of the reaction temperature.

Catalysts

Catalysts are compounds that accelerate the rate at which a reaction reaches equilibrium. They can be classed as **heterogeneous** where the catalyst remains undissolved and provides a solid surface for the reaction (e.g. palladium charcoal used in hydrogenation), or they can be **homogeneous** where they dissolve in the reaction solvent and form complexes with the reactants (e.g. rhodium catalysts used in hydrogenation). Catalysts can have different activities and it is possible to alter the outcome of a reaction by 'poisoning' a catalyst so that it is less reactive. For example, the reduction of alkynes with hydrogen over an active palladium charcoal catalyst results in the formation of an alkane. Carrying out the same reaction with a less active catalyst results in the formation of a *Z*-alkene instead (*Fig. 4*).

If a variety of catalysts can be used for a particular reaction, it is often worth trying out each one to see if one catalyst gives a better result than another. For

Fig. 4. Reduction of an alkyne.

example, various Lewis acids such as $AlCl_3$ and $ZnCl_2$ can be tried as catalysts for the Friedal-Crafts acylation of aromatic rings (*Fig. 5*).

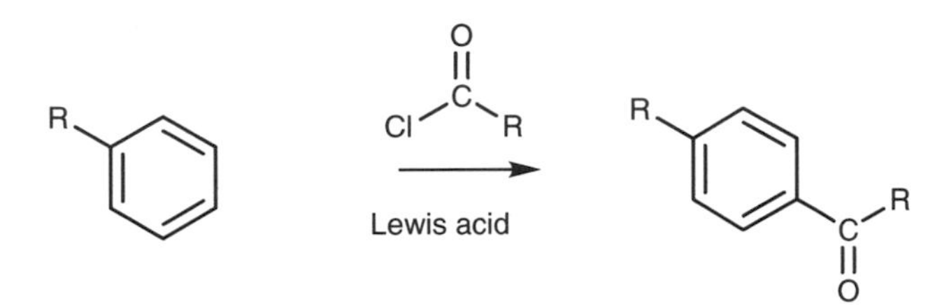

Fig. 5. Friedal-Crafts acylation.

Excess reactant

If the reaction in question is thermodynamically controlled (as opposed to kinetically controlled), there is an equilibrium between the reactants and products, and the reaction will proceed until that equilibrium is reached. The **equilibrium constant** (K_{equil}) for the reaction is defined as below:

$$K_{equil} = \frac{\text{Concentration of products}}{\text{Concentration of reactants}}$$

For a reaction involving two reactants and two products, the equilibrium constant is as follows:

$$A + B \rightleftharpoons C + D \qquad K_{equil} = \frac{[C]\ [D]}{[A]\ [B]}$$

In order to increase the yield, the equilibrium can be disturbed by increasing the concentration of one of the reactants so that it is in excess. This shifts the equilibrium to the right and more product is formed. The excess reactant should be cheap and readily available. For example, if reactant A is the third intermediate of a reaction sequence, and is being treated with a commercially available reactant B, then it is B that is used in excess. Another important factor to consider is whether the excess reactant can be easily removed from the reaction mixture. There is no point using excess reactant to force the equilibrium to products if a long purification procedure is required to isolate the desired product.

In some reactions, it is necessary to add an excess of a reagent or reactant to obtain the desired reaction product. For example, 1,2-diaminoethane can react with an acid anhydride to give a mono-acylated or a diacylated product (*Fig. 6*). If equal quantities of each reactant are used with the aim of obtaining the

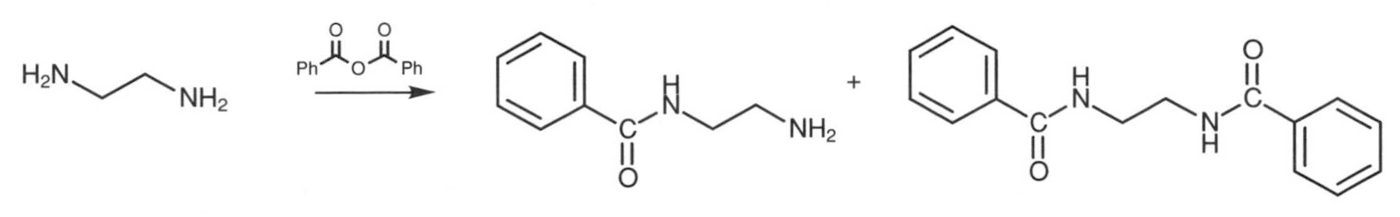

Fig. 6 Acylation of 1,2-diaminoethane.

monoacylated product, then a mixture of products is guaranteed since the monoacylated product will compete with 1,2-diaminoethane for the acid anhydride. Therefore, in order to increase the proportion of mono-acylated product, the 1,2-diaminoethane should be used in excess.

Removing a product

A different method of forcing an equilibrium reaction to the right is to remove one of the products from the reaction mixture. For example, if one of the products is volatile, the reaction mixture could be heated to distil off the product as it is formed. Alternatively, one of the products might be induced to crystallize out of solution as it is formed. The removal of water from the ketalization of ketones is an important method of forcing this reaction to completion (*Fig. 7*).

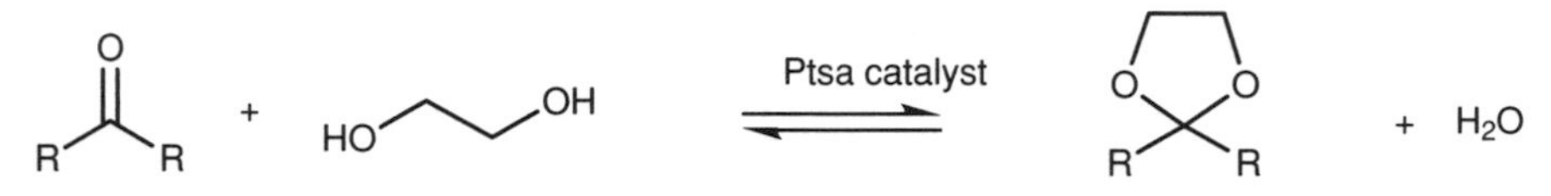

Fig. 7. Ketalization of ketones.

Methods of addition

The method and order by which reactants are physically added to the reaction vessel can often affect the outcome of a reaction. If the reaction is fast and exothermic, it is best to add one of the reactants slowly to the other so that the temperature of the reaction can be kept under control. However, problems can occur due to a temporary high concentration of the reactant at the point of addition. This can be minimized by diluting the reactant in solvent before it is added, and using vigorous stirring.

Sometimes the outcome and yield of a reaction can be affected by reversing the addition of the reactants. For example, in the Horner–Emmons reaction shown below (*Fig. 8*), butyl lithium was added to the phosphonate to form the phosphonate anion, then reacted with an aldehyde to form the desired alkene. However, a methylated impurity was also present. This was formed by the phosphonate anion reacting with unreacted methyl phosphonate. The side reaction could be avoided by adding the phosphonate to the butyl lithium instead.

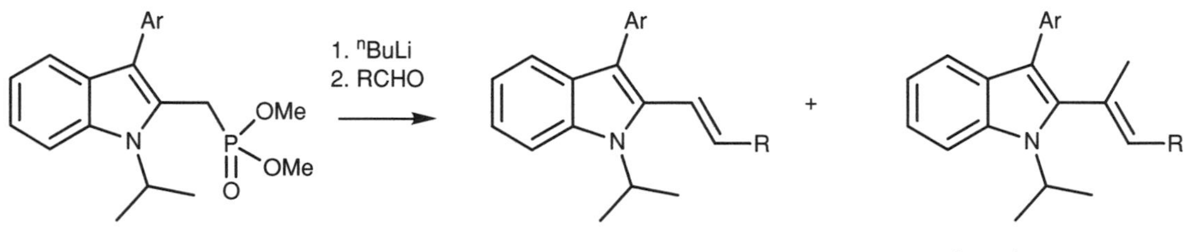

Fig. 8. Horner–Emmons reaction.

Reactivity of reagents/reactants

Sometimes there is an advantage in varying the reactivity of a reagent. For example, the reaction between benzoyl chloride and 1,2-ethanediamine gave a 1:1 mixture of the mono- and diacylated products, even when the acid chloride was diluted in solvent and added slowly to an excess of the diamine with vigorous stirring (*Fig. 9*). By replacing benzoyl chloride with the less reactive benzoic anhydride, it was possible to get a better ratio of mono- to diacylated products of 1.86 : 0.14.

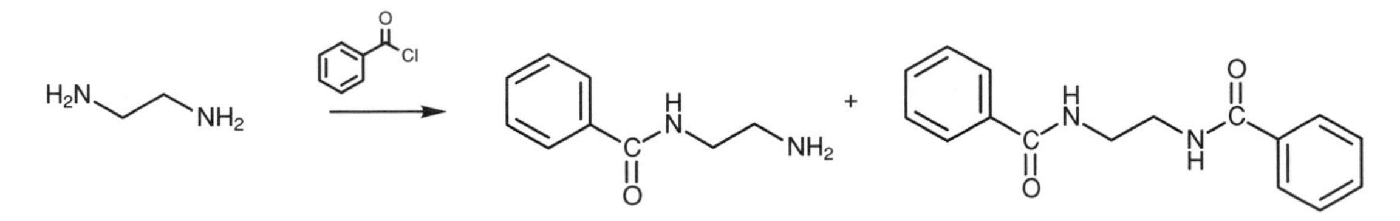

Fig. 9. Acylation of 1,2-ethanediamine.

J3 SCALE-UP ISSUES

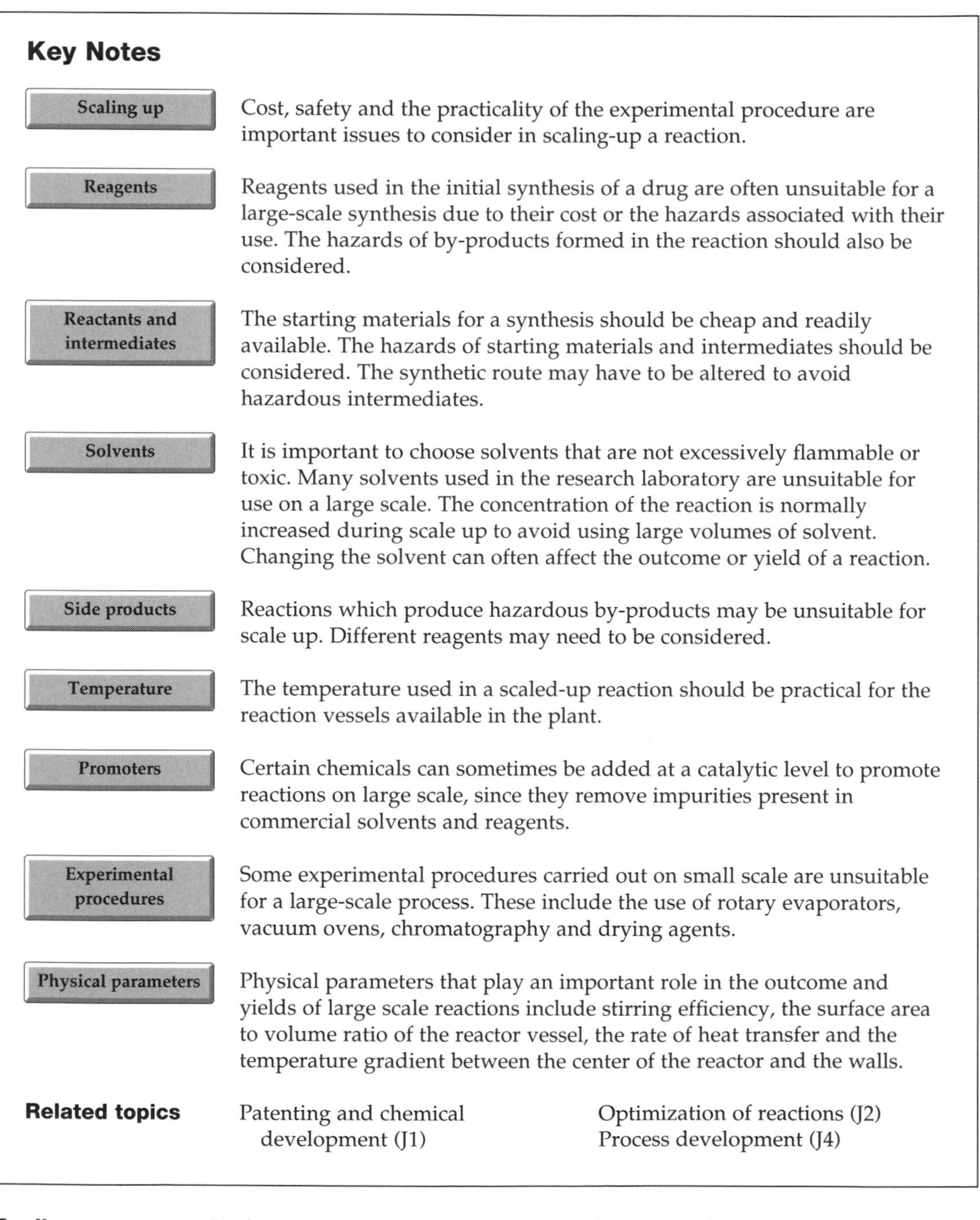

Key Notes

Scaling up — Cost, safety and the practicality of the experimental procedure are important issues to consider in scaling-up a reaction.

Reagents — Reagents used in the initial synthesis of a drug are often unsuitable for a large-scale synthesis due to their cost or the hazards associated with their use. The hazards of by-products formed in the reaction should also be considered.

Reactants and intermediates — The starting materials for a synthesis should be cheap and readily available. The hazards of starting materials and intermediates should be considered. The synthetic route may have to be altered to avoid hazardous intermediates.

Solvents — It is important to choose solvents that are not excessively flammable or toxic. Many solvents used in the research laboratory are unsuitable for use on a large scale. The concentration of the reaction is normally increased during scale up to avoid using large volumes of solvent. Changing the solvent can often affect the outcome or yield of a reaction.

Side products — Reactions which produce hazardous by-products may be unsuitable for scale up. Different reagents may need to be considered.

Temperature — The temperature used in a scaled-up reaction should be practical for the reaction vessels available in the plant.

Promoters — Certain chemicals can sometimes be added at a catalytic level to promote reactions on large scale, since they remove impurities present in commercial solvents and reagents.

Experimental procedures — Some experimental procedures carried out on small scale are unsuitable for a large-scale process. These include the use of rotary evaporators, vacuum ovens, chromatography and drying agents.

Physical parameters — Physical parameters that play an important role in the outcome and yields of large scale reactions include stirring efficiency, the surface area to volume ratio of the reactor vessel, the rate of heat transfer and the temperature gradient between the center of the reactor and the walls.

Related topics

Patenting and chemical development (J1)
Optimization of reactions (J2)
Process development (J4)

Scaling up

In the previous section, we discussed the various factors that are involved in optimizing a reaction. However, it is a waste of time optimizing a reaction if the reagents, reactants, solvents or experimental procedures are incompatible with scale up. The main factors to consider here are cost, safety, environmental issues and the practicality of the experimental procedure.

Reagents

Some **reagents** used in the initial small-scale synthesis of a drug are too expensive to be used in large-scale reactions, and it may be necessary to replace an expensive reagent with something cheaper. For example, zinc-copper amalgam is used in cyclopropanation reactions (*Fig. 1*), but is too expensive for a large-scale synthesis. Fortunately, it can be replaced with cheaper zinc powder.

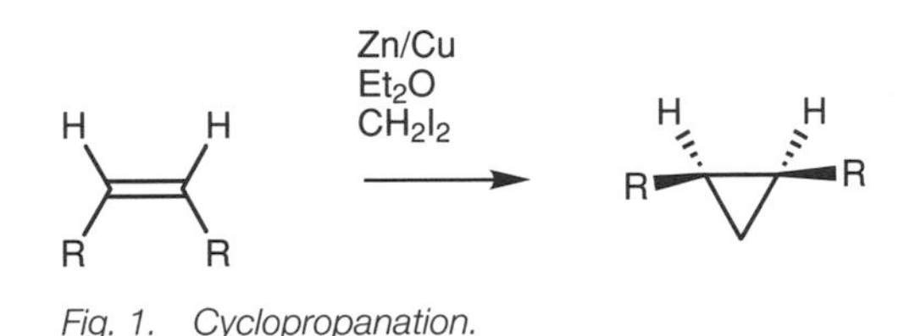

Fig. 1. Cyclopropanation.

The commercial availability of reagents also has to be taken into account. If the reagent has to be prepared 'in-house', it will add to the cost of the process.

Some reagents are unsuitable for large-scale synthesis on safety grounds. Highly toxic, carcinogenic or explosive reagents should be avoided if possible. For example, the reagents palladium chloride (used for the deoxygenation of *N*-oxides) and pyridinium chlorochromate (used for the oxidation of alcohols to aldehydes) are both carcinogenic (*Fig. 2*). If a hazardous reagent has to be used because there is no other alternative, then it is more acceptable if it is used at an early stage of a synthesis than at a later stage, since there is less chance of the reagent contaminating the final product. If a hazardous reagent is used late in a synthesis, the risks may become unacceptable. For example, a synthesis involving a carcinogenic reagent in the final step would be rejected by regulatory bodies. If a safer reagent cannot be found, the synthetic route itself would have to be changed.

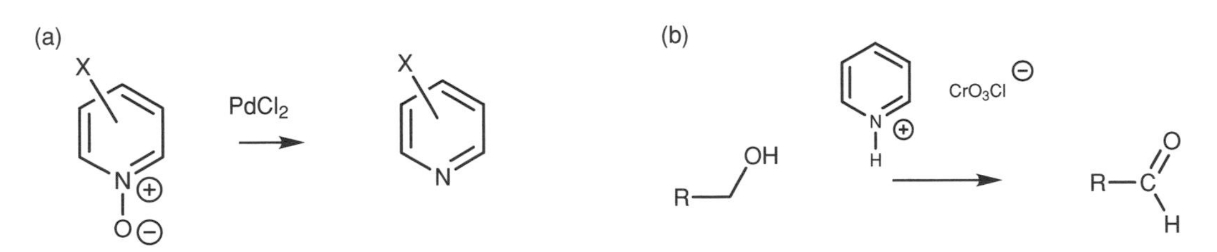

Fig. 2. Examples of carcinogenic reagents.

Some reagents may be unsuitable on environmental grounds since they are too volatile or odorous. Others may be difficult to handle, being hygroscopic or lachrymatory. It is also important to consider the hazards associated with the by-products formed from reagents. For example, the use of mercuric acetate can result in the formation of mercury.

Sometimes a choice has to be made between cost and safety. For example, *m*-chloroperbenzoic acid is often used for the Baeyer-Villiger oxidation (*Fig. 3*). Cheaper peroxide reagents are available, but are more hazardous. Therefore, despite its cost, *m*-chloroperbenzoic acid is preferred since it is a relatively safe peracid with a relatively high decomposition temperature.

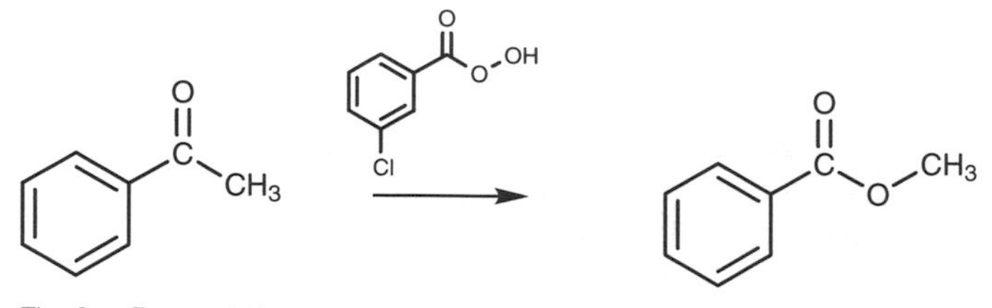

Fig. 3. Baeyer-Villiger oxidation.

Reactants and intermediates

Many of the potential problems described above also apply to **reactants** and **intermediates**. The starting material for the synthetic route may be expensive or it may not be commercially available, in which case an alternative synthetic route that makes use of a cheap, readily available starting material should be devised. Failing that, the starting material may have to be synthesized in house.

The potential hazards of reactants and intermediates in the synthetic route should be considered. For example, diazonium salts are potentially explosive and are best avoided on large scale.

Solvents

The nature and quantity of the **solvent** used in a reaction is one of the most common issues that needs to be addressed in chemical development. There are two major issues here – the **safety** and the **volume** of the solvent.

As far as volume is concerned, small-scale reactions carried out in the research laboratory are relatively dilute. A straight scale-up would result in massive volumes of solvent being used, which would be wasteful, hazardous and costly. Therefore, it is normal to reduce the relative volume of solvent as the reaction is scaled up. Typically, the concentration may be increased 100-fold and volumes reduced to a solvent : solute ratio of 5 : 1 or less. There are potential advantages in running reactions at higher concentration since the reaction may go faster and be completed sooner. Some reactions can even be carried out without solvent.

The nature of the solvent is also important. Factors such as cost, possible hazards and the ease with which the solvent can be recycled need to be considered. Some common solvents that are commonly used in the research laboratory are avoided on scale-up. For example, diethyl ether is considered too hazardous for large-scale use due to its low boiling point and flammability. Solvents such as chloroform, carbon tetrachloride, dioxane, benzene, and hexamethylphosphoric triamide are not used since they are carcinogenic.

When choosing a suitable solvent for scale up, it must be suitable for the reaction, but its physical properties must also be considered. It is necessary to know the solvent's **ignition temperature** (the temperature at which the compound ignites), its **flash point** (the temperature at which vapors of the solvent ignite in the presence of an ignition source, like a spark or flame), its **vapor pressure** (a measure of its volatility), **vapor density** (a measure of whether vapors of the solvent will rise or creep along the floor), and the mixture range of solvent and air that is flammable. Solvents that are flammable at a low solvent/air mixture and over a wide range of solvent/air mixtures are particularly hazardous. For example, diethyl ether has a flammable solvent/air range of 2–36%.

In general, one would not work with solvents having a flash point less than –18°C. This rules out diethyl ether and carbon disulfide, but not acetone. Diethyl ether is particularly hazardous since it is heavier than air and can creep across a laboratory or plant floor, allowing the possibility of ignition on any hot pipes that might be present.

Therefore, many solvents used in the research laboratory are changed for scale up. Diethyl ether can be replaced with dimethoxyethane, which has a higher boiling point and is not as flammable. Dimethoxyethane offers additional benefits in that it has a higher heat capacity and allows higher reaction temperatures to be used. Another option is to use t-butyl methyl ether, which is cheap, relatively safe and does not form peroxides.

Flammable solvents such as pentane and hexane can be replaced with heptane. Toxic solvents such as chloroform, dichloromethane and carbon tetrachloride are commonly replaced with ethyl acetate. Benzene, which is carcinogenic, is usually replaced with toluene or xylene. Dioxane is another carcinogenic solvent, which can be replaced with tetrahydrofuran. Of course it should be remembered that replacing solvents can have an effect on the rate and the outcome of reactions. Therefore, the chemical development laboratories need to experiment with different solvents to test what effect they have on the reaction.

Finally, it should be noted that it is not feasible to carry out elaborate purification procedures on solvents once the reaction has been scaled-up to a production scale. Therefore, experiments at the development stage should be carried out using commercial solvents without purification.

Side products

Some reactions may be unsuitable for scale up if they produce dangerous **side products**. For example, the preparation of a phosphonate by the Arbusov reaction produces methyl chloride, which is gaseous, an alkylating agent and toxic (*Fig. 4*). This can be avoided by switching from trimethyl phosphite (which also stinks) to sodium dimethyl phosphonate. The side product formed is sodium chloride.

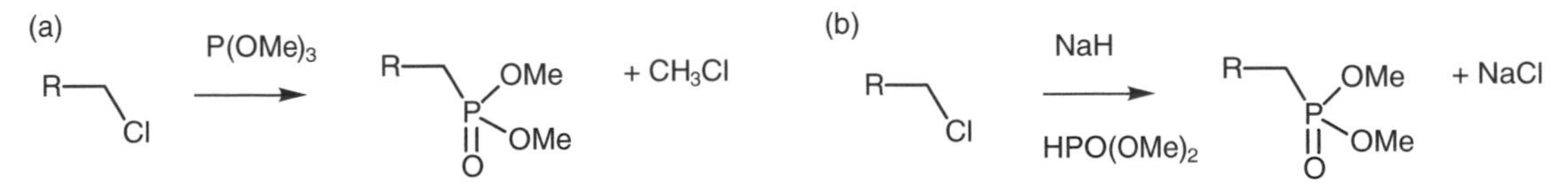

Fig. 4. Preparation of a phosphonate.

Temperature

As mentioned previously, the **temperature** of a reaction should be optimized for each reaction. However, it is also important to consider what temperatures are practical for the reaction vessels available in the plant. Typical coolants for reaction vessels are glycol (–20°C) and water (15°C).

Promoters

The addition of a catalytic amount of a **promoter** can have a beneficial effect on a scaled-up reaction, especially when commercial grade solvents and reagents are being used. For example, a catalytic amount of RedAl can be used in the cyclopropanation reaction with zinc mentioned previously. The catalyst prevents a delay to the start of the reaction by removing zinc oxides from the surface of the zinc, removing water from the solvent and removing peroxides from the solvent.

Adding a promoter to a Grignard reaction in the form of methyl magnesium iodide is also commonly used.

Experimental procedures

Some **experimental procedures** are not conducive to scale up. For example, although it is possible to scrape a solid out of a round-bottomed flask on small scale, it is not possible to carry out this operation on a reaction vessel in a production plant. There are many other examples of procedures commonly used in research laboratories that cannot be used on the plant. The following are a selection.

- Adding drying agents such as sodium sulfate to an organic solution is impractical on large scale. Instead, one could add a suitable organic solvent and azeotrope the water off by distillation. Alternatively, the organic solution can be extracted with brine.
- Using a rotary evaporator to concentrate solutions is not possible on a large scale and a conventional distillation would have to be carried out. It is therefore important to see whether this would affect the outcome and yield of the reaction before attempting the scale-up.
- Concentrating solutions to dryness should be avoided since dry solids cannot be removed from the reaction vessel. It is better to keep intermediates in solution, so that they can be easily transferred from one reaction vessel to another.
- Oily products are often dried in the research laboratory by heating them in a vacuum oven, but this cannot be done as effectively on a large scale. Sometimes, adding solvent for the next reaction can flush out excess solvent from the previous step.
- Chromatography is commonly used to purify compounds in the research laboratory and might be used after each stage of the synthesis. Although chromatography is possible on a large scale, it is preferable to purify compounds by crystallization.
- The scale-up process has an inevitable effect on the length of time taken over certain operations. Distillations and the addition of reagents take longer. For example, it is possible to add a reagent to a small-scale reaction in one portion. On scale-up, this addition may take 15 minutes.
- On a small scale, the work-up procedure may involve several washings and extractions using a separating funnel. Shaking separation funnels in a production site is not possible and solutions are washed/extracted by stirring the solvent phases in large reaction vessels. This takes time and so the number of such operations should be cut to a minimum. Countercurrent extraction is an alternative.

Physical parameters

During chemical development, it is important to study all the parameters that are likely to have an effect on a reaction. However, there are some parameters that are unique to the scale-up process and cannot be effectively modeled with smaller scale experiments in the laboratory. These parameters include the stirring efficiency of the reaction tank, the surface area to volume ratio of the reactor vessel in comparison with the laboratory vessel, the rate of heat transfer and the temperature gradient between the center of the reactor and the walls, all of which will be different on scale-up.

J4 PROCESS DEVELOPMENT

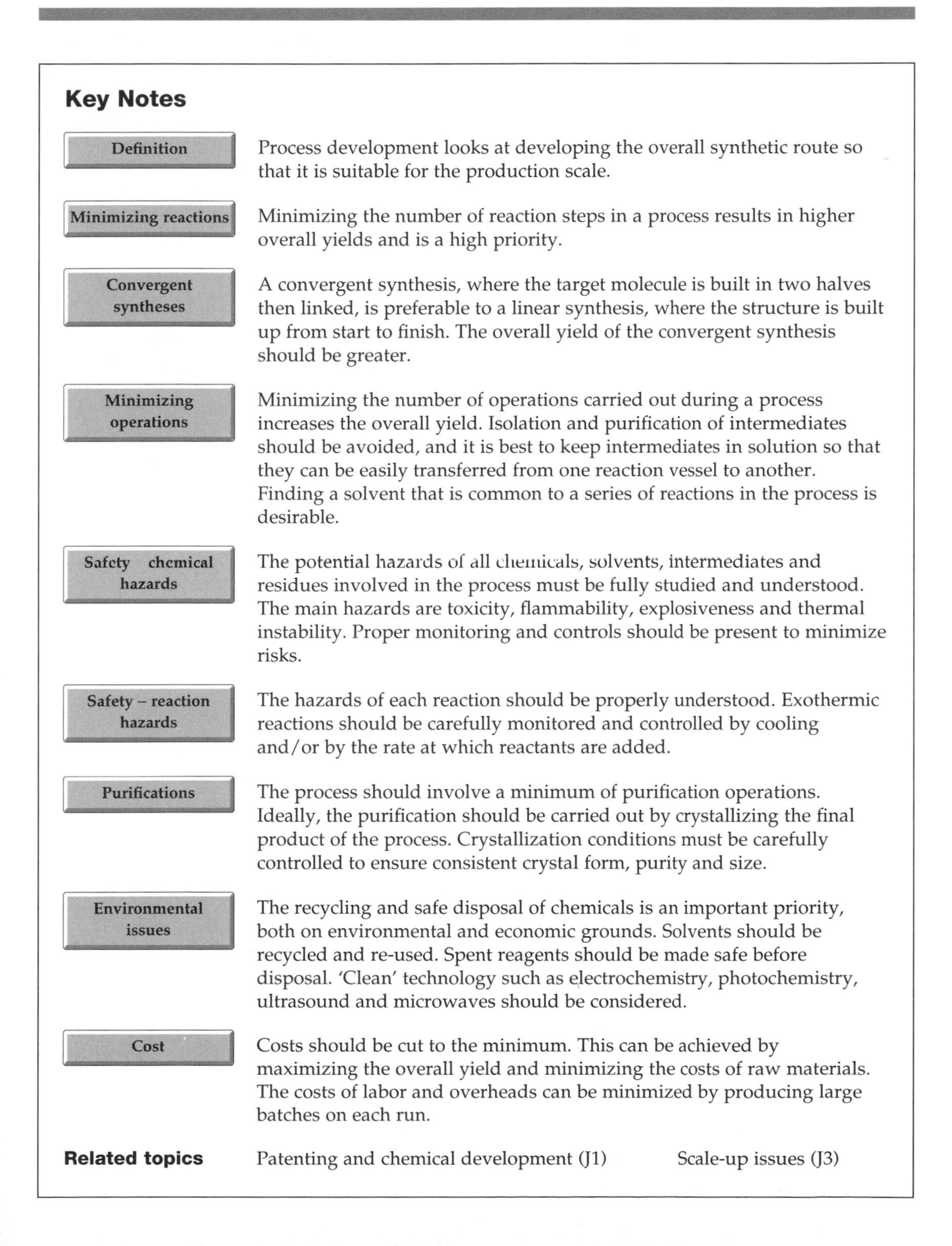

Key Notes

Definition	Process development looks at developing the overall synthetic route so that it is suitable for the production scale.
Minimizing reactions	Minimizing the number of reaction steps in a process results in higher overall yields and is a high priority.
Convergent syntheses	A convergent synthesis, where the target molecule is built in two halves then linked, is preferable to a linear synthesis, where the structure is built up from start to finish. The overall yield of the convergent synthesis should be greater.
Minimizing operations	Minimizing the number of operations carried out during a process increases the overall yield. Isolation and purification of intermediates should be avoided, and it is best to keep intermediates in solution so that they can be easily transferred from one reaction vessel to another. Finding a solvent that is common to a series of reactions in the process is desirable.
Safety chemical hazards	The potential hazards of all chemicals, solvents, intermediates and residues involved in the process must be fully studied and understood. The main hazards are toxicity, flammability, explosiveness and thermal instability. Proper monitoring and controls should be present to minimize risks.
Safety – reaction hazards	The hazards of each reaction should be properly understood. Exothermic reactions should be carefully monitored and controlled by cooling and/or by the rate at which reactants are added.
Purifications	The process should involve a minimum of purification operations. Ideally, the purification should be carried out by crystallizing the final product of the process. Crystallization conditions must be carefully controlled to ensure consistent crystal form, purity and size.
Environmental issues	The recycling and safe disposal of chemicals is an important priority, both on environmental and economic grounds. Solvents should be recycled and re-used. Spent reagents should be made safe before disposal. 'Clean' technology such as electrochemistry, photochemistry, ultrasound and microwaves should be considered.
Cost	Costs should be cut to the minimum. This can be achieved by maximizing the overall yield and minimizing the costs of raw materials. The costs of labor and overheads can be minimized by producing large batches on each run.

Related topics Patenting and chemical development (J1) Scale-up issues (J3)

Definition

In previous sections, we saw that development chemists are responsible for optimizing and scaling up reactions. **Process development** looks not only at how individual reactions can be adapted for full scale production, but also at how the overall synthesis can be integrated so that it is as efficient and manageable as possible. There are several important principles to bear in mind.

Minimizing reactions

One of the major priorities in process development is to reduce the number of individual reactions in the overall synthesis. This can often have a dramatic effect on the overall yield. For example, a synthetic route that involves ten reactions (each of which goes in 80% yield) will lead to a 10% overall yield. If the number of reactions can be reduced to five having the same level of yield, the overall yield would be 33%. Achieving this goal involves a good understanding of organic chemistry in order to identify alternative reactions or reagents that can be used to cut out a step in the synthesis.

Convergent syntheses

Sometimes it may be necessary to alter the whole synthetic strategy in order to achieve a good overall yield. For example, the initial synthesis might involve 10 steps (*Fig. 1*). If each step went in 80% yield, the overall yield would be 10.7%. This kind of approach is known as a **linear approach** to synthesis. A more effective approach is to devise a **convergent synthesis**. A convergent synthesis involves synthesizing two halves of the final product separately, then linking them together. The total number of reactions might be the same, but the overall yield would be greatly increased. In this example, the yield would be 26.2% from L and 32.8% from R, assuming 80% yields at each step.

Minimizing operations

Once a suitable synthetic route has been identified, it is possible to increase the overall yield by reducing the number of operations that have to be carried out. One of the best ways of doing this is to develop the synthesis so that intermediates do not need to be isolated and purified. For example, it is possible to convert an alcohol to an alkyl chloride using thionyl chloride. Rather than isolate the alkyl halide intermediate, it could be treated immediately with triphenylphosphine to make a Wittig reagent.

On large scale, it is preferable to form one intermediate and keep it in solution for the next stage without attempting to isolate it. Therefore, using a solvent that is common to a series of reactions is preferable to using a different solvent for each reaction.

Safety – chemical hazards

Safety is a major issue in any process. The development process already described will have replaced particularly hazardous chemicals and solvents with less hazardous ones. However, hazards can only be minimized, they cannot be removed entirely. Therefore, the properties and reactivities of the

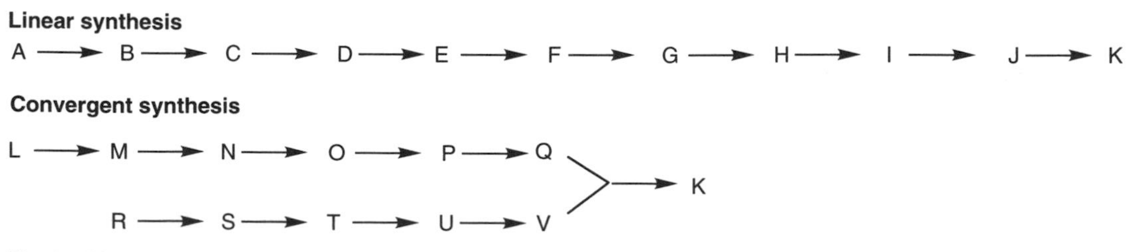

Fig. 1. Linear and convergent syntheses.

chemicals and reagents used in the synthetic route must be properly understood and suitable precautions taken.

Hazards can be divided into four categories.

- **Thermal instability**. The decomposition temperature of all chemicals, solvents and distillation residues must be established so that one stays well below these temperatures in the reaction process.
- **Toxicity**. In general, compounds with a LD_{50} of less than 100 mg kg^{-1} are not used in production. This corresponds to a lethal dose of about a teaspoon.
- **Flammability**. Solvents with a high risk of flammability should be avoided on large scale. Solvents with moderate hazards of flammability require precautions such as the grounding of drums and reactors to prevent any build up of electrostatic electricity.
- **Explosiveness**. The explosive properties of chemicals can be tested in various ways. The **dust explosion test** determines whether a spark can ignite a dust cloud of the compound. The **hammer test** involves dropping a weight on the sample and observing any sound or light that might be produced, which might suggest an explosive nature. It is also possible to predict the probability of explosiveness from the structure of the compound by looking for functional groups that are known to be unstable (e.g. peroxides and hydrazines). Compounds that can eliminate small stable molecules (e.g. N_2, O_2, NO, NO_2) are potentially explosive because such decompositions tend to be exothermic.

Safety – reaction hazards

The **hazards** associated with particular reactions should be fully understood. In particular, care must be taken with exothermic reactions at high concentration. With little solvent present, there is less scope for heat to be absorbed by the solvent, and it is important to be certain that such reactions can be controlled in the plant. Controlling the rate at which reactants are added is a good way of doing this. Nevertheless, adequate cooling should be provided to prevent such reactions getting out of control. The rate of stirring in such reactions can also have a vital role to play and should be carefully monitored. Particular care must be taken with autocatalytic reactions as these are potentially dangerous.

Purifications

At the research level the product from each reaction step may be purified by chromatography or crystallization. This is impractical on a large scale. Ideally, a synthetic process used in production would have no chromatographic purifications and would rely instead on **crystallization** of the final product. The crystallization process needs to be carefully studied so that the crystals obtained are consistent not only in terms of purity, but in terms of crystal size and form. Therefore, such aspects as the cooling and stirring rates of the hot solution must be specified so that the correct crystals are obtained and the mixture does not become too thick once crystallization takes place. The crystallization should be controlled in such a way that fine crystals are not formed since this can plug up the filter when the crystals are filtered. If the hot solution has to be filtered before crystallization takes place, it should be performed at least 15°C above the crystallization temperature to prevent crystallization during the filtration process.

Environmental issues

The chemical industry has a bad public image for polluting the world so it is important that a production plant takes care of its local **environment**. The recycling of chemicals avoids disposal problems and is a high priority, both on environmental and economic grounds. This is particularly true for **solvents**. A good choice of solvent allows easier recovery and re-use. For this reason, it is better to design processes that avoid the use of mixed solvents as these are more difficult to recycle. Solvents with low boiling points should be avoided since they are more likely to escape into the atmosphere on distillation. Water is a preferred solvent on environmental grounds. If chemicals *do* have to be disposed of, they should be treated so that they are in the least harmful or toxic form possible.

Other methods by which a process can be made cleaner and more environmentally friendly are the use of catalysts, or experimental procedures such as electrochemistry, photochemistry, ultrasound or microwaves.

Cost

The **cost** of the process by which a compound is produced is clearly crucial. Costs can be cut by increasing the overall yield of the process. Other methods of reducing costs are finding cheaper suppliers of raw materials, and the reduction of labor and overhead costs. The latter is best achieved by running the process such that large batches are produced on each run, rather than preparing a large number of small batches.

J5 Specifications

Key Notes

Specifications — The specifications of a product define its properties and purity. All batches must pass predetermined limits.

Impurities — Impurities in the final product must be identified and their source determined. Work should then be carried out to minimize or eliminate any impurities.

Purifications — Impurities can often be removed by precipitation, crystallization or distillation.

Impure reagents and reactants — Commercially available reagents and reactants may contain impurities that will affect the purity of the final product. Such chemicals should be analyzed for purity before they are used in the process.

Reaction conditions — Varying the reaction conditions may have an influence on the formation of impurities. There are many variables to consider, such as the nature of the solvent, the reaction temperature, the rate of addition, and the stirring rate. Some reagents are susceptible to oxidation if they are exposed to air. Fresh batches of the reagent should be used and the reaction may need to be carried out under nitrogen.

Order of addition — Some side reactions can be avoided by reversing the addition of the reagents.

Troublesome by-products — It may be necessary to change a reagent in order to avoid the formation of troublesome by-products.

Changing a synthesis — If it proves impossible to remove a troublesome impurity, it may be necessary to change the synthetic route.

Inorganic impurities — Inorganic impurities are also unacceptable and it may be necessary to use deionized water in some processes.

Related topic — Regulatory affairs (K5)

Specifications

The **specifications** of a product refer to its properties and its purity. Before a drug can be tested or sold, it must meet product quality specifications. These define predetermined limits for a range of properties such as melting point, color of solution, particle size, polymorphism and pH. The product's chemical and stereochemical purities must also be defined, and the presence of any impurities or solvent quantified. Acceptable limits for different compounds are proportional to their toxicity. For example, the specifications for ethanol, methanol, mercury, sodium and lead are 2%, 0.05%, 1 p.p.m., 300 p.p.m. and

2 p.p.m., respectively. Carcinogenic compounds such as benzene or chloroform should be completely absent, which means in practice that they must not be used in the final stages of the synthesis.

Impurities

One of the most challenging aspects of chemical development is dealing with **impurities** observed in the final product. If an impurity is detected, it has to be isolated, purified, and identified. Isolation and purification are carried out by high performance liquid chromatography (HPLC). Identification is carried out by NMR and/or mass spectroscopy. Once the structure of the impurity has been identified, it is then a case of working out how the impurity was formed and at which stage of the process it was formed. Sometimes the origin of the impurity is obvious. At other times, tracking it down may involve quite a bit of detective work. Having discovered the source of the impurity, the development chemist must study the reaction concerned and find a method of eliminating or minimizing the impurity.

Purifications

The obvious way of removing an impurity is to carry out a **purification**, either following the offending reaction or at the end of the overall synthesis. There are various methods that one can try. For example, it might be possible to precipitate the impurity from solution by adding a suitable solvent. Alternatively, the product could be precipitated to leave the impurity in solution. Crystallization and distillation are other common purification methods that could be tried.

Impure reagents and reactants

Sometimes the impurity is introduced by a reagent used in the synthetic route. It is important to realize that commercially purchased chemicals are not 100% pure and will contain small levels of impurities. Such an impurity may survive the synthetic route and remain to contaminate the final product. Alternatively, the impurity itself may undergo reactions during the synthetic route and contaminate the final product with an unexpected structure.

An example of this occurred during the large-scale synthesis of **fluvostatin** (*Fig. 1*). Analysis of the final product revealed the presence of an impurity that was identified as the *N*-ethyl analog (*Fig. 2*). The source of this impurity was tracked down to the batch of *N*-isopropylaniline used in the second step. It was discovered that the batch contained a small amount of *N*-ethylaniline. Since this compound was so closely related to the authentic starting material, it under-

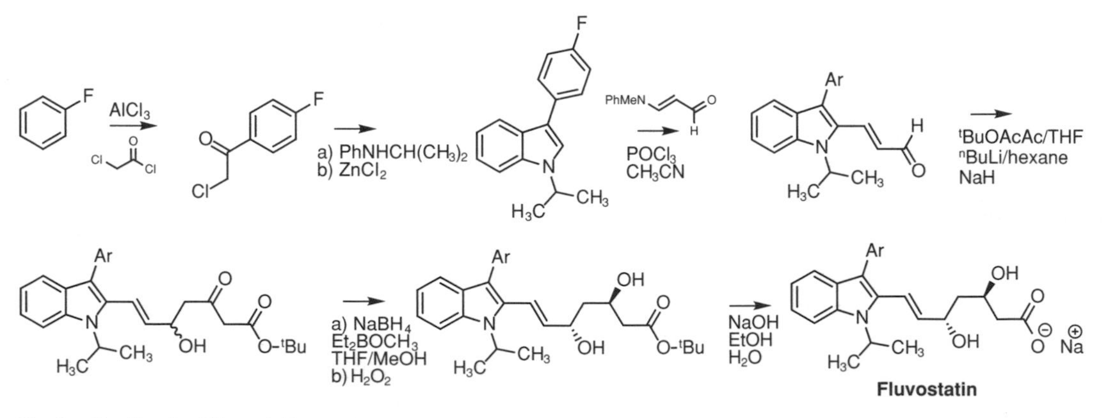

Fig. 1. Synthesis of fluvostatin.

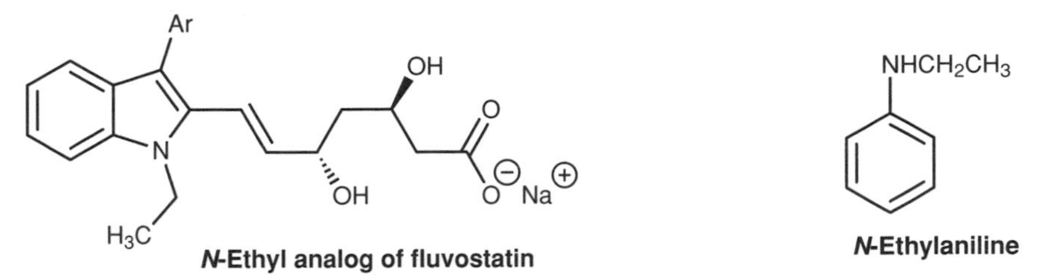

Fig. 2. N-Ethyl analog of fluvostatin and N-ethylaniline.

went the same reactions throughout the entire synthetic route to give the observed analog.

It is therefore important to analyze the purity of all the starting materials and reagents used in the synthetic route. Care should also be taken if it is decided to change the supplier of a particular chemical, since batches from different suppliers may have significantly different impurities.

Reaction conditions

If an impurity is known to occur as a result of a side reaction, it may be possible to minimize its occurrence by varying the **reaction conditions**. There are many variables to consider. For example, varying the polarity of the solvent may affect the reactivity of the reagent or reactant and have an effect on the formation of impurities. The formation of impurities may also be affected by varying any catalyst or base used in the reaction, or by varying the ratio of reactants and reagents.

Varying the reaction temperature is important if one compound is formed under kinetic control and another is formed by thermodynamic control. If the desired product is kinetically controlled and the impurity is a thermodynamic product, then heating will result in more of the impurity. Kinetic control can be helped by rapid addition of the reactant.

Some reagents are air sensitive. For example, if *N*-butyllithium is left exposed to air, oxidation occurs and lithium butoxide is formed. Similarly, if benzaldehyde is left exposed to air, oxidation can lead to the formation of benzoic acid. The use of fresh reagent minimizes this problem and the reaction can be carried out under a nitrogen atmosphere.

Order of addition

Impurities sometimes occur depending on the order in which reagents are added to the reaction vessel. For example, in the bromination of alcohols with PBr_3, *O*-alkylation is a possible side reaction (*Fig. 3*). This can arise since the bromide product formed can react with any unreacted alcohol. This is more likely to arise when the PBr_3 is added to the alcohol since there will be an excess of alcohol present during the initial stages of the addition. Reversing the

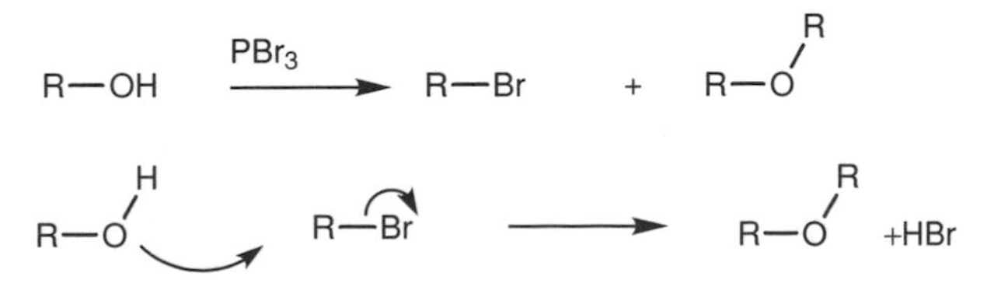

Fig. 3. Bromination and O-alkylation of an alcohol.

addition so that the alcohol is added to the PBr_3 gets round this problem and no impurity is formed. The same situation is found for the conversion of alcohols to alkyl chlorides using thionyl chloride.

Troublesome by-products

In some reactions, impurities are formed as a natural result of the reaction mechanism. If these are difficult to remove, it may be necessary to change the reaction or the reagent such that less troublesome by-products are formed. For example, the **Wittig reaction** carried out with triphenylphosphine results in triphenylphosphine oxide as a by-product, which usually has to be removed by chromatography (*Fig. 4*). Chromatography is not ideal for scale up and so one could try switching to the **Horner–Emmons reaction** instead (*Fig. 5*). In this reaction, there are no difficult by-products to remove. The phosphonate ester by-product is water soluble and can easily be removed by washing.

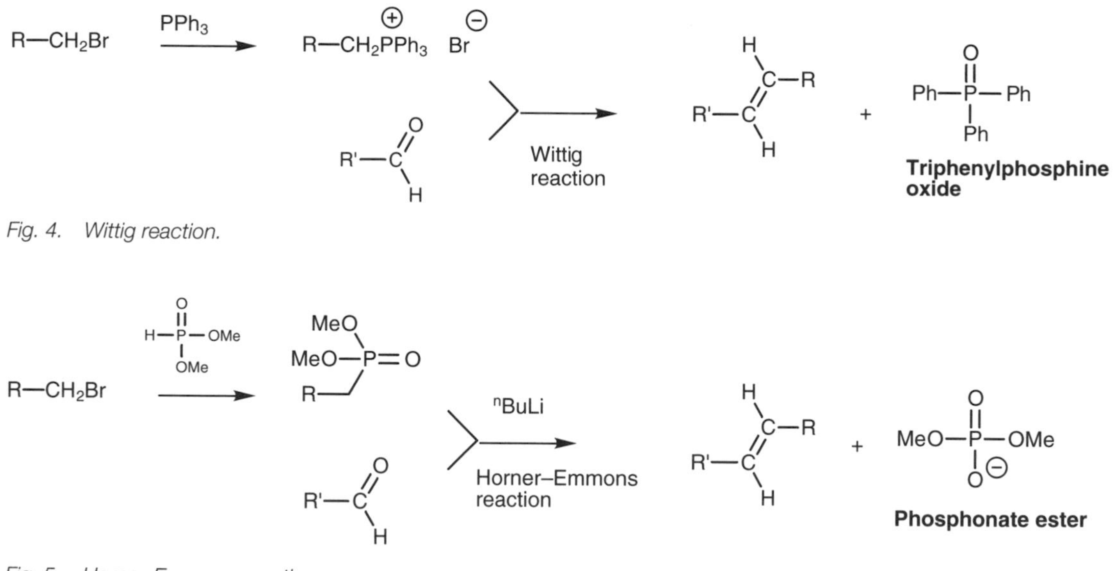

Fig. 4. Wittig reaction.

Fig. 5. Horner-Emmons reaction.

Changing a synthesis

Sometimes an impurity may prove so troublesome that it may be necessary to develop a completely different route. For example, the Grignard reaction below (*Fig. 6*) suffered from an ester impurity, which arose from oxidation of the

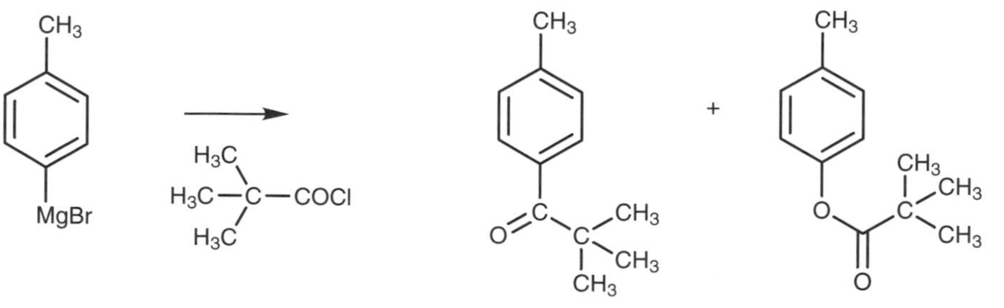

Fig. 6. Grignard reaction.

Grignard reagent to a phenol. The phenol then reacted with the acid chloride to give the ester observed.

The reaction could be improved by reversing the addition such that the Grignard reagent was added to the acid chloride. However, this was not easy on large scale since the Grignard reagent was air sensitive and had to be kept warm to prevent it solidifying. Therefore, a different synthetic approach was sought. Several different approaches were tried as shown below (*Fig. 7*).

However, none of these methods was particularly effective. Success was finally achieved by carrying out a Grignard reaction on an aromatic nitrile (*Fig. 8*).

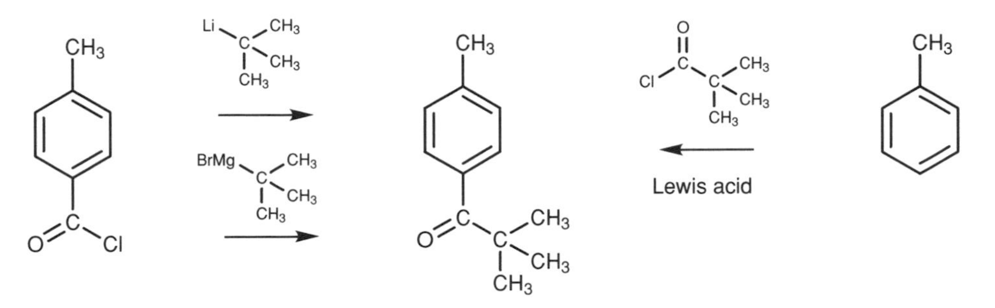

Fig. 7. Alternative syntheses.

Inorganic impurities

Products should be checked for **inorganic impurities** such as metal salts. In some cases, deionized water may need to be used in the reaction process if the desired compounds are metal ion chelators or are isolated from water.

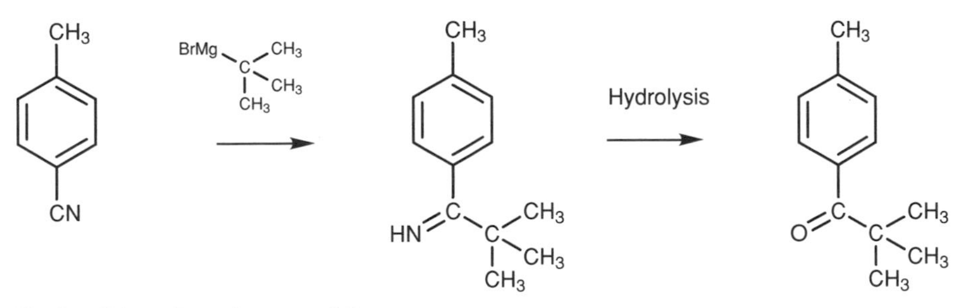

Fig. 8. Grignard reaction on a nitrile.

K1 TOXICOLOGY

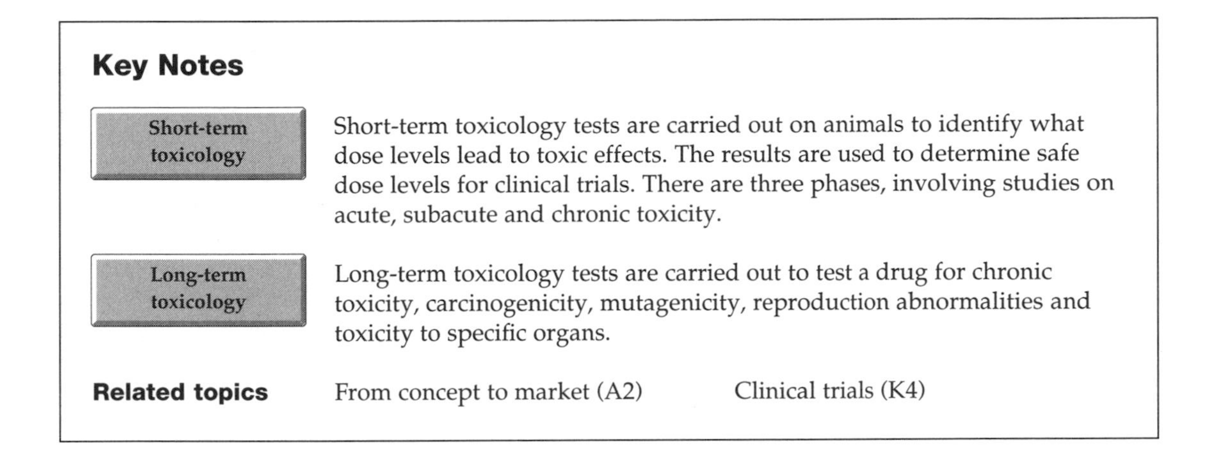

Key Notes

Short-term toxicology	Short-term toxicology tests are carried out on animals to identify what dose levels lead to toxic effects. The results are used to determine safe dose levels for clinical trials. There are three phases, involving studies on acute, subacute and chronic toxicity.
Long-term toxicology	Long-term toxicology tests are carried out to test a drug for chronic toxicity, carcinogenicity, mutagenicity, reproduction abnormalities and toxicity to specific organs.
Related topics	From concept to market (A2) — Clinical trials (K4)

Short-term toxicology

Short-term toxicology studies are carried out to test whether a drug is safe enough to be used for phase I clinical trials. The drug is given to animals at sufficiently high doses to ensure a toxic effect, and the results of these tests are used to determine suitable dose levels for clinical trials. Various kinds of tests are carried out.

Acute toxicity studies determine the short-term effects of a drug when it is administered as a single dose, or as several doses over a period of 24 hours or less. These studies provide insights into the drug's activity and its obvious toxic effects. Several species of animal are used and the drug is administered in the same way as it will be administered clinically. The animals are observed for at least a week, in order to detect any delayed toxic effect. The results from these studies can identify particular organs that might be at more risk than others.

Subacute toxicity studies are carried out to observe any toxic effects that arise from administering several doses of the drug at levels expected to cause toxicity, but not death. Studies last 1–6 months and are carried out on rats or mice as well as another species (e.g. dogs). Again, the drug is administered as it will be administered clinically. The animals are observed for signs of toxicity, and blood and urine samples are taken regularly to monitor any changes that might suggest toxicity. At the end of the study, the animals are killed and key tissues and organs are examined both visually and under the microscope for signs of damage. These studies determine the doses that can be used for carcinogenicity tests as well as determining what doses will be safe for clinical trials.

Chronic toxicity studies lasting 6–12 months are carried out on drugs that are intended to be taken orally by patients over a long time scale. The tests are done in rats or mice plus a nonrodent species. These studies also include ophthalmological tests.

Long-term toxicology

Long-term toxicology tests are carried out to test the drug for chronic toxicology, carcinogenicity, special toxicology, mutagenicity, and reproduction abnormalities.

Carcinogenicity tests must be carried out on any drug that is intended to be used on patients over long time periods, as well as any drug whose structure suggests that it might be carcinogenic. The tests are normally carried out over a two-year period on rats or mice – the equivalent of a lifetime – and dose levels are chosen such that they are below toxic levels to ensure that the animals survive the trial. The animals are then killed and their tissues are studied for any signs of tumors. A positive control may often be used, where some of the animals receive a known carcinogen.

It is possible that a drug could damage DNA without producing any visible tumors, so **mutagenicity tests** are carried out both *in vivo* and *in vitro* to see if any genes are damaged or whether there is any chromosomal damage.

Reproduction studies have to be carried out if the drug is likely to be taken by women of child bearing potential. The drug is studied for its effects on male and female fertility, as well as any toxic effects that it might have on the embryo, fetus or newborn. There are three stages to these studies. Studies are first carried out on male and female rats to detect any effects on spermatogenesis and egg formation respectively. The tests on female rats are continued through gestation and weaning. In the second stage, studies are carried out on two different animal species, usually involving mice, rats or rabbits. The drug is given to pregnant animals to measure any embryo toxicity and teratogenic effects. In the final stage, tests are carried out to detect any effects the drug may have on fetal development, labor and delivery, and growth of the newborn.

Special toxicity tests may be carried out to study how a drug affects the function of any organ that is particularly susceptible to toxicity, and to study whether the toxic effects are reversible. The animal may also be killed so that the tissue can be studied under the electron microscope.

K2 PHARMACOLOGY AND PHARMACEUTICAL CHEMISTRY

Key Notes

Pharmacology

Pharmacology tests are carried out to determine a drug's mechanism of action at the intended target, and its breadth of pharmacological activity against other targets.

Formulation

Formulation studies are carried out to establish a dosage preparation, such as a pill or capsule, which will be consistent in its properties and will contain a constant level of the drug.

Stability

The stability of a drug preparation must be studied under various conditions of temperature, humidity and light. Containers must be used which do not interact with the preparation. Stability tests establish the shelf life of the preparation, and the storage conditions that should be used.

Related topic

Drug administration (C6)

Pharmacology

Pharmacology studies will have been carried out during the drug design process, but further studies may need to be carried out in order to define the activity of the drug and to provide a better insight into its mechanism of action. *In vivo* and *in vitro* studies are also carried out to see whether the drug has any other pharmacological activity apart from the one of interest. Further studies are then carried out to compare the drug's activity with known drugs. The studies determine a dose–response relationship, as well as the drug's duration and mechanism of action.

Formulation

Formulation involves the development of drug preparations that are stable and acceptable to the patient. A patient is more likely to accept a pill, than a spoonful of powder. However, a pill contains a variety of other substances apart from the drug itself, and studies have to be carried out to ensure that the drug is compatible with these other substances. **Preformulation** involves the characterization of a drug's physical, chemical and mechanical properties in order to choose what other ingredients should be used in the preparation.

Formulation studies start in advance of clinical trials, but are unlikely to be complete until after clinical trials have started. To begin with, simple preparations are developed for use in Phase I clinical trials. These typically consist of hand-filled capsules containing a small amount of the drug and a diluent. Proof of the long-term stability of these formulations is not required since they will be used in a matter of days. Consideration has to be given to what is called the **'drug load'** – the ratio of the active drug to the total contents of the dose. A low drug load may cause homogeneity problems. A high drug load may pose flow problems or require large capsules if the compound has a low bulk density.

By the time phase III clinical trials are reached, formulation of the drug should have been developed to be close to the preparation that will ultimately be used in the market. Knowledge of stability is essential and conditions must have been developed to maximize drug stability.

Formulation studies have to consider such factors as particle size, polymorphism, pH and solubility, since all of these can influence bioavailability and hence the activity of a drug. The drug must be combined with inactive additives by a method that ensures that the quantity of drug present is consistent in each dosage unit. The dosage should have a uniform appearance, with an acceptable taste, tablet hardness or capsule disintegration.

Stability

Tests must be carried out on a drug preparation to ensure that it is chemically stable. If the drug proves to be unstable, the results from clinical trials could be confusing and unreliable. Moreover, toxic degradation products might be formed. Safety data establishes the shelf life of the compound and allows an expiry date and the storage requirements to be defined.

The drug preparation is tested for stability when stored at room temperature and at various other temperatures (e.g. 5, 50 and 75°C). The effects of humidity, UV light and visible light are tested and the preparation is analyzed to identify any degradation products. Containers should be chosen that do not interact with the drug preparation either physically or chemically. For example, if a plastic container is used, it is important to check that plasticizers, lubricants, pigments and stabilizers do not leach out into the preparation. Conversely, it is important to ensure that the drug or other ingredients in the preparation do not bind to the plastic or become adsorbed. Similarly, it is important to ensure that interaction with the plastic does not result in any chemical reactions such as oxidation, degradation or precipitation. Any label adhesives should be checked to ensure that they do not leach through the container into the drug.

K3 Drug metabolism studies

Key Notes

Labeling studies
Radiolabeled drugs are used in drug metabolism studies to detect any drug metabolites that are formed.

Isotopes
The isotopes that are commonly used for labeling studies are heavy isotopes such as deuterium and carbon-13, and radioactive isotopes such as tritium and carbon-14. The radiochemical purity of a labeled compound is not the same as the chemical purity.

Synthesis
A labeling synthesis should be designed so that the isotopic label is incorporated as late as possible in the synthesis.

Incorporation of D or T
There are several methods of incorporating D or T, such as the exchange of acidic protons with labeled water, hydrogenation with labeled hydrogen gas, and reaction with labeled organic reagents such as iodomethane. It is important to ensure that the label is not easily exchanged with water, or lost as a result of drug metabolism.

Incorporation of carbon isotopes
Carbon isotopes are less likely to be lost from a molecule, but it is still important to choose positions for the label which are metabolically stable.

Related topics
From concept to market (A2)
Drug metabolism (C4)
Clinical trials (K4)
From lead compound to dianilino-phthalimides (L3)

Labeling studies

In order to study drug metabolism, it is necessary to incorporate an identifiable isotope into the drug. Both heavy isotopes and radioactive isotopes have been used, but radioactive isotopes are more common since the detection of radioactivity is easier and more sensitive. A radiolabeled drug can be administered to animals and human volunteers, then monitored in blood and urine, allowing studies of drug absorption, distribution and excretion. Blood and urine can also be studied by HPLC to detect any drug metabolites by the appearance of radioactive peaks other than those of the drug itself. These can be collected and the metabolites identified.

Isotopes

There are various **isotopes** that can be used to label drugs. These include stable isotopes such as deuterium (^{2}H or D) and carbon-13 (^{13}C), as well as radioactive isotopes like tritium (^{3}H or T) or carbon-14 (^{14}C). Radioisotopes such as ^{3}H and ^{14}C can be detected at small levels by measuring their β radiation. Tritium has a maximum specific activity of 29.1 Ci mmol^{-1} and a half-life of 12.3 years, whereas ^{14}C has a specific activity of 62.4 mCi mmol^{-1} and a half-life of 5730 years. Stable heavy isotopes can be detected by mass spectroscopy (e.g. D) or in some cases by NMR (e.g. ^{13}C).

Synthesis

When labeling a drug, not every atom in the molecule has to be labeled. Indeed, it is preferable to label only one position in the molecule. Nor is it necessary for every molecule in the sample to be labeled. Detection methods are accurate enough to detect labeled molecules even if they represent only a fraction of the total molecules present. The measure of how significantly a compound is radioactively labeled is given by the **specific activity** of the compound, given in mCi $mmol^{-1}$. This can be used to follow the **radiochemical purity** of the intermediates throughout a radiolabeled synthesis. Assuming the initial radiolabeled compound is radiochemically pure, the intermediates and product formed should have the same specific activity. It is also important to realize that a compound could be chemically pure, but radiochemically impure (or *vice versa*).

The method used to incorporate a label into a drug structure is determined by the labeled compounds that are commercially available. Typically, these are simple compounds such as labeled water, iodomethane, cyanide ion and labeled hydrogen gas. However, it is also possible to purchase labeled amino acids.

Introduction of the label should be as late as possible in the synthesis to reduce the number of steps involving labeled compounds. This is easier when introducing D or T, but not so easy for ^{13}C or ^{14}C. The latter have to be incorporated into the drug's skeleton, which demands a longer synthesis.

Incorporation of D or T

There are various methods by which deuterium or tritium can be incorporated into a drug. The easiest method is to identify an exchangeable proton in the drug structure and to replace it with D or T. For example, the protons of alcohols, phenols and carboxylic acids can be replaced with D or T by merely shaking a solution of the drug with D_2O or T_2O (*Fig. 1*). However, since it is easy to incorporate the label, it is also easy to lose the label by exchange with water when the drug is administered to a test animal. Therefore, the label has to be put into a more stable position than any of these functional groups. A better strategy would be to identify an acidic proton in the structure that is stable at a pH of 7.4, but which could be removed under more basic conditions. Protons that are alpha to a carbonyl group can be exchanged in this way (*Fig. 2*) and it is more likely that a D or T label will remain attached to the molecule when it is administered to a test animal. However, it is important to ensure that the basic conditions used to introduce the label do not racemize any asymmetric centers in the molecule.

Fig. 1. Labeling of an alcohol, phenol or carboxylic acid.

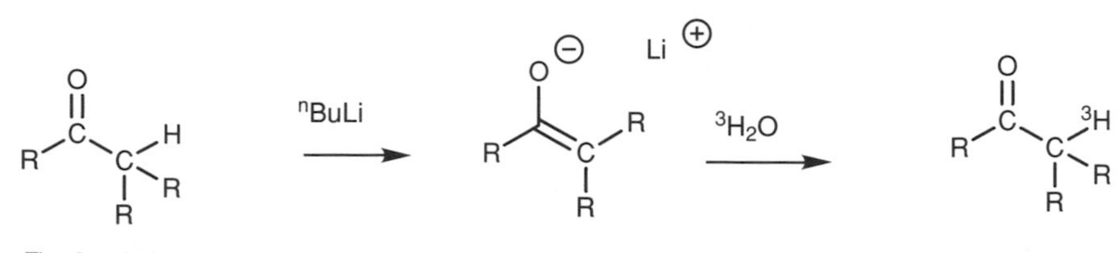

Fig. 2. Labeling alpha to a carbonyl group.

If the drug does not contain a suitably acidic proton, then there may be a synthetic intermediate that does. The label could be introduced at that stage and the labeled intermediate converted to the final product. However, it is important to ensure that the label remains attached throughout the synthesis and does not become 'scrambled' (i.e. moved to different positions due to isomerizations).

Other common methods of introducing a tritium label are shown below (*Fig. 3*). These methods alter the structure, so these methods would have to be carried out on a suitable synthetic intermediate that could then be carried through to the final product.

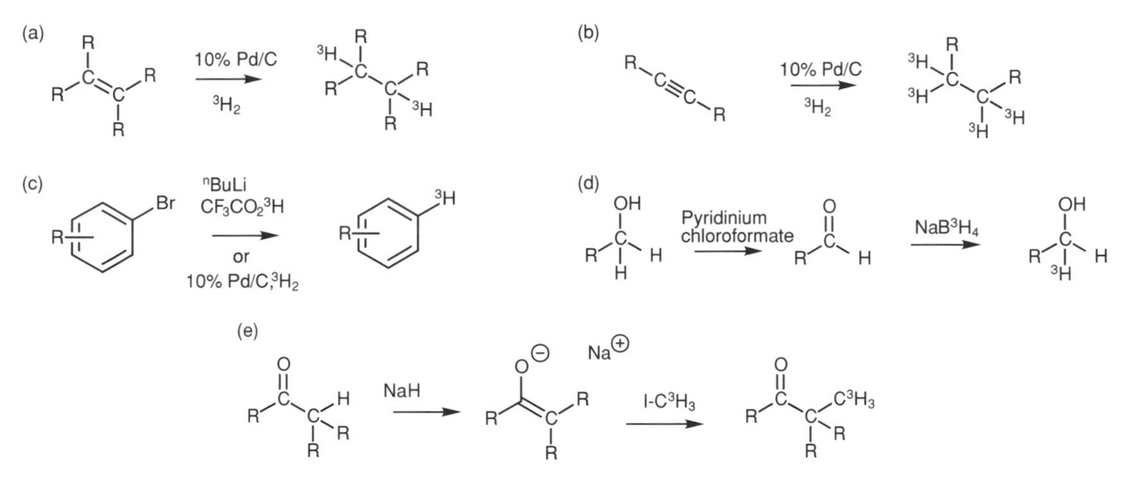

Fig. 3. Common methods of introducing a tritium label.

Although the above methods can introduce deuterium or tritium to a chemically stable position in the drug, there is always the possibility that a metabolic reaction may occur that results in their loss. As a result, that metabolite would not be detected. Another problem is a significant **isotope effect**. Ideally, the labeled molecule should behave like the unlabeled molecule. However, this is not always the case, especially with tritium. For example, it is sometimes possible to separate a tritium labeled compound from an unlabeled compound by chromatography.

Incorporation of carbon isotopes

There are several advantages in using **carbon isotopes** for drug metabolism studies, rather than deuterium or tritium. Primarily, carbon labels incorporated into the drug's skeleton do not run the risk of being exchanged or 'scrambled', and there is less of an isotope effect. The disadvantage of using a carbon label is the difficulties that might arise in incorporating it into the molecule. Indeed, a completely new synthesis may need to be devised in order to be successful.

When designing a labeled synthesis, it is important to identify the labeled compounds that are commercially available, and then determine how these can be incorporated into the desired structure. This will often restrict the positions of the drug that can be labeled. However, it is also important to find a biologically stable position for the label. For example, consider a drug that contains an *N*-methyl group. Labeled iodomethane is commercially available and so it would be tempting to label the drug at that position. The reaction would be easy to carry out and could be done as the final stage in a synthesis (*Fig. 4*). However, it is known that *N*-methyl groups are susceptible to drug metabolism. If this

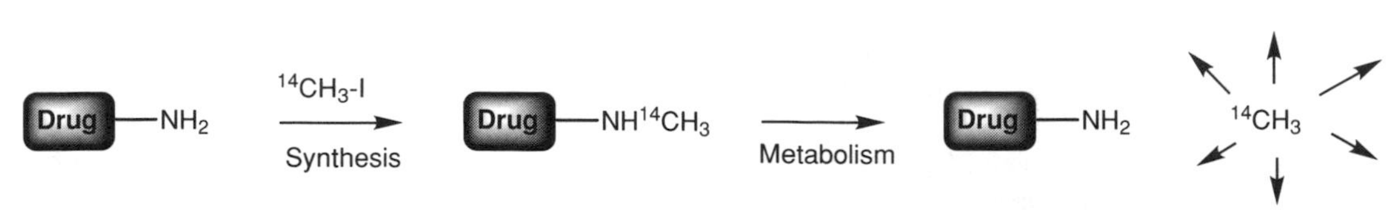

Fig. 4. Labeling a N-methyl group.

happened to the drug, the labeled methyl group would be lost and would enter the cell's general biosynthetic pool, resulting in a large number of radiolabeled compounds totally unrelated to the drug.

A significant synthetic effort may be required to incorporate a carbon isotope at a position that is both synthetically feasible and biologically stable. For example, the synthesis of ^{14}C radiolabeled **pseudomonic acid** was carried out as shown below (*Fig. 5*). ^{14}C radiolabeled ethyl bromoacetate is commercially available and was converted to a phosphonoacetic acid, which could be esterified to attach one half of the pseudomonic acid skeleton. A Wittig–Horner reaction was then carried out with ketone (B) to splice on the other half of the pseudomonic acid skeleton. This gave methyl pseudomonate along with the *E*-isomer. Methyl pseudomonate was purified by chromatography, then treated with yeast to remove the methyl ester and give radiolabeled pseudomonic acid.

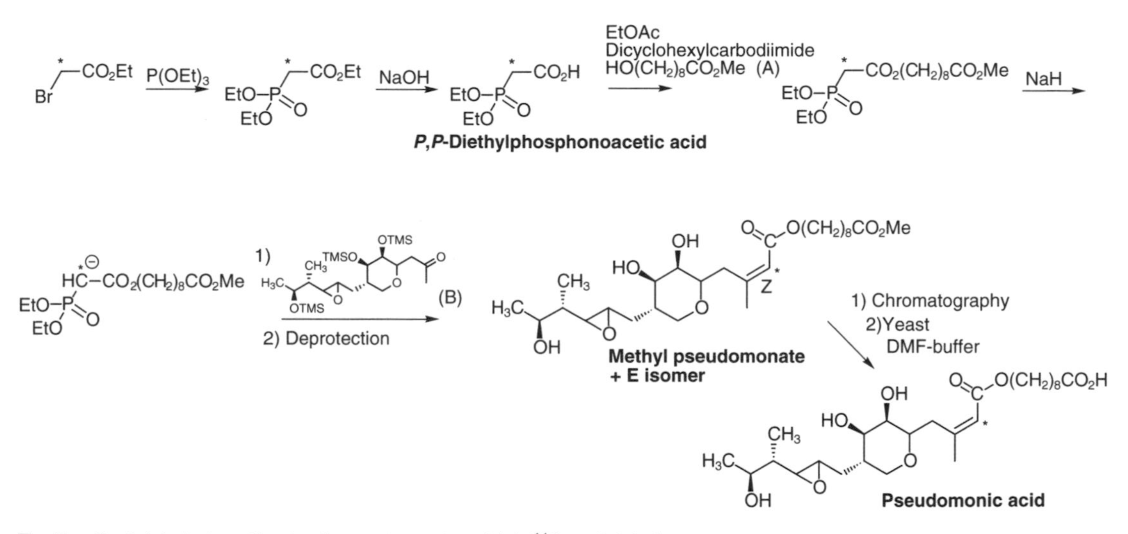

*Fig. 5. Radiolabeled synthesis of pseudomonic acid. *, ^{14}C radiolabel.*

Ketone (B) was obtained from pseudomonic acid itself, by ozonolysis, followed by protection with trimethylsilyl groups (*Fig. 6*).

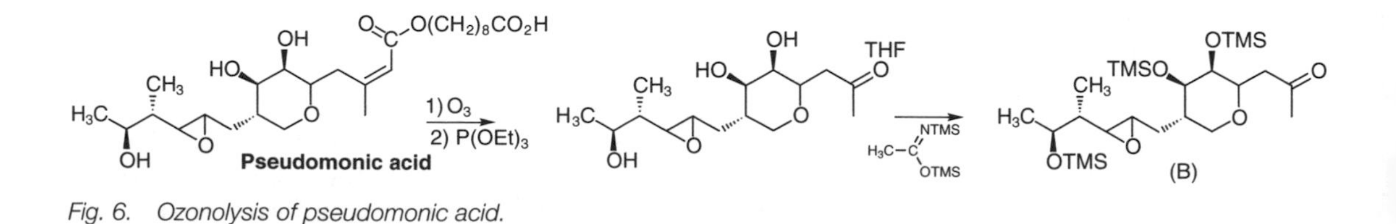

Fig. 6. Ozonolysis of pseudomonic acid.

Alcohol (A) was obtained from oleic acid by ozonolysis, followed by reduction and esterification (*Fig. 7*).

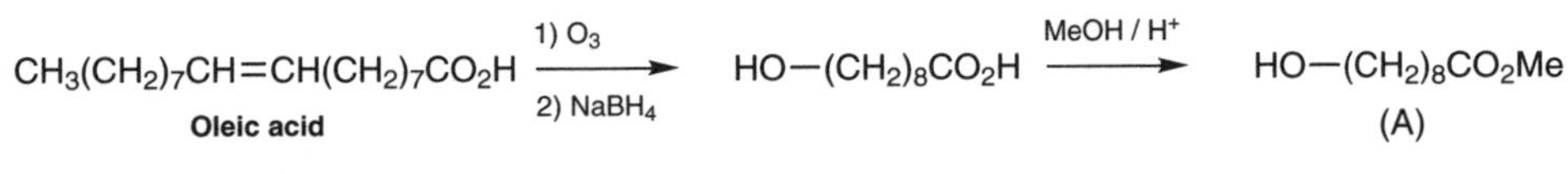

Fig. 7. Ozonolysis of oleic acid.

K4 CLINICAL TRIALS

Key Notes

Definition

Clinical trials are carried out to test the therapeutic effects of new drugs and to ensure that they have no unacceptable side effects. There are four phases.

Phase I

Phase I trials are normally carried out on healthy volunteers to establish dosing levels, and to carry out pharmacokinetic studies. The therapeutic effect is not tested. Special groups of volunteers may be tested if the drug is likely to be targeted at these groups.

Phase II

Phase II studies are carried out on patients suffering from the disease that the drug is intended to treat. One group of patients receives the drug and another group receives a placebo or a conventional drug. Neither patient nor doctor knows which patient receives placebo or drug. Different dose levels and regimes are used on different groups to establish the best dosing regime. The studies demonstrate whether the drug is therapeutically useful and whether it has any side effects.

Phase III

Phase III studies are carried out in a similar fashion to phase II studies, but on a larger number of patients in order to get statistical proof, both of the drug's efficacy and its safety.

Phase IV

Phase IV studies continue after the drug has gone onto the market. They are designed to study the effects of long-term use and to identify any rare side effects that may arise.

Related topics

From concept to market (A2)
Testing drugs *in vivo* (D3)
Toxicology (K1)
Regulatory affairs (K5)
Drug metabolism studies (K3)

Definition

Clinical trials must be carried out on all drugs before they are marketed to ensure that they have the therapeutic effect claimed, and do not have unacceptable side effects. On average, clinical trials take 5–7 years and do not necessarily have to take place in one laboratory or even in one country. It is important to ensure that the clinical trials satisfy all the requirements of the different regulatory bodies involved if the drug is to be marketed worldwide. Clinical trials can be divided into four phases.

Phase I

Phase I clinical trials are usually carried out on healthy human volunteers to provide a preliminary evaluation of the drug's safety, its pharmacokinetics and the dose levels that can be administered, but they are not intended to demonstrate whether the drug is effective or not. During the study, the volunteers do not take medication, caffeine, alcohol or cigarettes. This is to avoid any complications that might arise due to the drug–drug interactions (see Topic C4). As a result, adverse effects that arise due to such interactions will only appear at a later stage when the drug is tested in real life situations.

The **dose levels** that can be tolerated are usually studied first. For each dose level, 6–12 subjects are given the active drug and 2–4 subjects are given a placebo. The initial dose administered is based on the toxicology results from animal testing. For example, if the highest dose found to have no toxic effect in animals is 50 mg kg^{-1}, then the initial dose level will be a tenth of that value (i.e. 5 mg kg^{-1}). Once the initial dose is administered, extensive pharmacokinetic studies are carried out in order to follow the drug and its metabolites. After a full safety assessment has been made, a higher dose will be given, and this will be continued until mild adverse effects are observed. This will be taken as the **maximum tolerated dose**, and further studies will then concentrate on smaller doses.

Studies are carried out early on to determine whether there are any interactions between the drug and food. This is essential in order to establish when the dose should be taken relative to meals.

Studies are also carried out to test whether there are any interactions with other drugs that may affect the absorption and metabolism of either drug. These will concentrate on drugs that are most likely to be taken alongside the new drug. For example, a drug for Alzheimer's disease will be used mostly on elderly patients who are likely to be taking drugs such as diuretics, anticoagulants or aspirin.

Another study involving a radiolabeled drug is carried out on 4–8 healthy volunteers in order to follow the absorption, distribution, and excretion of the drug. These studies will also determine how the drug is metabolized in humans.

Studies may be carried out on special age groups of volunteers. For example, drugs intended for Alzheimer's disease are tested on healthy elderly volunteers to test the drug's pharmacokinetics in that particular population. Special studies may be carried out on volunteers who are not particularly healthy, but who have conditions that will affect the pharmacokinetics of the drug. For example, the rate of drug excretion and drug metabolism will be affected in patients with renal or hepatic impairment. The rate of drug absorption will be affected in patients with inflammatory bowel disease or other gastrointestinal diseases. There are also groups of individuals that metabolize drugs more slowly than the majority of the population, and a study involving them may be relevant if such groups are likely to be treated with the new drug.

Bio-equivalence studies are required when different dosage forms are to be used in the early and late phases of clinical trials. Powder-filled capsules are frequently used in phase I, while tablets are used in phases II and III. Therefore, it may be necessary to establish that these formulations show bio-equivalence in healthy volunteers. In addition, it has to be demonstrated that dissolution of both formulations is similar.

Phase I takes about a year and involves 100–200 volunteers at a cost of 2.5–3.0 million dollars. In some cases, phase I studies are not carried out on healthy volunteers but on volunteer patients. This is when the drug is potentially toxic, and is to be used against life threatening diseases such as AIDS or cancer.

The decision whether to go to phase II can be difficult since only a limited amount of safety data is available. Any adverse effects observed may or may not be due to the drug. For example, abnormal liver function in a healthy patient may be due to the drug or to alcohol. However, evidence of a serious adverse effect will usually result in clinical trials being terminated.

Phase II

Phase II studies generally last about 2 years and may start before phase I studies are complete. Phase II studies are carried out on patients suffering from the condition that the drug is meant to treat. The studies aim to establish the therapeutic value of the new drug, and to study the pharmacokinetics and short-term safety of the drug. They are also used to establish the best dosing regimes. Phase II trials can be divided into early studies (IIa) and late studies (IIb).

Initial trials (phase IIa) involve a limited number of patients to see if the drug has any therapeutic value at all and to see if there are any obvious side effects. If the results are disappointing, clinical trials may be terminated at this stage.

Later studies (IIb) involve a larger numbers of patients. They are usually carried out as **double blind, placebo controlled studies**. This means that patients are split into two groups, where one group receives the drug, while the other group receives a placebo. A double-blind study means that neither doctor nor the patient knows whether a placebo or drug is being administered.

The studies demonstrate whether the health of the patient group receiving the drug is improved relative to the patient group receiving the placebo. The studies also explore different dosing levels and regimes to try and clarify which regimes are the most effective. Most phase II trials require 20–80 patients per dose group to demonstrate efficacy.

Some form of **rescue medication** in placebo controlled trials may be necessary. For example, it would be unethical to continue asthmatic patients on a placebo if they suffer a severe asthmatic attack. A conventional drug would be given and its use documented. The study would then compare how frequently the placebo group need to use the rescue medicine compared to those taking the new drug. In some cases, such as the treatment of Aids or cancer, the use of a placebo is not ethical, so an established drug is used as a standard comparison.

The **endpoint** is the measure that is used to determine whether a drug is successful, and can be any factor that is relevant, measurable, sensitive and ethically acceptable. Examples of endpoints include blood assays, blood pressure, tumor regression, the disappearance of an invading pathogen from tissues or blood, etc. Less defined endpoints include perception of pain, use of rescue medications, joint stiffness, etc.

Phase III

Phase III studies can be divided into phases IIIa and IIIb, and may begin before the end of phase II studies. The basic design for phase III studies is similar to phase II, but usually includes fewer dose levels and more patients per dose in order to show statistically significant efficacy and safety. Therefore, the studies are carried out on a double-blind basis with a control group who take a placebo or an established drug. Phase IIIa studies are carried out prior to the drug's registration with the regulatory authority, and assess the therapeutic value of the drug, as well as the most safe and effective way in which it can be used.

Phase IIIb trials are carried out after the drug has been registered, but prior to approval. These studies are carried out in order to compare the new drug with other drugs already established in the field, or to provide further relevant information. If a patient responds well to the trial drug, they may be permitted to continue with it and this allows the researchers to evaluate long-term safety and side effects, including any interactions that might take place with other drugs.

Phase III studies normally take about 3 years but can vary, depending on the drug and the type of therapy.

Phase IV

Phase IV studies are performed after the regulatory authorities have granted market approval for the drug. These studies document the safety of the drug and identify rare side effects or unexpected interactions with other drugs. They also assess the drug's therapeutic value in special populations.

On average, for every 10 000 structures synthesized during drug design, 500 will reach animal testing, 10 will reach phase I clinical trials and one will reach the market place. The average overall development cost of a new drug was estimated as being 114 million dollars in 1987.

K5 Regulatory affairs

Key Notes

Regulatory bodies
Regulatory bodies are responsible for assessing the scientific and clinical data, and approving whether a drug can proceed to clinical trials or the market place.

Investigational Exemption to a New Drug Application (IND)
In the USA, pharmaceutical companies submit an IND to the FDA when they wish to start clinical trials on a drug. The IND must contain all scientific evidence regarding the production and testing of the drug, and must be approved by the FDA before clinical trials begin.

New Drug Application (NDA)
A New Drug Application has to be submitted to the FDA if a drug is to be marketed in the USA. This contains all the scientific and clinical information regarding the drug's manufacture and testing.

Fast tracking
Fast tracking is permissible for drugs that are effective against life threatening diseases where existing drugs are either lacking or inferior.

'Orphan' drugs
Orphan drugs are drugs that are useful against rare diseases and are therefore used on small populations of patients. Pharmaceutical companies will make little profit out of such drugs, so financial and marketing incentives are provided to encourage their development.

Labeling
Labeling must be approved by the regulatory authority and should give full details on the drug, its side effects and its administration.

Good laboratory, manufacturing and clinical practice
Good laboratory, manufacturing and clinical practice (GLP, GMP and GCP) are regulations designed to ensure high professional, scientific and clinical standards in the laboratory, manufacturing plant and clinic. Detailed documentation must be kept to prove that the company is adhering to these standards.

Marketing
Regulatory bodies must approve any marketing claims made about new drugs.

Related topics

From concept to market (A2)
Specifications (J5)
Clinical trials (K4)
A fledgling science (1900–1930) (M3)
The dawn of the antibacterial age (1930–1945) (M4)
The antibiotic age (1945–1970s) (M5)

Regulatory bodies

Regulatory bodies such as the **Food and Drugs Administration** (FDA) in the USA are responsible for approving whether a drug can proceed to clinical trials and whether it should be allowed on the market. The regulatory body has to evaluate the scientific and clinical data to ensure that the drug can be produced with consistently high purity, that it has the clinical effect claimed, and that it does not have unacceptable side effects. It must also approve the labeling of the drug and the directions for its use. In general, the regulatory body is interested in all aspects of a drug once it has been identified as a potentially useful medicine.

Investigational Exemption to a New Drug Application (IND)

Once preclinical studies have been completed on a drug, the pharmaceutical company will assess the scientific evidence and decide whether it wants to proceed to clinical trials. Usually, this will happen if the drug has the desired effect in animal tests, demonstrates a distinct advantage over established therapies, has acceptable pharmacokinetics, few metabolites, a reasonable half-life and no serious side effects. If the company decides to go ahead, the reports are submitted to the regulatory authority. In the USA, this takes the form of an **Investigational Exemption to a New Drug Application (IND)** which is submitted to the FDA. The IND should contain information regarding the chemistry, manufacture and quality control of the drug. It should also include information regarding the drug's pharmacology, distribution and toxicology in animals, and any clinical information that may be known about the drug. The IND is a confidential document and is not released to the public.

The FDA is responsible for assessing this information and approving the start of clinical trials. Dialog then continues between the FDA and the company as the clinical trials are carried out. Any adverse results must be reported to the FDA who will discuss with the company whether the trials should be stopped. If the clinical trials proceed smoothly, the FDA advises the company how to present its application for the drug's approval.

New Drug Application (NDA)

In order to get a drug onto the market in America, a pharmaceutical company has to submit a **New Drug Application** (NDA) to the FDA. An NDA is typically 400–700 volumes in size with each volume containing 400 pages! The application has to state what the drug is intended to do, along with scientific and clinical evidence for its efficacy and safety. It should also give details of the chemistry and manufacture of the drug, as well as the controls and analysis that will be in place to ensure that the drug has a consistent quality. Any advertising material must be submitted to ensure that it makes accurate claims and that the drug is being promoted for its intended use.

The FDA has inspectors who will visit clinical investigators to ensure that their records are consistent with those provided in the NDA, and that patients have been adequately protected. An approval letter is finally given to the company and the product can be launched, but the FDA will continue to monitor the promotion of the product as well as further information regarding any unusual side effects.

Once an NDA is approved, any modifications to a drug's manufacturing synthesis or analysis must be approved. In practice, this means that the manufacturer will stick with the manufacturing route described in the NDA and perfect that, rather than consider better alternatives.

Abbreviated NDAs can be filed by manufacturers who wish to market a generic variation of an approved drug once its patent life has expired. The manufacturer is only required to submit chemistry and manufacturing information, and demonstrate that the product is comparable with the product already approved.

Fast tracking

The regulations of many regulatory bodies include the possibility of '**fast tracking**' certain types of drug, so that they can reach the market as quickly as possible. Fast tracking is made possible by demanding a smaller number of Phase II and Phase III clinical trials than normal before the drug is put forward for approval. Fast tracking would be carried out for drugs that show promise for diseases where no current therapy exists, and for drugs that show distinct advantages over existing drugs in the treatment of life-threatening diseases such as cancer.

'Orphan' drugs

Orphan drugs are drugs that are effective against relatively rare medical problems. Since there is not a large market for such drugs, pharmaceutical companies are unlikely to reap a great reward and may decide not to develop and market the drug. In the USA, an orphan drug is defined as one that is used by less than 200 000 people. Financial and commercial incentives are given to firms to encourage the development and marketing of such drugs.

Labeling

The **labeling** of a drug preparation must be approved by the regulatory authority. Correct labeling is extremely important because it instructs physicians about the mechanism of action of the drug, the medical situations for which it should be used and the correct dosing levels and frequency. Possible side effects, toxicity or addictive effects should be detailed, as well as special precautions that might need to be taken (e.g. avoiding drugs that interact with the preparation).

Good laboratory, manufacturing and clinical practice

Good Laboratory Practice (GLP) and **Good Manufacturing Practice** (GMP) refer to scientific codes of conduct that must be adhered to in a pharmaceutical company's laboratories and production plants. They detail the scientific standards that are necessary in a pharmaceutical company, and the company must not only adhere to these standards, but prove that it is doing so.

GLP regulations cover the various laboratories involved in testing a drug, including pharmacology, drug metabolism, and toxicology laboratories.

GMP regulations are relevant to the production plant and chemical development laboratories. They encompass the various manufacturing procedures used in the production of the drug, as well as the checks and balances that must be in place to ensure that the production process is reliable and gives a product that is of consistently high quality. As part of the GMP regulations, the pharmaceutical company is required to set up an independent quality control unit to monitor a wide range of factors, including employee training, the working environment, operational procedures, instrument calibration, batch storage, labeling, and the quality control of all solvents, intermediates and reagents used in the process. The analytical procedures used to test the final product must be defined, as well as the specifications that have to be met. Each batch of drug that is produced must be sampled to ensure that it passes those specifications. Written operational instructions must be in place for all special equipment (e.g. freeze dryers), and **Standard Operating Procedures** (SOPs) must be written for the calibration and maintenance of equipment.

Implicit in all the above is the need for detailed and accurate paperwork, which must be properly filed and open to inspection by the regulatory bodies. All manner of records need to be maintained – calibration records, maintenance records, production reviews, batch records, master production records, inventories, analytical reports and records, equipment cleaning logs, batch recalls and customer complaints. Clearly, GMP involves a lot of paperwork and there is a danger that applying the rules too rigidly may prove a barrier to any changes that might improve the process. Indeed, it is quite possible that an improvement may not be adopted if the benefits are outweighed by the amount of paperwork involved.

For clinical trials, investigators must be experienced in clinical research and demonstrate that they can carry out the work according to **Good Clinical Practice** (GCP) regulations. The regulations require proper staffing, facilities and equipment for the required work. There must also be evidence that a patient's rights and wellbeing are properly protected. Each test site involved in clinical trials must be approved. In the USA, approval is given by the **Institutional Review Board** (IRB). While the work is in progress, regulatory authorities may carry out data audits to ensure that no research misconduct is taking place (e.g. plagiarism, falsification of data, poor research procedures, etc.). In Britain, the **General Medical Council** or the **Association of British Pharmaceutical Industry** can discipline unethical researchers.

Problems can arise during clinical trials due to the pressures that are often placed on researchers to obtain their results as speedily as possible. This can lead to hasty decisions, resulting in mistakes and poorly thought out procedures. However, there have also been individuals who have deliberately falsified results or have cut corners to speed up results. Sometimes, personal relationships can prove to be a problem. The investigator can be faced with a difficult dilemma between doing the best for his or her patient, and maintaining good research procedures. Patients may also mislead clinicians if they are desperate for a new cure, and falsify their actual condition in order to take part in the trial. Other patients have been known to have their drugs analyzed to see whether they are getting placebo or drug.

Marketing

The **marketing** of a drug has to be approved by regulatory authorities to ensure that the claims made for the product are accurate. This avoids the possibility of pharmaceutical companies making exaggerated claims for their product in an increasingly difficult market. This has arisen since health authorities across the world have attempted to cut the costs of medical health care by focusing on cheap generic drugs. As a result, profits have shrunk for the pharmaceutical companies and a promising new drug may not progress to the market if it offers no clear advantage over competitors' drugs. Nowadays, the strategy is to concentrate on drugs that can be marketed globally and offer some advantage over existing therapies.

L1 EPIDERMAL GROWTH FACTOR RECEPTOR

Key Notes

Introduction	The following case study looks at the design of inhibitors against a tyrosine kinase-linked receptor called the epidermal growth factor receptor.
Target choice	Kinase-linked receptors are found to be overexpressed in many cancers, so inhibitors may be potential anticancer agents.
Epidermal growth factor receptor	The epidermal growth factor (EGF) is both a receptor and an enzyme. Activation of the receptor opens up an active site on the intracellular region of the protein, triggering a signaling cascade within the cell.
Kinase-catalyzed phosphorylation	The reaction catalyzed by the EGF-receptor involves the phosphorylation of protein tyrosine residues by adenosine triphosphate (ATP). The enzyme has two binding sites – one for the tyrosine residue and one for ATP. The enzyme could be inhibited by designing inhibitors that bind to either site.
Related topic	Receptors (B2)

Introduction

In this section, we will look at a case study that illustrates many of the general points made in previous sections. The case study in question covers a specific project carried out by researchers at **Ciba Pharmaceuticals**, which looks at the design of inhibitors against a tyrosine kinase-linked receptor called the **epidermal growth factor receptor** (EGF-receptor) (Topic B2).

Target choice

Tyrosine kinase-linked receptors are important to cell growth and cell division. They have a dual action in that they act both as a receptor and as an enzyme (tyrosine kinase). In the resting state, the active site of the enzyme is closed, but when the receptor is activated the active site is opened, triggering a signaling cascade within the cell, which leads to gene activation, protein biosynthesis, and subsequent cell growth. When these receptors become deregulated or overexpressed, it is found that certain diseases such as cancer can result. The overexpression of the receptors is caused by overexpression of the genes that code for them. For example, overexpression of a gene called ***erbB1*** results in excess biosynthesis of the EGF-receptor, which is then incorporated into the cell membrane in excess quantities. As a result, the cell becomes supersensitive to the receptor's natural ligand, resulting in excess growth and division. Therefore, it makes sense to design inhibitors of the receptor kinase in order to prevent the start of the signal cascade. Such inhibitors may prove useful anticancer drugs.

Epidermal growth factor receptor

Like other kinase linked receptors, the **EGF-receptor** is located in the cell membrane and doubles up as a receptor and as an enzyme (*Fig. 1*). The receptor binding site is located in the extracellular portion of the receptor, while the enzyme active site is on the intracellular region and is normally closed. The ligand or messenger for this receptor is a protein called **epidermal growth factor** (EGF). When EGF binds, it activates the receptor causing it to change shape. This opens the enzyme's active site in the intracellular region of the protein and a catalytic reaction takes place, triggering the start of a signaling cascade which ultimately leads to increased cell growth and cell division (Topic B2).

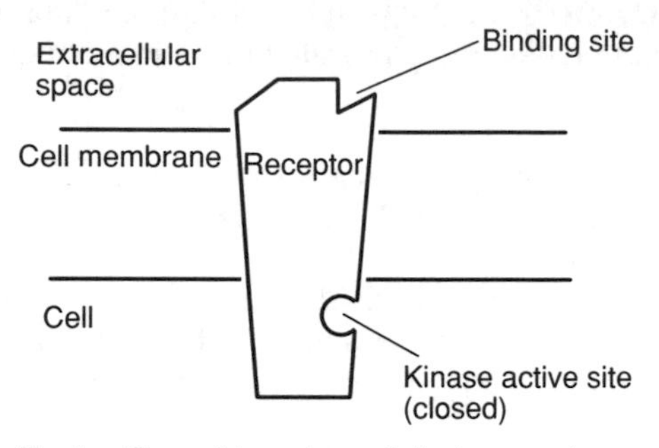

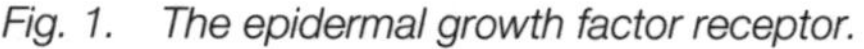

Fig. 1. The epidermal growth factor receptor.

Overexpression of this receptor kinase has been observed in a variety of epithelial type cancers, such as squamous cell carcinoma, breast cancer, and ovarian cancer. Therefore, inhibitors that bind to the active site and prevent it from functioning as an enzyme could be useful in treating these diseases.

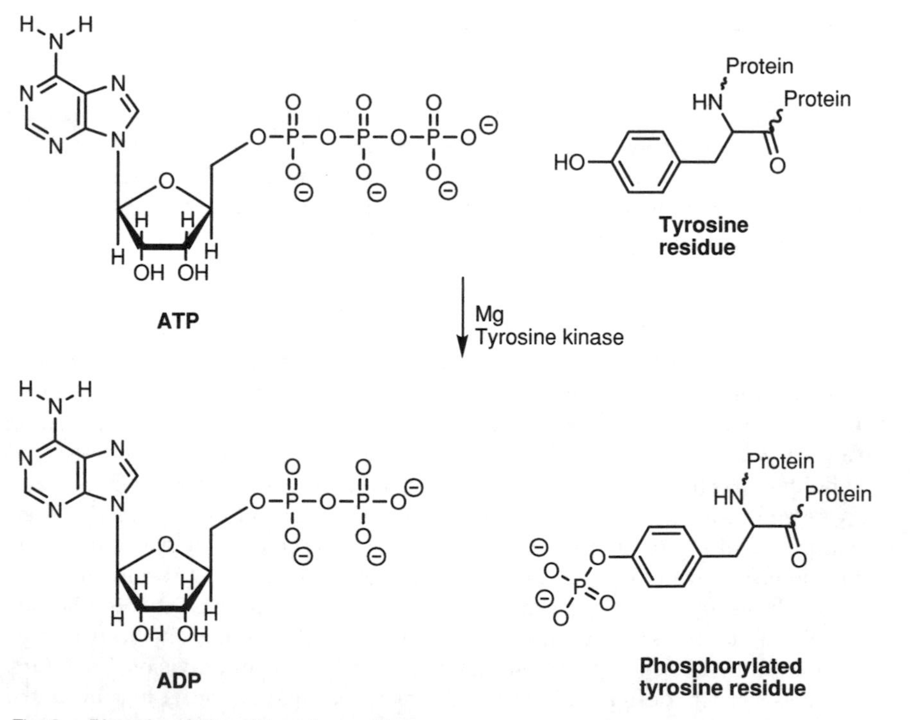

Fig. 2. Phosphorylation of tyrosine residues.

Kinase-catalyzed phosphorylation

The EGF-receptor is called a tyrosine kinase-linked receptor since its enzyme active site catalyzes the **phosphorylation** of tyrosine residues on another protein. An understanding of this reaction and the substrates involved is essential if effective inhibitors are to be designed. There are two substrates involved. One is the **tyrosine** residue of another protein while the other is **adenosine triphosphate** (ATP), which acts as the phosphorylating agent (*Fig. 2*). Magnesium is also important in binding ATP to the active site. The reaction involves a phosphate group being transferred from ATP to the tyrosine residue.

Clearly, the active site of tyrosine kinase enzymes must contain two binding sites – one for the tyrosine residue and one for ATP. Therefore, one could design inhibitors to fit the binding site for the tyrosine residue or design them to fit the binding site for ATP. In this project, the Ciba workers designed inhibitors that bind to the ATP binding site.

L2 TESTING PROCEDURES

Key Notes

Testing procedures — Bioassays had to be established to determine the activity and selectivity of kinase inhibitors. Various *in vitro* and *in vivo* tests were also established to demonstrate whether that inhibition resulted in the prevention of signal transduction, cell growth and tumor growth.

Enzyme assay — The kinase portion of the EGF-receptor was produced by recombinant DNA techniques and used in solution to test the ability of compounds to inhibit a standard phosphorylation reaction. This assay can be used to compare the inhibitory activities of different compounds without the complications of the compounds having to cross cell membranes.

Cell assays — Three kinds of cell assay were carried out. One established whether the inhibitor could cross the cell membrane and prevent the phosphorylation of tyrosine residues in the cell. Another tested whether kinase inhibition prevented the transcription of an mRNA molecule essential for cell growth. The final test measured whether kinase inhibition prevented the proliferation of cells.

***In vivo* assays** — Only compounds that showed *in vitro* activity were tested *in vivo*. Human tumor cells were grafted onto the backs of mice and the compounds injected into the mice to see whether they inhibited tumor growth.

Selectivity assays — Similar assays were developed to check whether test compounds showed any activity against other kinase enzymes. To be an effective agent, the inhibitor would have to show high selectivity for the EGF-receptor kinase.

Related topics Testing drugs (D1) Testing drugs *in vitro* (D2) Testing drugs *in vivo* (D3)

Testing procedures

A lead compound was required which had some kinase inhibitory effect, and which could be modified to improve its activity and selectivity. However, in order to achieve these goals it was first necessary to establish suitable **bioassays** to determine activity and selectivity. In this project, various tests were developed which demonstrated whether a test compound could cross the cell membrane and inhibit the kinase enzyme, and whether that inhibition prevented signal transduction, transcription and cell division both *in vitro* and *in vivo*.

Enzyme assay

An **enzyme assay** was required to test whether compounds could inhibit the kinase enzyme. However, the situation was complicated by the fact that the EGF-receptor is membrane bound with the enzyme region located within the cell. Inhibitors would have to cross the cell membrane to reach the enzyme, and so it would be difficult to rationalize whether a variation in activity was due to the inhibitor binding more strongly to the enzyme or crossing the cell membrane more efficiently. To simplify the situation, **genetic engineering** (or recombinant DNA technology) was used to produce the enzyme portion alone of the EGF-receptor. This proved to be soluble in water and so a simple enzyme assay was now possible. The enzyme was used to catalyze a standard reaction (the phosphorylation of the hormone **angiotensin II**), and inhibitors were tested by measuring how effectively they inhibited this reaction.

The enzyme assay made it possible to rationalize the activities of different inhibitors, but it also increased the chances of finding a **lead compound**. This is because it allowed the detection of inhibitors that would be too polar to cross the cell membrane. One might ask why one would want to detect a lead compound that could not reach the enzyme in the 'real' situation. The answer is that a good lead compound could always be modified to be less polar to enable it to cross cell membranes.

Cell assays

Cell assays were developed to test whether the inhibitors that worked on the isolated enzyme could also act on whole cells. These would test whether the compounds could cross the cell membrane in order to reach the enzyme, and whether they inhibited the **signal transduction** events resulting from receptor (and enzyme) activation. For these tests, cancerous human epithelial cells were used that overexpressed the EGF-receptor in their cell membranes, so they were sensitive to EGF and grew rapidly in its presence. Since the kinase activity of the EGF-receptor leads to the phosphorylation of tyrosine residues within the cell, a high throughput ELISA assay was developed which measured the total **tyrosine phosphorylation** resulting from exposure to EGF and how these levels were affected when an inhibitor was present. This particular test demonstrated whether an inhibitor could cross the cell membrane and inhibit the kinase enzyme.

Another assay was developed that tested whether the test compounds could actually prevent signal transduction as a result of kinase inhibition. This involved an assay for a specific mRNA molecule that was normally transcribed when the EGF-receptor was activated. If this molecule failed to appear, it could be taken as evidence that kinase inhibition also prevented signal transduction.

Finally, another cell assay was carried out on mice cells that divided rapidly in the presence of EGF. The ability of inhibitors to prevent that cell division could thus be measured.

In vivo assays

***In vivo* assays** were developed to study whether the inhibitors could be administered to a live animal and have antitumor activity. In order to carry out these tests, human tumor cells were grafted onto the backs of mice and the test compounds were then administered by injection. The ability of the compounds to inhibit these tumors could then be measured. These tests demonstrated whether the compounds had satisfactory pharmacokinetic properties and could reach the tumor, whether they were active in an *in vivo* situation, and whether they had any toxic effects. Only drugs that proved satisfactory in the *in vitro* tests were tested *in vivo*.

Selectivity assays

Kinase inhibitors for the EGF-receptor must be highly selective, because there are many other kinds of kinase enzymes present in the cell. These include other tyrosine kinases, as well as kinases that phosphorylate the serine and threonine residues of proteins. A battery of similar *in vitro* and *in vivo* tests were developed to establish whether the test compounds inhibited any of these enzymes.

L3 FROM LEAD COMPOUND TO DIANILINO-PHTHALIMIDES

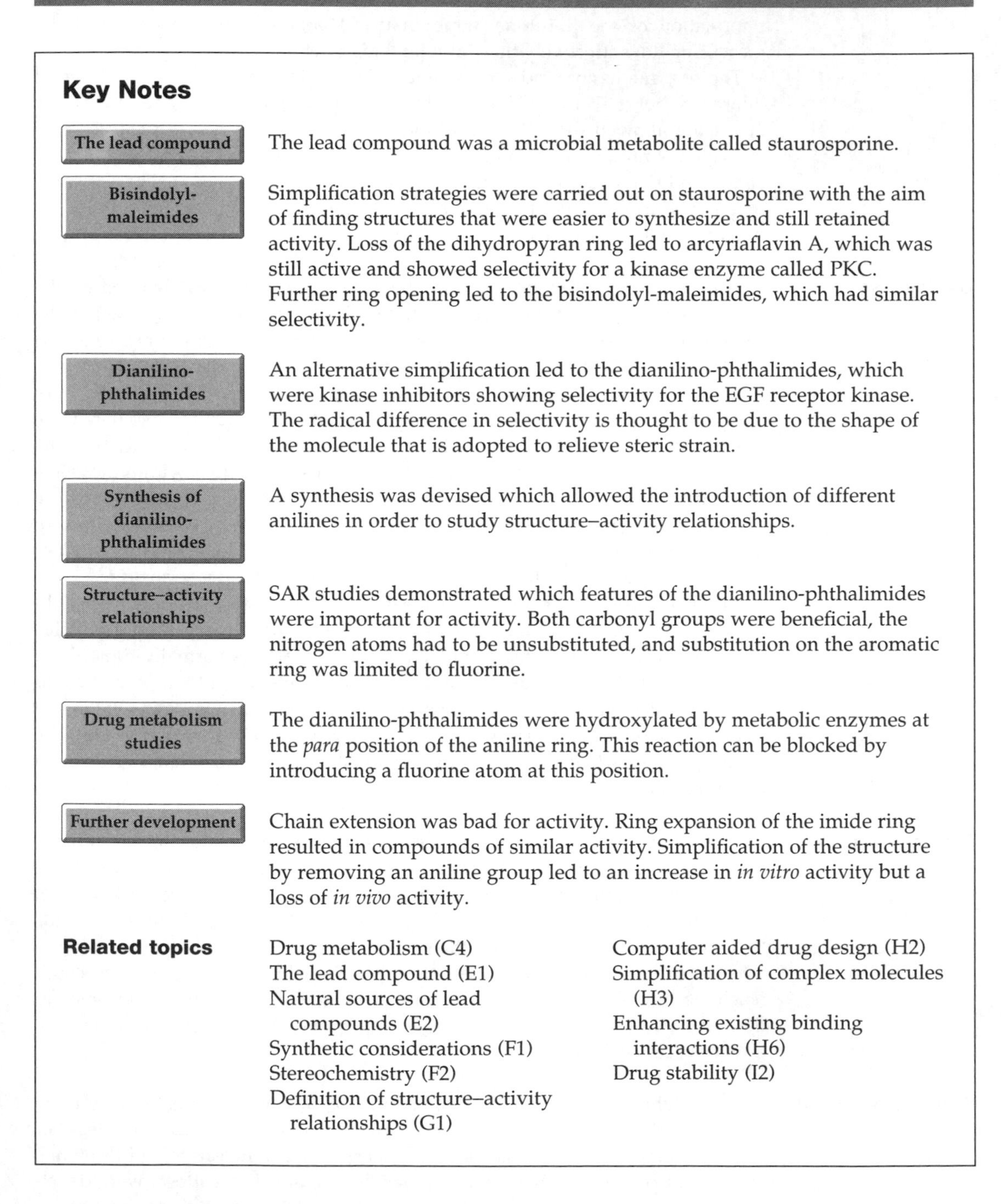

Key Notes

The lead compound

The lead compound was a microbial metabolite called staurosporine.

Bisindolyl-maleimides

Simplification strategies were carried out on staurosporine with the aim of finding structures that were easier to synthesize and still retained activity. Loss of the dihydropyran ring led to arcyriaflavin A, which was still active and showed selectivity for a kinase enzyme called PKC. Further ring opening led to the bisindolyl-maleimides, which had similar selectivity.

Dianilino-phthalimides

An alternative simplification led to the dianilino-phthalimides, which were kinase inhibitors showing selectivity for the EGF receptor kinase. The radical difference in selectivity is thought to be due to the shape of the molecule that is adopted to relieve steric strain.

Synthesis of dianilino-phthalimides

A synthesis was devised which allowed the introduction of different anilines in order to study structure–activity relationships.

Structure–activity relationships

SAR studies demonstrated which features of the dianilino-phthalimides were important for activity. Both carbonyl groups were beneficial, the nitrogen atoms had to be unsubstituted, and substitution on the aromatic ring was limited to fluorine.

Drug metabolism studies

The dianilino-phthalimides were hydroxylated by metabolic enzymes at the *para* position of the aniline ring. This reaction can be blocked by introducing a fluorine atom at this position.

Further development

Chain extension was bad for activity. Ring expansion of the imide ring resulted in compounds of similar activity. Simplification of the structure by removing an aniline group led to an increase in *in vitro* activity but a loss of *in vivo* activity.

Related topics

Drug metabolism (C4)
The lead compound (E1)
Natural sources of lead compounds (E2)
Synthetic considerations (F1)
Stereochemistry (F2)
Definition of structure–activity relationships (G1)
Computer aided drug design (H2)
Simplification of complex molecules (H3)
Enhancing existing binding interactions (H6)
Drug stability (I2)

The lead compound

Every medicinal chemistry project requires a **lead compound** – a compound that either shows some useful biological activity or has some affinity for the target receptor or enzyme under study. Often lead compounds are obtained from natural products, a source of highly diverse chemical structures that often have biological potency of one form or another. The lead compound for this project was one such structure, a compound called **staurosporine** (*Fig. 1*). Staurosporine is a microbial metabolite that was found to be a highly potent protein kinase inhibitor. Moreover, it was discovered that it competed with ATP for the ATP binding site. Unfortunately, staurosporine cannot be used in medicine since it is unselective in its action, inhibiting both serine-threonine kinases and tyrosine kinases alike.

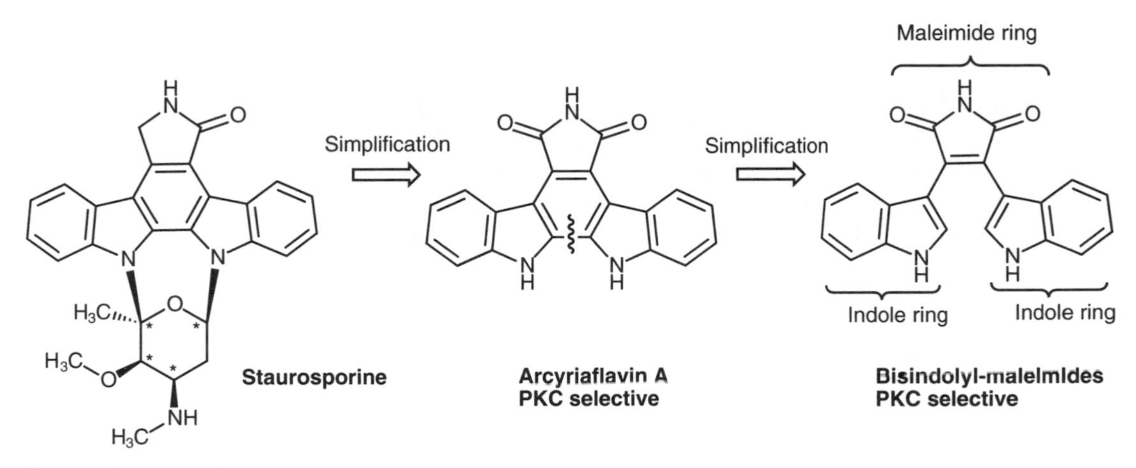

*Fig. 1. Kinase inhibitors. *, asymmetric centers.*

Bisindolyl-maleimides

Staurosporine has a complex polycyclic structure, which includes a dihydropyran ring with four asymmetric centers. Such a compound is not easily synthesized and so it is difficult to make a series of analogs. **Simplification** is a tactic that is often tried on complex lead compounds with the aim of finding a simpler structure that is easier to synthesize and retains the important structural features associated with activity. In this case, an obvious structure to try is one lacking the dihydropyran ring system since this simplifies the polycyclic system and also removes the four asymmetric centers. If possible, drugs should have a minimum number of asymmetric centers to make their synthesis easier and to avoid the possibility of different isomers.

Arcyriaflavin A (*Fig. 1*) is a suitable structure lacking the dihydropyran ring. It should be noted that an extra carbonyl group is present, but this could be considered a form of simplification since it results in a symmetrical molecule. In general, symmetrical molecules are easier to synthesize. Although the activity of arcyriaflavin A is less than that of staurosporine, it is still a kinase inhibitor, demonstrating that the dihydropyran ring is not essential to activity. Moreover, the molecule shows selectivity against a specific kinase enzyme called **protein kinase C** (PKC). There is poor activity against the EGF-receptor kinase and at this stage the project showed more promise in developing a selective inhibitor against PKC.

Further simplification involved the synthesis of structures that lacked the central aromatic ring, leading to a series of compounds known as the bisindolyl-maleimides (*Fig. 1*). These were also found to be potent PKC inhibitors.

Dianilino-phthalimides

An alternative simplification strategy was tried where the central phthalimide ring structure of arcyriaflavin was retained but both indolyl rings were opened to give a different series of symmetrical molecules called the **dianilino-phthalimides** (*Fig. 2*). These were expected to have similar activity and selectivity to the previous structures. The structures were certainly kinase inhibitors, but surprisingly, they proved to be selective inhibitors of the EGF-receptor kinase and were inactive against PKC and other kinases.

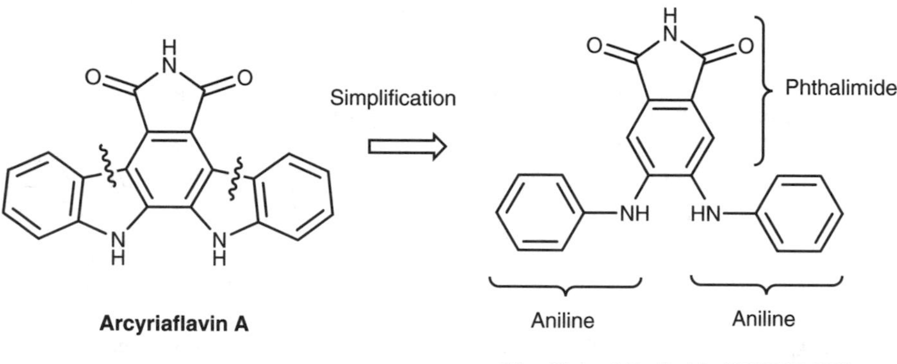

Fig. 2. Dianilino-phthalimides.

The question then arose as to why there should be such a radical difference in selectivity between arcyriaflavin A and the bisindolyl-maleimides on one hand, and the dianilino-phthalimides on the other. The structures of the three compounds were studied by **X-ray crystallography** and the results showed that there was a significant difference in shape between them all (*Fig. 3*). Arcyriaflavin A is planar, while the bisindolyl-maleimides adopt a symmetrical bowl shape. However, the dianilino-phthalimides, although symmetrical looking, actually adopt an asymmetric propeller shaped conformation.

The propeller-shaped conformation of the dianilino-phthalimides is formed in order to relieve a steric clash between protons which would occur if the structure was planar (*Fig. 4*). This clash does not occur in either arcyriaflavin or the bisindolyl-maleimides since the protons involved are not present. It was proposed that the differences in shape between the three molecules were important in determining their selectivity of action.

Synthesis of dianilino-phthalimides

A **synthesis** was devised to make a series of **dianilino-phthalimides**, so that their structure–activity relationships could be studied (*Fig. 5*). The synthesis involved a Diels-Alder reaction to form a cyclohexadiene, to which a range of different anilines could be added. A cyclic anhydride was then formed which was heated strongly with ammonia, resulting in aromatization and formation of the phthalimide ring structure.

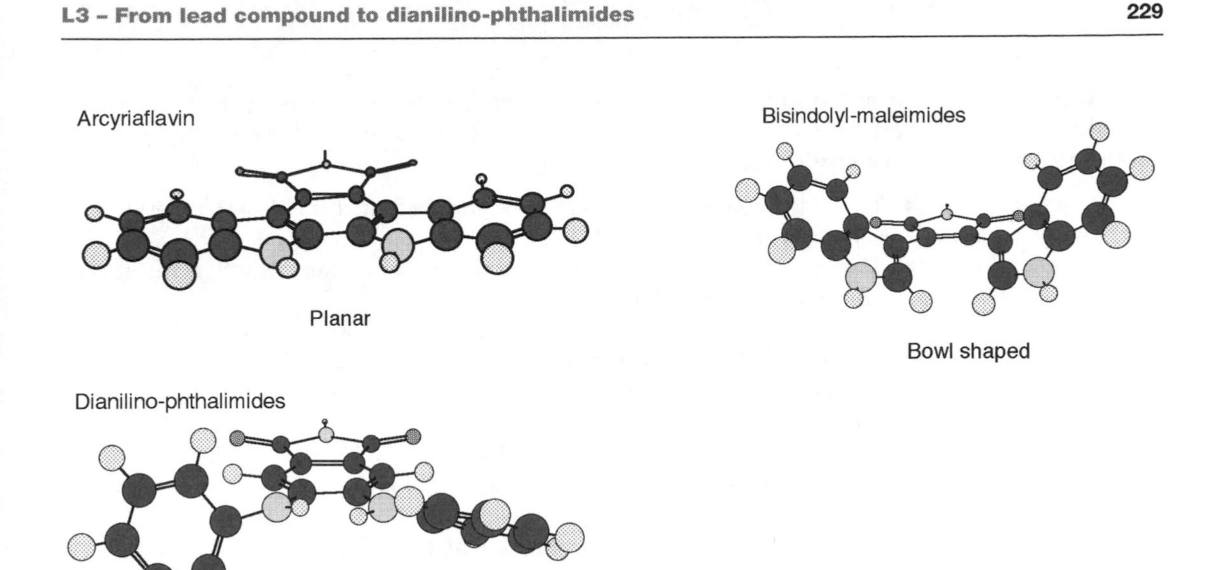

Fig. 3. Comparison of conformations for kinase inhibitors.

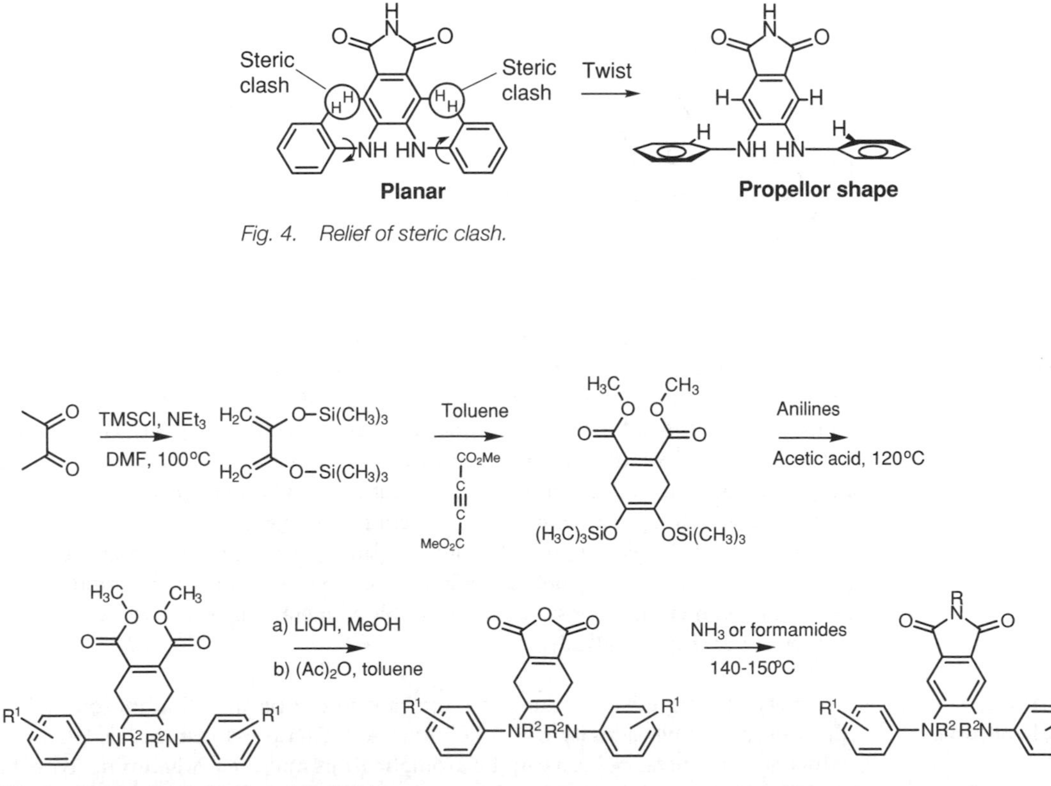

Fig. 4. Relief of steric clash.

Fig. 5. Synthesis of dianilino-phthalimides.

Structure–activity relationships

Over 250 various analogs were synthesized by the above synthetic route, or variations of it. They were then tested for activity resulting in the following **SAR results** (*Fig. 6*).

- The phthalimide nitrogen has to be unsubstituted (i.e. R=H), otherwise all activity is lost.
- The aromatic rings of the anilino groups are essential. Activity is lost if they are replaced with cyclohexane rings.
- The aniline substituents (R^1) must be small or activity drops. In fact, the only feasible substituent is fluorine, which is approximately the same size as hydrogen. Adding methyl groups decreases activity, while adding ethyl groups eliminates it.
- The aniline nitrogen has to be unsubstituted (i.e. R^2 must be H). If one of the R^2 substituents is a methyl group, activity falls. If both R^2 substituents are methyl groups, activity is lost completely.
- The aniline nitrogen is essential for activity. Replacement of the aniline nitrogen (NR^2) with S eliminates activity altogether.
- Both carbonyl groups are important to activity. Loss of a single carbonyl oxygen to form a lactam rather than an imide causes a large drop in activity.

Of the 250 structures synthesized, the parent structure **CGP52411** was chosen for preclinical trials. It proved to be a highly selective inhibitor for the EGF-receptor kinase, with an IC_{50} value of 0.7 μM, and may prove useful in the treatment of **psoriasis** (a proliferative disease of the epidermis).

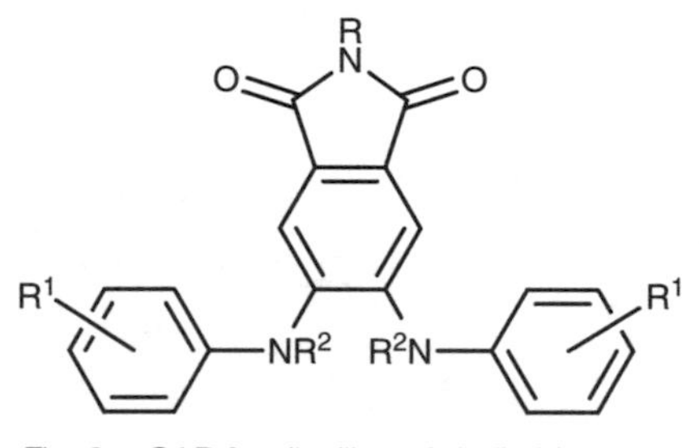

Fig. 6. SAR for dianilino-phthalimides.

Drug metabolism studies

Drug metabolism studies were carried out on **CGP52411** to see whether it was stable to metabolism in humans and various test animals. It was well absorbed orally, but was also rapidly metabolized in several species (including man) by *para*-hydroxylation of one of the phenylamino moieties (*Fig. 7*). The phenol group was then conjugated by glucuronylation and excreted. In monkeys, the drug was metabolized differently. Here, both phenylamino groups were *para*-hydroxylated before conjugation and excretion took place.

In order to prevent drug metabolism, fluorine substituents were placed at the *para* positions to act as metabolic blockers. The resulting compound (**CGP 53353**) had similar activity, and proved stable to metabolism (*Fig. 8*). It, too, was put forward for clinical trials.

Further development

A variety of drug design strategies were carried out on the dianilino-phthalimides to see whether activity could be improved. **Chain extension** was tried out to increase the distance between the aromatic rings and the phthalimide 'core' to give **CGP58109** (*Fig. 9*). The idea of this strategy was to test whether the

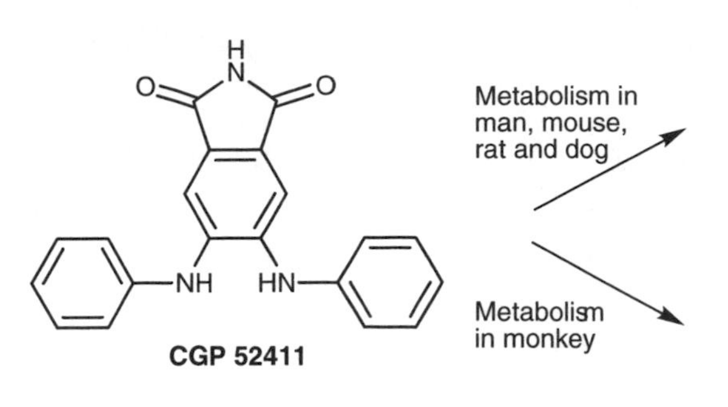

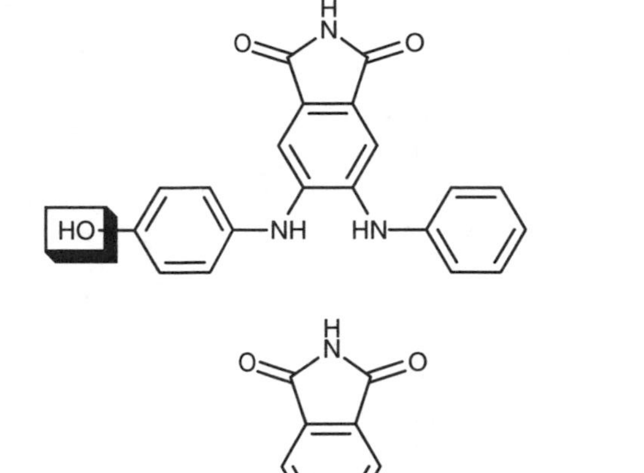

Fig. 7. Drug metabolism of CGP52411.

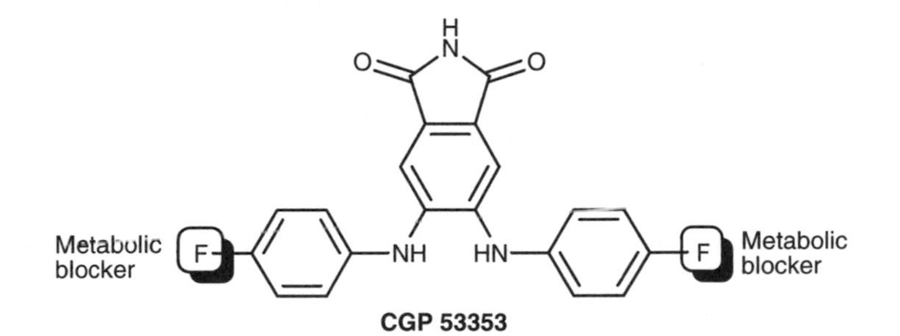

Fig. 8. Use of metabolic blockers.

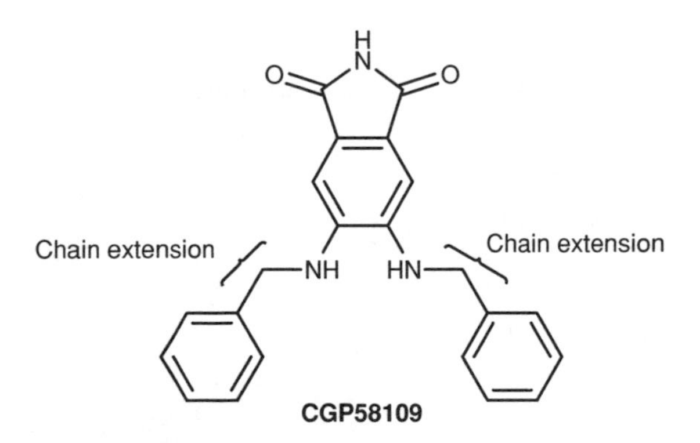

Fig. 9. Chain extension.

aromatic rings and phthalimide core needed to be further apart in order to have better binding interactions with the binding site. However, chain extension led to a large drop in activity, which demonstrated that the separation was now too great.

The strategy of **ring expansion** was then carried out to see whether it was possible to expand the five-membered imide ring to a six-membered ring (*Fig. 10*).

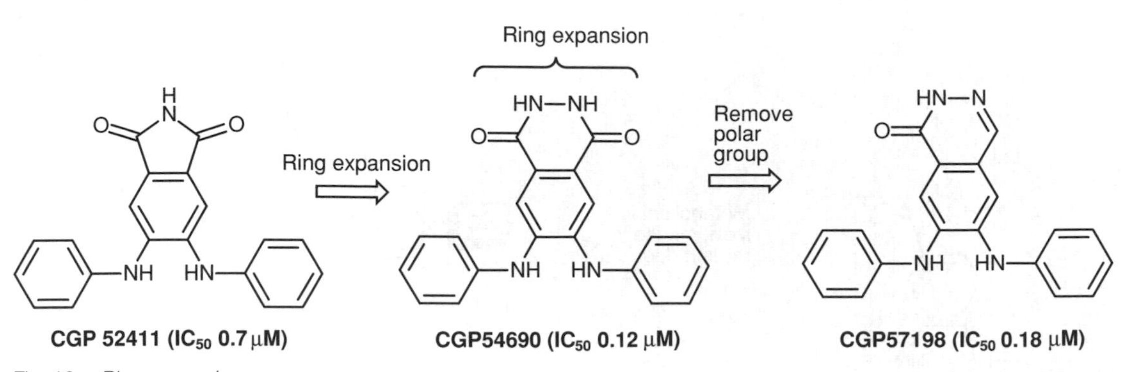

Fig. 10. Ring expansion.

Introduction of an extra NH group expanded the imide ring to the hydrazone structure (**CGP54690**). This compound had comparable activity to CGP52411 in enzyme inhibition assays, but lacked activity in cellular assays, demonstrating that it was unable to cross cell membranes. This can be explained by the fact that an extra polar group (NH) has been introduced. In order to reduce the polarity of the compound, structure **CGP57198** was synthesized. This compound is less polar since it lacks one of the carbonyl oxygens. Moreover, the extra nitrogen that was introduced can no longer act as a hydrogen bond donor. The modifications proved successful since CGP57198 showed good activity in both *in vitro* and *in vivo* tests.

Further simplification of CGP52411 was tried by removing one of the aniline groups (*Fig. 11*). This gave **CGP58522**, which was slightly more active than CGP52411 *in vitro*, but was inactive in cellular tests.

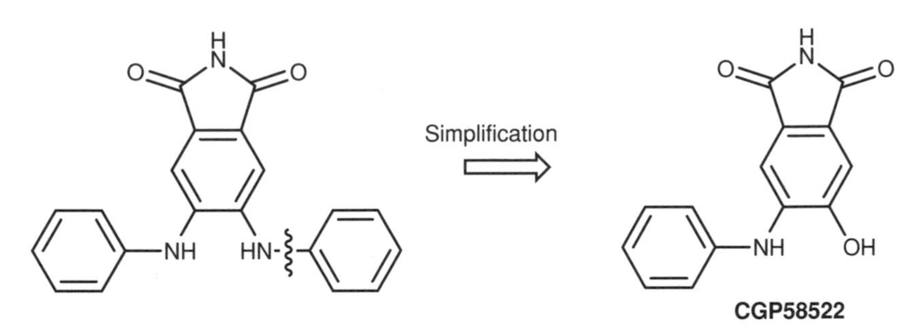

Fig. 11. Simplification.

L4 Modeling studies

Key Notes

Model binding site

The structure of the ATP binding site for the EGF-receptor kinase was modeled on the known structure of a related kinase. ATP binds into a cleft with the adenine buried deep into the cleft and interacting with the binding site by two hydrogen bonds. The ribose sugar of ATP fits into a small pocket and the triphosphate chain points towards the opening of the cleft. There is a large unoccupied pocket in the binding site.

Binding model of dianilino-phthalimides

The dianilino-phthalimides can bind to the ATP binding site so that the imide moiety mimics the adenine ring of ATP and participates in three hydrogen-bonding interactions. One of the aniline aromatic rings fits into the small ribose pocket. The binding model explains the SAR results for these compounds. The bisindolyl-maleimides can bind through their imide group, but do not bind to the ribose pocket.

Selectivity of action

Selectivity against the EGF-receptor kinase is due to the presence of a cysteine residue in the ribose pocket. The cysteine residue increases the hydrophobic character of the pocket and binds to the aromatic ring of the inhibitor.

Pharmacophore

The pharmacophore for EGF-receptor kinase inhibitors consists of a hydrogen bond donor, hydrogen bond acceptor and an aromatic ring at the correct relative positions in space.

Related topics

The pharmacophore (G4)   Computer aided drug design (H2)

Model binding site

The design of enzyme inhibitors is made easier if the shape and format of the enzyme's active site is known. Ideally, this involves crystallizing the enzyme with an inhibitor bound to the active site then carrying out an X-ray crystallographic analysis. The structure can then be determined and the position of the inhibitor reveals the location of the active site. The protein–inhibitor complex can then be fed into a computer and molecular modeling software can be used to study how the inhibitor is bound to the active site (i.e. to which amino acids it is bound and by which types of interaction). Areas of space within the active site that are not occupied can be identified, which allows the possibility of designing further inhibitors that will fit more snugly and which may bind more effectively (drug extension).

Unfortunately, not all proteins can be obtained or crystallized for such a study, and this is particularly the case for membrane-bound receptors such as the EGF-receptor. However, the fact that most receptors belong to families of proteins having similar overall structure means that it is possible to model receptors or enzymes based on a closely similar protein whose structure has been determined.

An X-ray crystal structure has not been obtained for the EGF-receptor, so it was not possible to study the structure of the EGF receptor directly. However, the X-ray crystal structure of a related protein kinase, **cyclic AMP-dependent protein kinase**, has been worked out, with the enzyme complexed to an inhibitor, Mg, and ATP. This complex showed that ATP fitted into a cleft in the enzyme with the adenine portion buried more deeply than the rest of the molecule so that it was close to the predominantly hydrophobic bottom of the cleft. The rest of the molecule, consisting of the ribose sugar and the triphosphate moiety, extended out towards the opening of the cleft. Molecular modeling studies made it possible to identify the binding interactions. From SAR studies it was known that the adenine portion of the molecule was crucial in binding ATP to kinases, and the molecular modeling studies revealed hydrogen bond interactions involving the 6-amino group and N1 nitrogen of adenine. The former group acted as a hydrogen bond donor and the latter as a hydrogen bond acceptor.

Based on these results, a **model ATP binding site** for the EGF-receptor kinase was constructed as in *Fig. 1*, where the comparable amino acids for the EGF-receptor kinase were introduced in place of those identified for cyclic AMP-dependent protein kinase. This model revealed that the important hydrogen bonding interactions were between the adenine group and the peptide bonds connecting amino acid residues Gln-767, Leu-768 and Met-769. The ribose ring fitted into a small pocket in the binding site, which was named the 'ribose' pocket. It was also observed that there was a large pocket in the binding site that was unoccupied by any part of ATP.

Binding model of dianilino-phthalimides

Using molecular modeling, ATP was removed from the binding site and CGP52411 was fitted into the binding site in its place, then docked in order to get the strongest binding interactions (*Fig. 2*). The bidentate binding interaction observed for the pyrimidine ring of ATP is mimicked by the imide moiety of

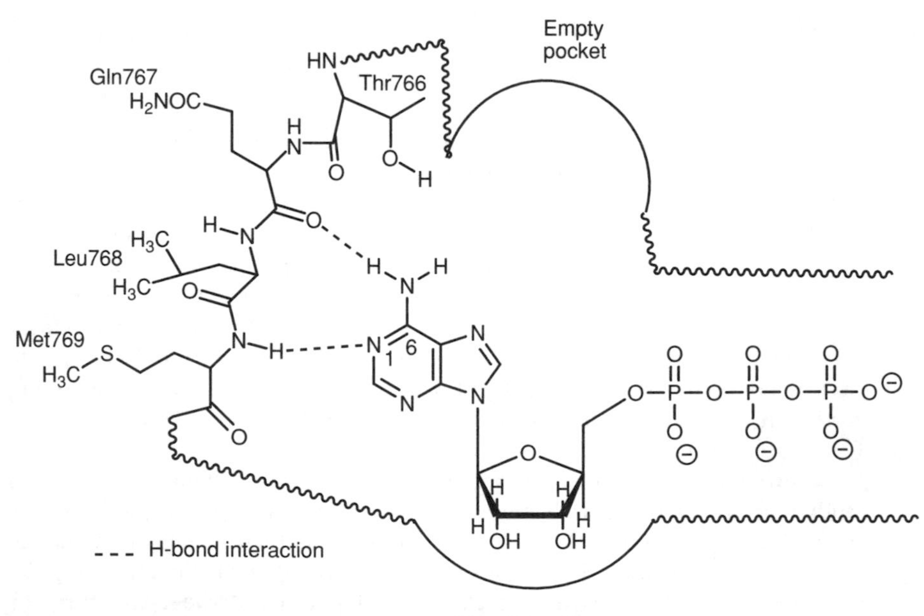

Fig. 1. Binding model of ATP to EGF-receptor kinase active site.

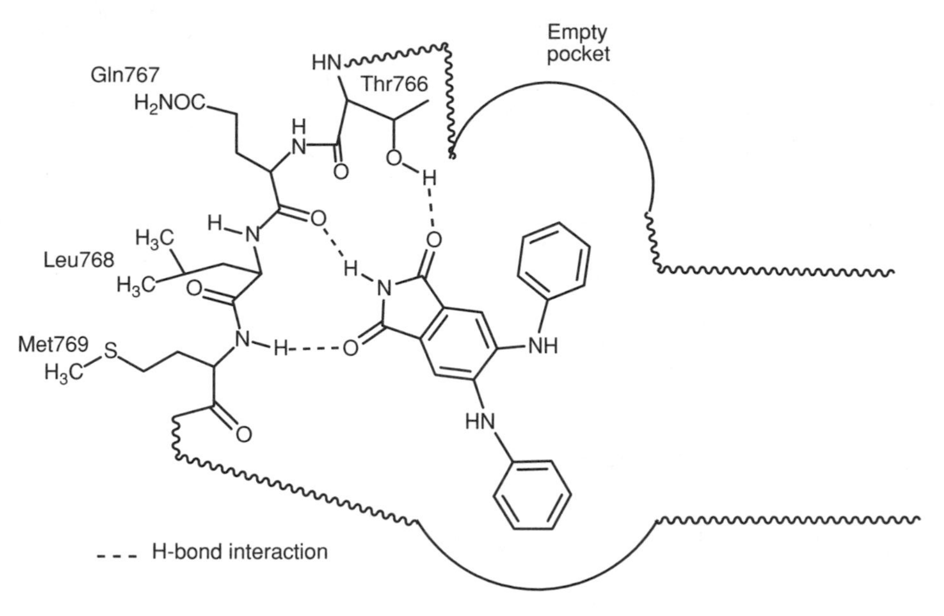

Fig. 2. Binding interactions of CGP52411 in model binding site.

CGP52411, where NH acts as a hydrogen bond donor and a carbonyl oxygen acts as a hydrogen bond acceptor. An additional hydrogen bonding interaction could also take place between the second imide carbonyl and threonine 766.

These interactions explain some of the **SAR results** obtained for the dianilino-phthalimides. For example, substitution of the imide NH would prevent this group acting as a hydrogen bond donor, while removal of either carbonyl oxygen would remove an important hydrogen bonding interaction.

The model also reveals that one of the aniline aromatic rings fits into the small 'ribose pocket'. This explains several other SAR results. Since the pocket is small, any substituents on the aromatic ring make this group too large for the pocket. Similarly, the chain extension strategy that led to CGP58109 was unsuccessful since this pushed the aromatic ring too far into the pocket.

The **ring expansion** strategy was successful since the six-membered hydrazone ring that resulted could still participate in the three hydrogen bonding interactions identified for CGP52411 (*Fig. 3*). As a comparison, the bisindolylmaleimide structure was docked into the model binding site (*Fig. 4*). The imide ring of the structure could bind using the same by hydrogen bonding interactions, but the aromatic ring of the indole moiety did not fit into the ribose pocket.

Selectivity of action

The **selectivity** of the dianilino-phthalimides for the EGF-receptor kinase appears to be due to the aromatic ring, which can fit into the ribose pocket, implying that this is a favorable interaction. However, this raises a couple of questions. First, the ribose pocket normally accepts a highly polar ribose group and interacts favorably with it. Therefore, how could a hydrophobic phenyl group fit into a polar binding pocket and take part in any binding interactions? Even if this was possible, why should this lead to selectivity? The ATP binding sites in other kinases are very similar, so there seems no reason why the

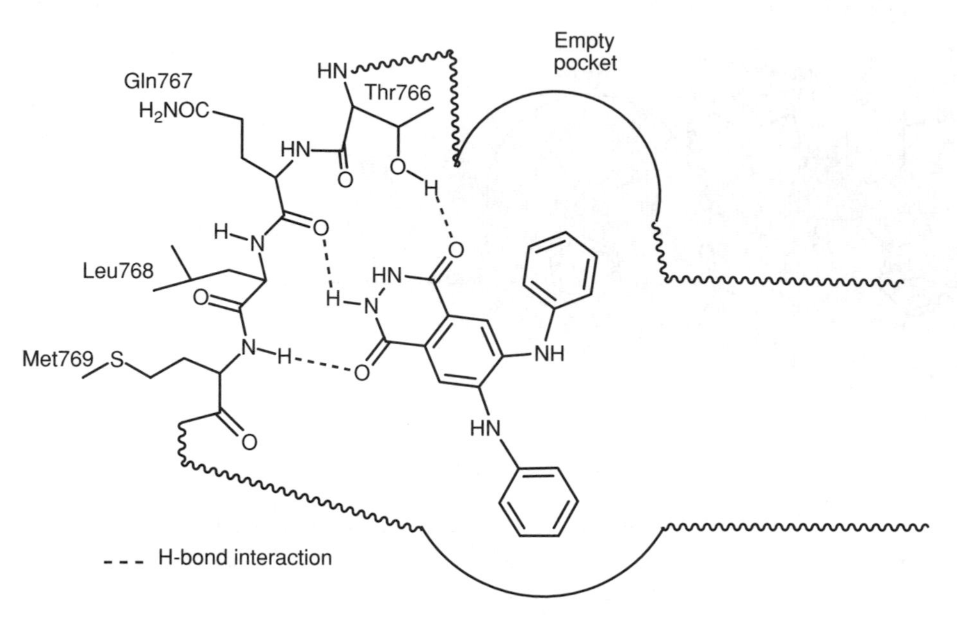

Fig. 3. Binding interactions for the hydrazone analog with the model binding site.

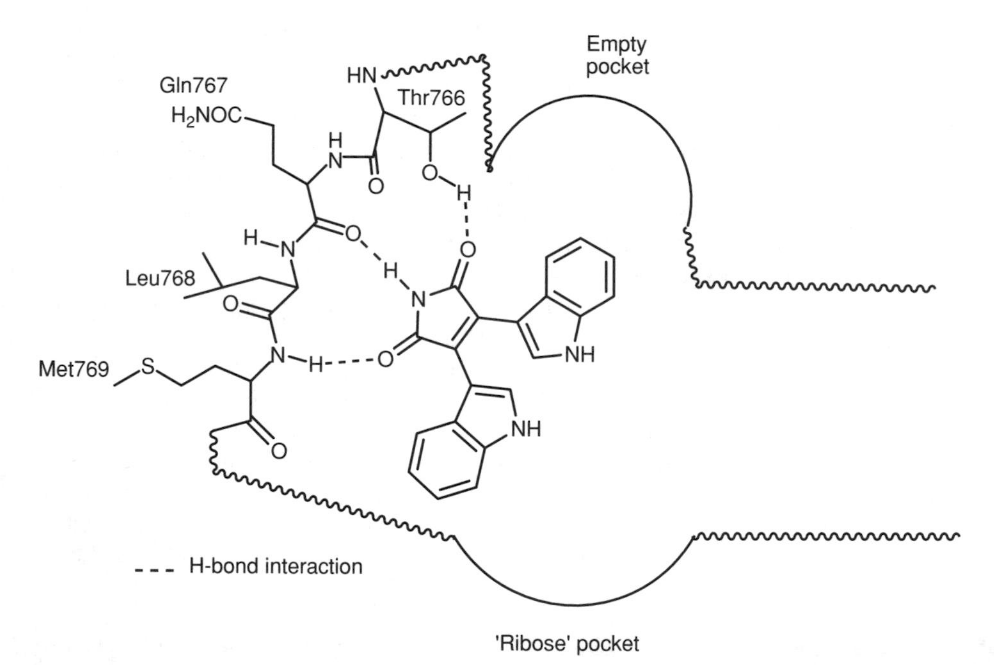

Fig. 4. Binding interactions of the bisindolyl-maleimides with the model binding site.

dianilino-phthalimides should not bind to other kinases in the same way. Indeed, it was long assumed that the ATP binding sites in different kinases were so similar that it would not be possible to design selective inhibitors for different kinases. This was because amino acid sequencing had showed a high conservation of amino acids in the different kinases. Nevertheless, the differences that are present can be decisive.

The amino acids present in the **ribose pocket** of **protein kinase A** consist of four hydrophobic amino acids (leucine, glycine, valine and leucine) plus four polar amino acids (two glutamic acids, asparagine and threonine). The two glutamic acid residues are important in binding the ribose group through hydrogen bonding interactions.

In the **EGF-receptor kinase**, the corresponding amino acids are the same, but with two very important exceptions, mainly the two glutamic acids have been replaced with cysteine and arginine. The replacement of a glutamic acid with cysteine is of particular importance. Cysteine is a sulfur-containing amino acid, which is more hydrophobic than glutamic acid, and therefore makes the ribose pocket itself more hydrophobic. Moreover, it has been found that the sulfur atom of cysteine can interact favorably with aromatic rings. These two factors help to explain why an aromatic ring can be accepted into the ribose pocket and be stabilized (*Fig. 5*). Since the cysteine residue is not present in the ribose pockets of other kinases, the presence of the aromatic ring in the ribose pocket would not be favored.

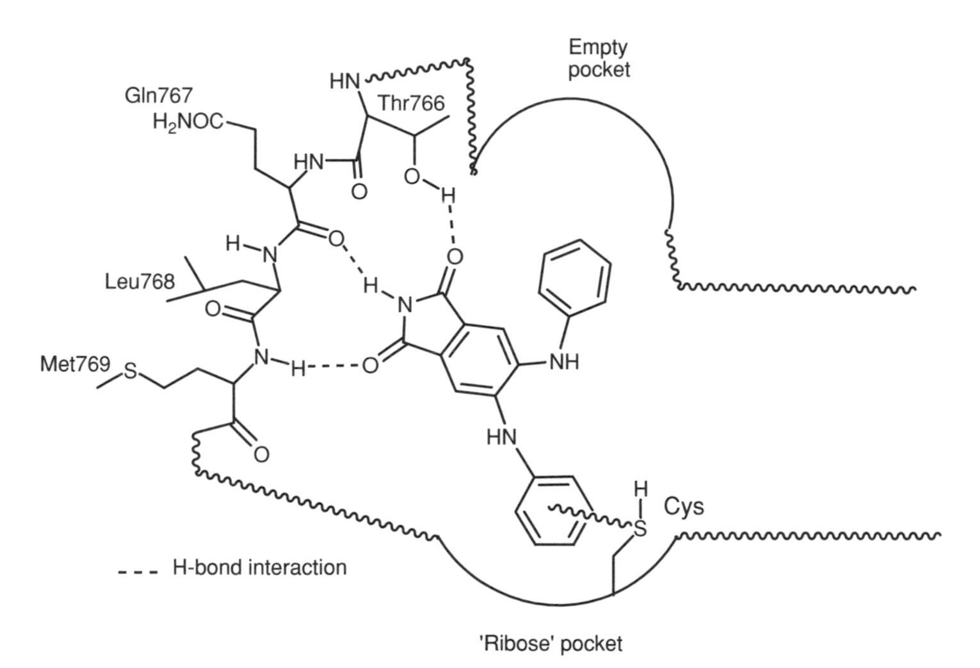

Fig. 5. Interaction of an aromatic ring with a cysteine residue.

Pharmacophore

The molecular modeling studies and SAR results for the dianilino-phthalimides made it possible to define a pharmacophore for kinase inhibitors that should be selective for the EGF-receptor kinase (*Fig. 6*). An important bidentate hydrogen bonding interaction was required where the inhibitor provided one hydrogen bond donor and one hydrogen bond acceptor. An aromatic ring was also required which could fit into the ribose pocket and interact with cysteine. Such a group would be required both for activity and selectivity.

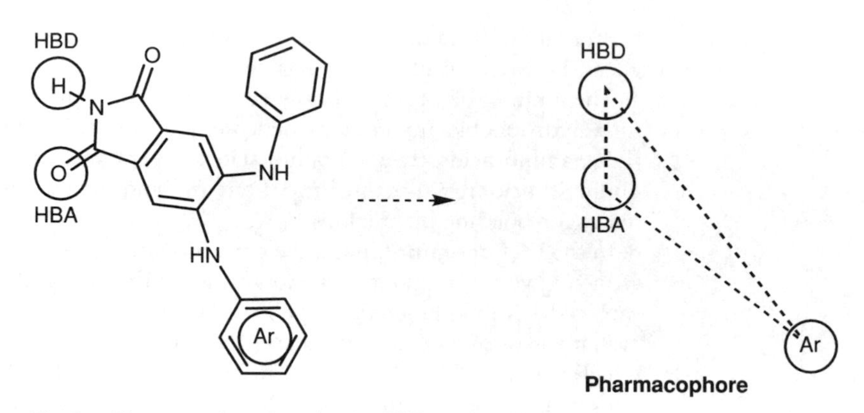

Fig. 6. Pharmacophore for selective EGF-receptor kinase inhibitors.

L5 4-(PHENYLAMINO)-PYRROLOPYRIMIDINES

Key Notes

Introduction

Molecular modeling can be used to identify potential inhibitors that are not structurally related to the lead compound. Target structures can be designed, then fitted into the binding site to see if they are worth synthesizing.

4-(Phenylamino) pyrrolopyrimidines

4-(Phenylamino)pyrrolopyrimidines are potent kinase inhibitors with selectivity for the EGF-receptor kinase. Two different modes of binding are possible for these structures. The most likely mode is the one that fits an aromatic ring into the ribose pocket.

Related topic

Computer aided drug design (H2)

Introduction

The advantage of having a model binding site, binding theory and pharmacophore, is that it is possible to design new inhibitors that are not structurally related to the original lead compound, but which are capable of interacting with the binding site in the same way. Using molecular modeling, target structures can be docked into the model binding site, tested to see if they fit, then synthesized. This helps to concentrate the synthetic work on those compounds that are most likely to be active.

4-(Phenylamino) pyrrolopyrimidines

CGP 59326 (*Fig. 1*) is an example of this class of compounds and shows good activity. It contains a pyrimidine ring like ATP and could certainly interact with

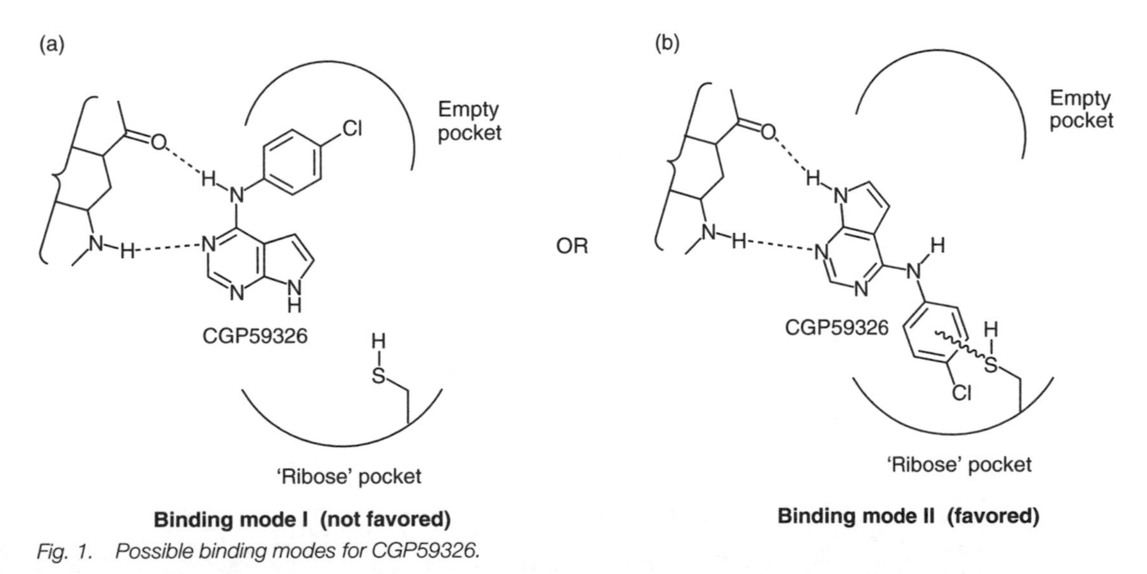

Fig. 1. Possible binding modes for CGP59326.

the binding site in the same way as ATP (binding mode I). However, it is more likely that it binds as shown in binding mode II. In this mode, the bidentate hydrogen bonding interaction is still possible using different atoms, but there is the added binding interaction between the aromatic ring and the ribose pocket. Without this extra interaction, it is difficult to explain why the compound is as active as it is. This illustrates the dangers in trying to design inhibitors based purely on the structure of a lead compound (in this case ATP) without understanding the binding interactions taking place.

Pharmacophore

The 4-(phenylamino)pyrrolopyrimidines contain the **pharmacophore** previously defined for the dianilino-phthalimides (*Fig. 2*).

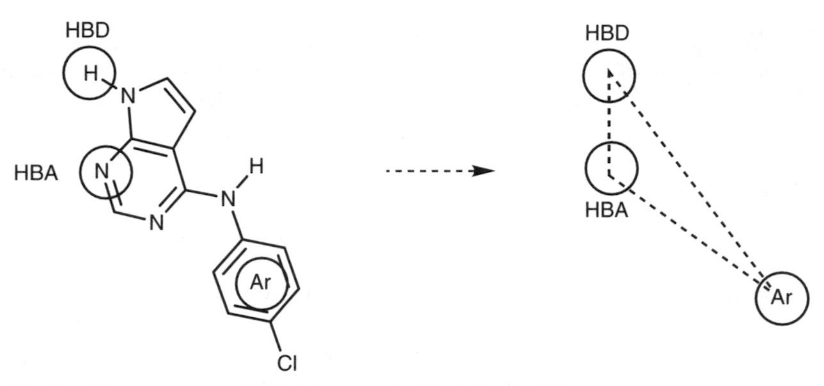

Fig. 2. Pharmacophore for 4-(phenylamino)pyrrolopyrimidines.

L6 Pyrazolopyrimidines

Key Notes

Lead compounds

A random screening of known compounds resulted in the discovery of two pyrazolopyrimidine structures, which acted as lead compounds for novel kinase inhibitors.

Binding models

Docking experiments suggested that the two lead compounds interacted with the binding site in two different binding modes. One structure mimicked ATP in its binding mode, while the other mimicked the 4-(phenylamino)pyrrolopyrimidine (CGP 59326).

Drug design

The lead compound that was proposed to bind like CGP59326 was developed further by removing a primary amino group then adding an aromatic ring to fit the ribose pocket. An extra hydrogen bonding interaction was established with the larger pocket. It was found that a larger variety of substituents was possible on the aromatic ring fitting the larger pocket than on the ring fitting the ribose pocket.

Related topics

The lead compound (E1)
Synthetic sources of lead compounds (E3)
Definition of structure–activity relationships (G1)
Computer aided drug design (H2)
Extra binding interactions (H5)

Lead compounds

Lead compounds can be obtained by **screening** synthetic compounds as well as natural ones. To this end, a random screening of chemicals produced by CIBA was carried out, revealing two chemicals (I and II) that acted as kinase inhibitors (*Fig. 1*).

Binding models

Since both structures are **pyrazolopyrimidines**, it is tempting to propose that they both bind to the model binding site in the same manner. However, it is

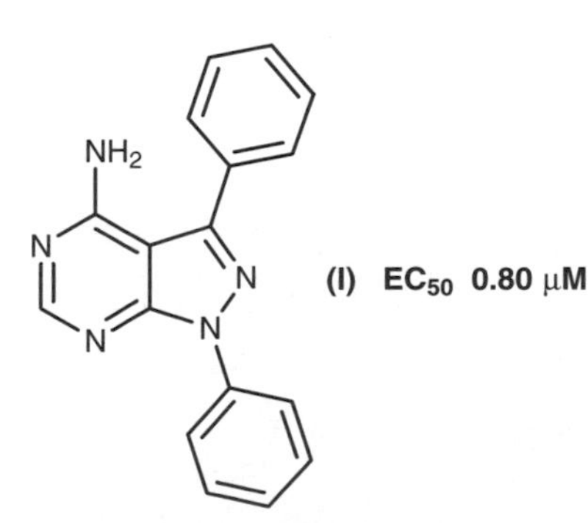

(I) EC_{50} **0.80 μM**

(II) EC_{50} **0.22 μM**

Fig. 1. Pyrazolopyrimidine lead compounds.

easy to be fooled. The **docking experiments** were carried out to see if the structures could form the important bidentate hydrogen bonding interaction as well as occupying the ribose binding site (*Fig. 2*). With structure I, this posed no problem. It was possible to dock the structure such that the pyrimidine ring interacted by hydrogen bonding in a similar fashion to ATP, allowing one of its aromatic substituents to occupy the ribose binding region. Moreover, it was found that the second aromatic substituent could occupy the empty pocket previously identified. The fact that the analog (III) (*Fig. 3*) lacked one of the aromatic rings and had less activity supported this theory.

When structure II was docked in the same way as I, it was found that the aromatic substituent could not fit the ribose pocket. Therefore, it was proposed that structure II might be interacting in the same way as the 4-(phenylamino)pyrrolopyrimidine (CGP 59326). By binding in this mode, it was possible for the aromatic ring to bind to the alternative pocket resulting in a stronger interaction overall.

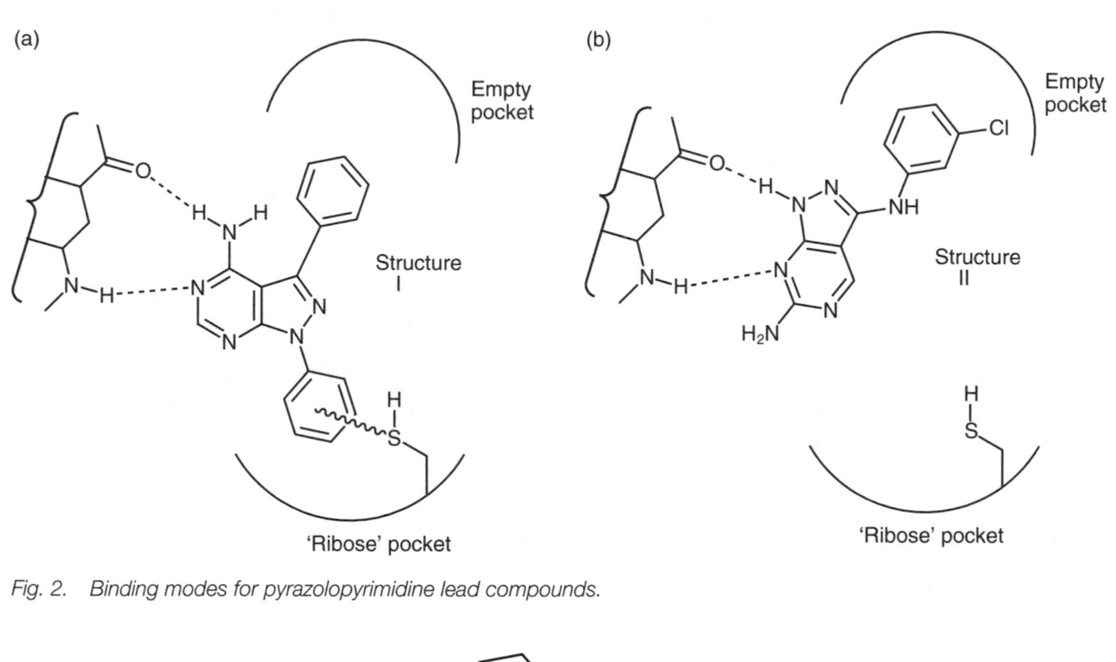

Fig. 2. Binding modes for pyrazolopyrimidine lead compounds.

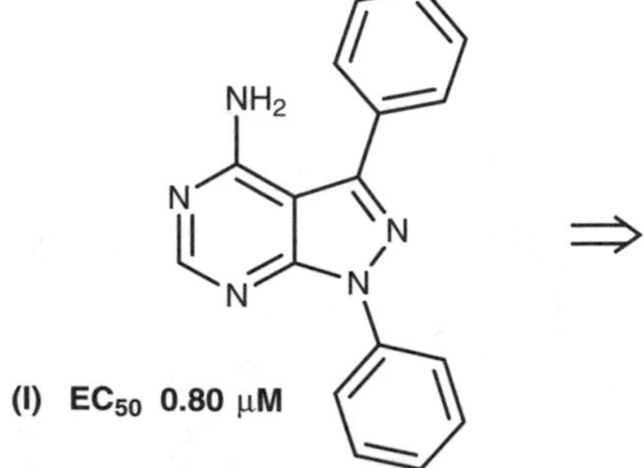

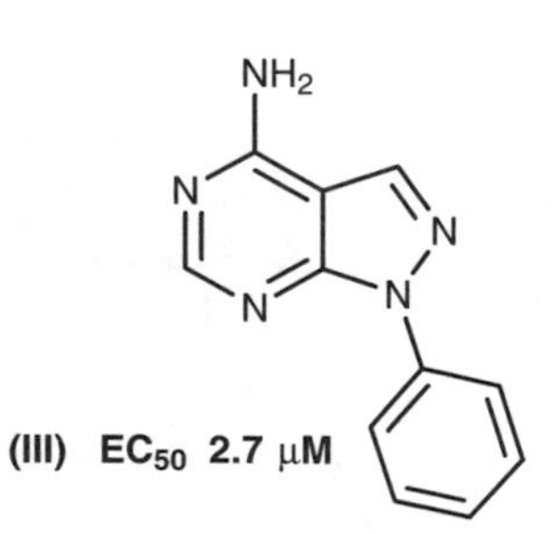

Fig. 3. Analog lacking an aromatic ring.

Drug design

If the second binding mode was the preferred interaction for structure II, then this meant that the primary amino group could not be involved in the bidentate hydrogen binding interaction. Therefore, it should be possible to remove this group without losing activity. Consequently, **simplification** of II led to analog IV, which was found to have improved activity, thus supporting the binding theory (*Fig. 4*). Having established this, it was now clear that the ribose pocket in the binding site must be empty when structure II or IV is bound. Therefore, it was reasoned that an **extension strategy** involving the addition of an aromatic substituent could position an aromatic ring into the ribose pocket. This led to structure V, which was found to have a ten-fold increase in activity over II.

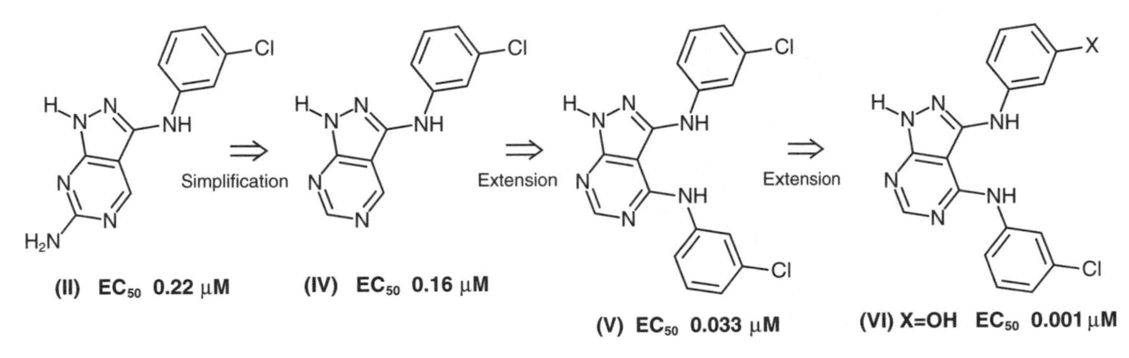

Fig. 4. Drug design of pyrazolopyrimidines.

A **SAR study** was now carried out, which concentrated on the aromatic group occupying the larger upper pocket. A range of different substituents was added to the ring and it was found that activity was improved by the introduction of an –OH or an NH_2 group at either the *meta* or *para* positions. Both these substituents are good hydrogen bonding groups, so the improved activity suggested that an extra hydrogen bonding interaction had been found with the binding site. This could be viewed as another example of the **extension strategy** (i.e. adding extra groups to find extra binding interactions).

Chain extensions and **contractions** were now tried out and it was found that the activity did not alter significantly (*Fig. 5*). This suggested that the pocket was

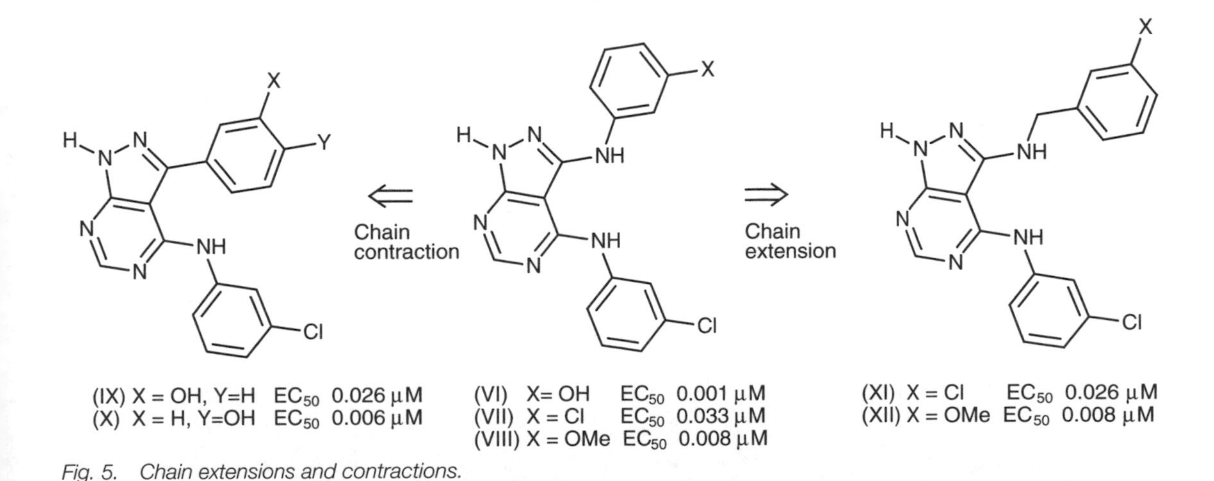

Fig. 5. Chain extensions and contractions.

large enough to accommodate a variety of different substituents of varying size and bulk. The results demonstrated that this pocket (unlike the ribose pocket) had a large capacity and that the activity only fell when extremely large substituents were added (e.g. structure XIII) (*Fig. 6*).

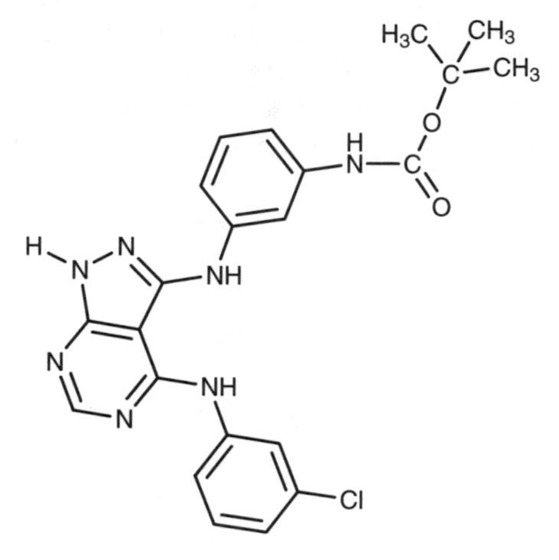

Fig. 6. Inactive compound (XIII).

M1 THE AGE OF HERBS, POTIONS AND MAGIC

Key Notes

Ancient remedies — Ancient civilizations around the world have used plant extracts as medicines. Many of these are merely placebos, but some ancient remedies are truly effective.

Ceremony and magic — In many cultures, the administration of remedies was accompanied by magic or religious ritual, which tended to emphasize the placebo effect.

Related topics — The lead compound (E1); Natural sources of lead compounds (E2); The nineteenth century (M2)

Ancient remedies

For centuries, various cultures around the world have used plants, bushes and trees as food sources, lethal poisons for hunting, and potions for the treatment of various human ailments. The discovery of useful plants relied on chance and good observation, a combination of events that is common in medicinal chemistry even to the present day.

A variety of **herbal recipes** were developed by the early herbalists, and different parts of each plant were recognized as being more useful than others. Thus, the leaves of one plant might be used for one remedy while the roots of another plant might be used for a different remedy. Invariably, the natural material would be crushed and boiled up in a 'witches brew' – a crude extraction process that would dissolve a concoction of different chemicals, one of which might actually be useful. In truth, many of these ancient remedies were no more than foul-tasting potions that had no real beneficial effect at all. The fact that many of them appeared to be effective was probably due to a **placebo effect**, where the patients improved in health through their belief in the potion.

Although many ancient potions have no proven pharmacological effect, there are several important ancient remedies that do, and these have been the basis for many of today's most important drugs. For example, **opium** from poppies is one of the world's most ancient remedies and was used as a sedative and analgesic both in Europe and Asia. In South America, the Incas chewed leaves from the **coca bush** when faced with physical hardship, since the leaves were known to increase stamina and decrease hunger. The bark from the **cinchona tree** has been used for centuries to treat fever and malaria, while ***Ipecacuanha*** is a root that was used to treat dysentery and fever.

Ceremony and magic

In many ancient civilizations, the preparation and administration of ancient remedies was undertaken by witch doctors, medicine men and shamans. These 'specialists' often kept their potions a closely guarded secret, only revealing them when it came time to pass on their knowledge to a successor. As a result, many of these medicine men were extremely powerful people and were viewed with respect and not a little fear. To emphasize their power, the medicine men would often include bizarre rituals, dances and ceremonies when administering their potions, thus introducing an element of magic or religion into the whole process. The importance of such rituals should not be ignored. The patient would be awestruck by the whole process, and the placebo effect would be strengthened. Who would dare to disobey the local witch doctor if he ordered you to get better?

M2 THE NINETEENTH CENTURY

Key Notes

The birth of chemistry

The nineteenth century saw the birth of chemistry as a science. Chemists developed the practical procedures required to separate and purify pure compounds from complex mixtures, allowing the isolation of active principles. Analytical and synthetic procedures were developed, allowing the structural identification of simple compounds and the ability to make synthetic analogs.

The active principle

The active principle is the chemical that is chiefly responsible for the biological effect observed in a plant or herbal extract, or other such mixture.

Structural analysis

Functional groups in a molecule were identified by characteristic reactions. The overall structure was determined by degradation to identifiable fragments, which were used to propose a structure. The structure was then synthesized to test whether it had the same properties as the unknown compound.

Semi-synthetic drugs

Semi-synthetic drugs are compounds that are synthesized from natural products.

Synthetic drugs

Early examples of synthetic drugs were gases and volatile liquids, which were used as general anesthetics.

The hypodermic syringe

The introduction of the hypodermic syringe allowed drugs to be administered directly into the blood supply and proved useful for drugs that were poorly absorbed from the digestive system.

The 'dark side'

Although the activity of active principles was greater than that of the original potion from which they were derived, their side effects were also enhanced. The dangers of drug addiction and tolerance began to be recognized by the end of the century.

Related topics

Natural sources of lead compounds (E2)
Synthetic considerations (F1)
The age of herbs and potions (M1)

The birth of chemistry

In the nineteenth century, **chemistry** came of age as a science, both in terms of experimental procedures and scientific theory. On the practical side, scientists developed techniques to isolate and purify single compounds from natural extracts using such techniques as crystallization, solvent–solvent extraction and distillation. Methods of analysis were developed which allowed the identification of molecular formulae, and even the molecular structures of simple compounds. With an understanding of molecular structure and functional groups, methods of organic synthesis were developed which allowed chemists to alter structures in a predictable way. These advances allowed chemists to study the traditional herbs and potions used in medicine from a scientific viewpoint. It was recognized that these ancient remedies were mixtures of different compounds, so chemists started to separate out the various components in order to discover whether a single compound was responsible for the medicinal activity – known as the active principle.

The active principle

The **active principle** is the compound that is chiefly responsible for the biological effect of a plant or herbal extract. For example, **morphine** is the active principle of opium, **cocaine** is the active principle from coca leaves and **quinine** is the active principle from cinchona bark (*Fig. 1*). Other active principles isolated in the nineteenth century included **nicotine**, **strychnine**, **caffeine**, **emetine**, **colchicine**, **codeine**, **atropine**, and **physostigmine**. Many of these active principles were found to be amines, and were therefore basic in character. As a result, they were called **alkaloids**. However, since structural determination was still in its infancy, the molecular structure of most of these compounds was unknown. Nevertheless, that did not stop them being tested as potential medicines, and it was soon discovered that the active principle had greater activity than the original potion. This is not too surprising since the active principle is diluted in the natural extract. Recognizing the commercial potential of active principles, 19th century scientists often set up small family businesses to market their product. From these humble origins, many of today's multinational pharmaceutical companies evolved.

Structural analysis

As active principles were isolated, chemists attempted to identify their **structures**. However, this was no easy task in the 19th century. It was possible to find out the **molecular weight** and the **molecular formula**, thus revealing the types of atoms present and their ratio. However, there was no easy method of finding out how these atoms were connected together. In those days, there was no X-ray crystallography or spectroscopy. The only way that a chemist could gain clues about a compound's structure was to carry out chemical reactions on it. Over time, it was recognized that certain arrangements of atoms (**functional groups**) could undergo characteristic reactions, and this allowed identification of functional groups in various active principles. For example, the presence of an alcohol or phenol could be tested by esterification, while the presence of an amine could be established if the compound dissolved in dilute hydrochloric acid. However, the only way to identify the full structure was to carry out reactions that would degrade the molecule into smaller identifiable molecules. It was then a case of working out how these various molecular components fitted together in the original molecule. A theoretical structure could then be proposed which would then be synthesized to see whether its properties matched those of the authentic compound. As an analogy, one could imagine a Scottish castle being dismantled stone by stone by a wealthy American millionaire, so that he can rebuild it in the USA. The only problem is that no record was kept of what the original castle looked like, so the architect employed to reconstruct the castle in the Arizona desert is faced with the challenge of visualizing what the castle must have looked like from the stones laid out on the ground.

Consequently, successes took many years. The structure of **cocaine** was determined by the end of the century but identifying the structure of **morphine** took over 100 years.

Semi-synthetic drugs

For the best part of the 19th century, medicines were derived from natural sources. However, as chemists carried out reactions on isolated active principles, a series of analogs were synthesized, some of which had useful pharmacological activity and which could also be used in medicine. Such analogs are defined as **semi-synthetic drugs** since they are synthesized from a naturally occurring chemical, rather than from simple starting materials. An early example of a semi-synthetic drug was **diacetylmorphine** (diamorphine or heroin) (*Fig. 1*), which was synthesized from morphine.

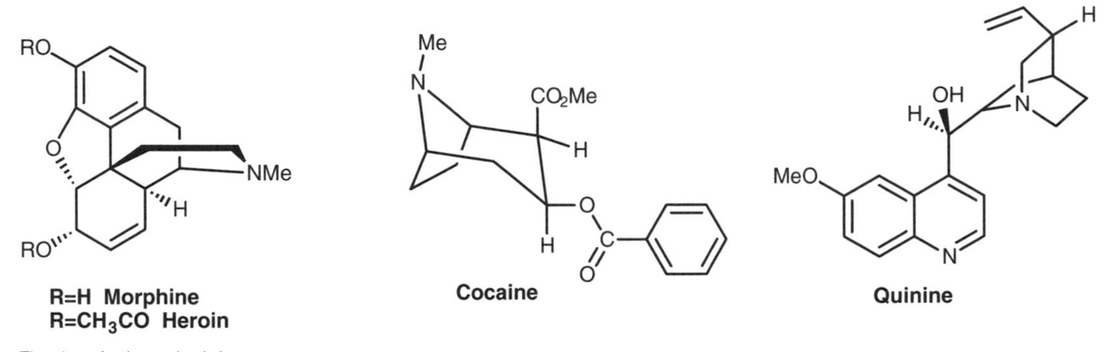

Fig. 1. Active principles.

Synthetic drugs

The 19th century saw the first **synthetic drugs** – the **general anesthetics**. These were simple chemicals such as diethyl ether and chloroform, which were used to induce unconsciousness. **Chloroform** was used on Queen Victoria during childbirth. However, these compounds were highly risky agents since many patients were put to sleep for good! Nowadays, concerns about the potential carcinogenicity of chloroform means that it is avoided even in the chemistry laboratory.

One common feature of general anesthetics is that they are insoluble in water and have a 'fatty' or hydrophobic character. This was the first recognition that the physical properties of a molecule can influence a particular pharmacological effect.

The hypodermic syringe

The **hypodermic syringe** was introduced in the mid-19th century and was used to inject drugs directly into the blood supply. This had important consequences for medicines such as **morphine**, which had been neglected up until then, since it only had weak oral activity. In contrast, morphine proved a powerful analgesic when it was administered by injection.

The 'dark side'

The advances of the 19th century did not come without a price. Although active principles were more active than the potions from which they were derived, they also demonstrated more serious **side effects**, some of which were not revealed until the drug had been taken for several days or weeks. No clinical trials were carried out in the 19th century, and no tests were carried out on potential long-term hazards, such as drug addiction or tolerance, partly because the existence of such hazards was not fully appreciated. Scientists often tested potential new drugs on themselves – for example, Sigmund Freud tested the effects of **cocaine** on himself. If the effects of a drug were found to be beneficial in the short term, then the compound was quickly publicized and introduced into medical practice. **Heroin** and cocaine are examples of drugs that were quickly introduced into medicine and whose long-term hazards were only recognized through bitter experience. Indeed, heroin was introduced onto the market at the end of the 19th century as the 'heroic' drug, which would banish pain for good. Five years later, the drug had to be withdrawn once its serious addictive effects were recognized.

M3 A FLEDGLING SCIENCE (1900–1930)

Key Notes

Chemotherapy	The principle of chemotherapy was established, whereby selective poisons were designed to act versus microorganisms, but not host cells (the magic bullet theory). The principle of selective action has been a mainstay in medicinal chemistry ever since. The dose in which a drug is taken determines whether it acts as a poison or as a medicine.
Synthetic drugs	The barbiturates and local anesthetics were developed at the start of the 20th century.
Endogenous compounds	Several of the body's own natural chemicals were isolated and identified. The role of neurotransmitters, hormones, and vitamins was investigated, leading to an increased understanding of the body's chemistry.
Regulation	Laws were first introduced to restrict the use of addictive and narcotic drugs. In America, regulations requiring the correct labeling of drugs were introduced.

Related topics Drug dosing (C7); Natural sources of lead compounds (E2); Regulatory affairs (K5)

Chemotherapy

By the beginning of the 20th century, it was recognized that many diseases were caused by microorganisms. Unfortunately, there were no effective drugs to treat these diseases. However, a scientist called Paul Ehrlich proposed the principle of **chemotherapy**, which stated that poisons might be developed that were selectively toxic to microorganisms rather than humans, and would thus prove useful medicines. This was the first application of the '**magic bullet**' principle whereby drugs are targeted against specific cells.

Ehrlich investigated arsenic-based structures that would be less toxic than arsenic itself, and eventually developed **salvarsan** (*Fig. 1*), which was introduced in 1907. The agent proved effective against protozoal infections, such as syphilis, and earned Ehrlich the Nobel prize for medicine in 1908.

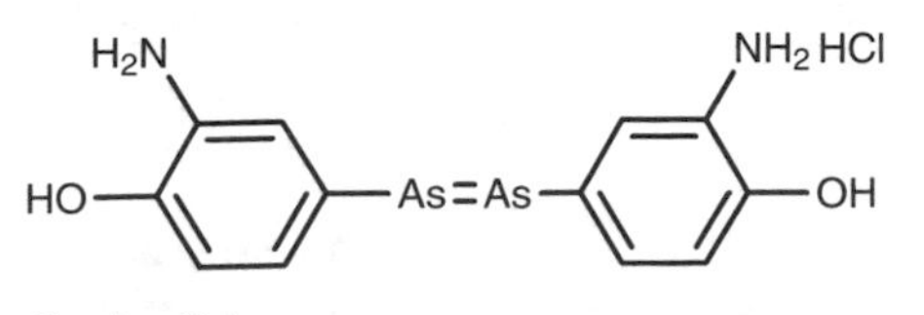

Fig. 1. Salvarsan.

The principle of selective toxicity, or selective action, has been the basis for drug design ever since. Ehrlich's research also demonstrated that there is often little difference between poisons and medicines, and that it is the dose administered that determines whether a drug acts as one or the other.

Synthetic drugs

At the beginning of the 20th century, two important groups of synthetic drugs were discovered – **barbiturates** and **local anesthetics**. Barbiturates [e.g. **phenobarbitone** and **pentobarbitone** (*Fig. 2*)] were used as hypnotics or sedatives, and remained unchallenged in this role until the 1950s. Local anesthetics were developed as simplified analogs of cocaine, following the observation that cocaine itself had local anesthetic activity. **Procaine** (*Fig. 2*) was the most important of these and became the local anesthetic of choice in dentistry for almost 50 years. The success of the local anesthetics also established one of the important strategies used in medicinal chemistry – the **simplification** of complex natural products.

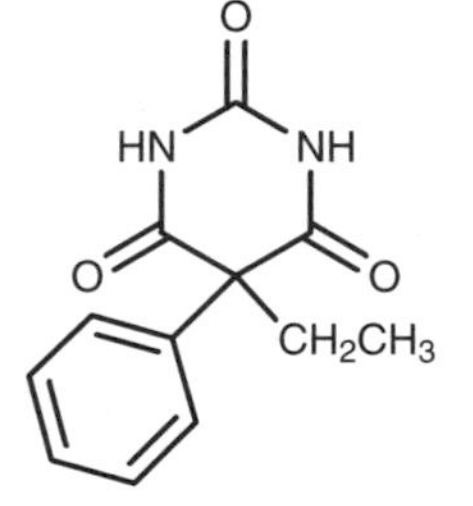

Phenobarbitone

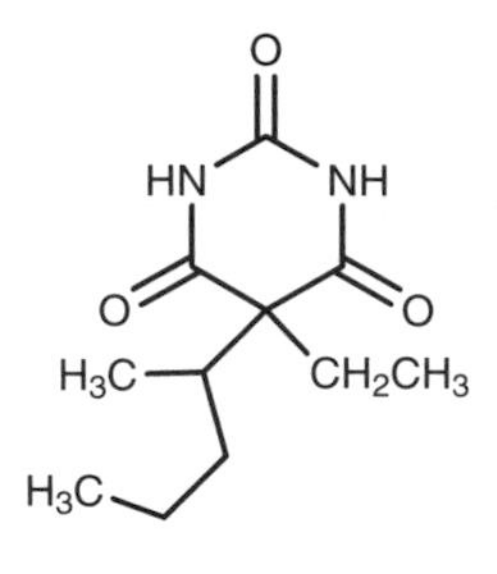

Pentobarbitone

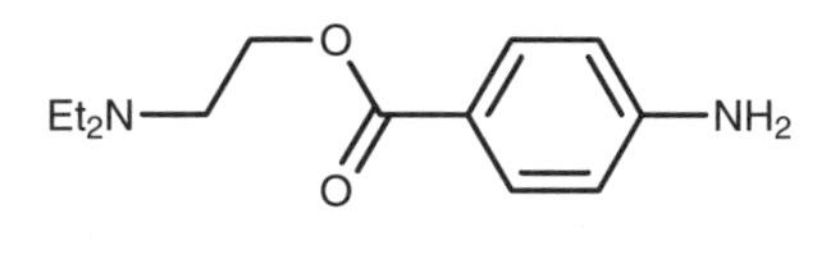

Procaine

Fig. 2. Synthetic drugs.

Endogenous compounds

Endogenous compounds are defined as the body's own natural chemicals. During the early part of the 20th century, a number of these agents were isolated, purified and tested for their pharmacological activity (*Fig. 3*). This provided

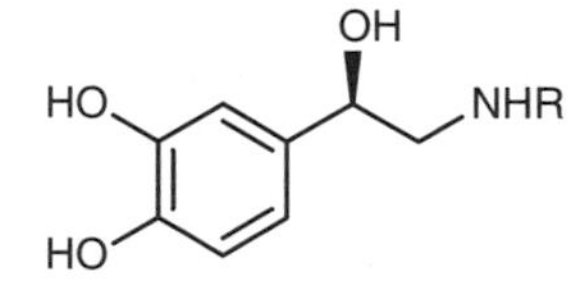

R= Me, Epinephrine (adrenaline)
= H, Norepinephrine (noradrenaline)

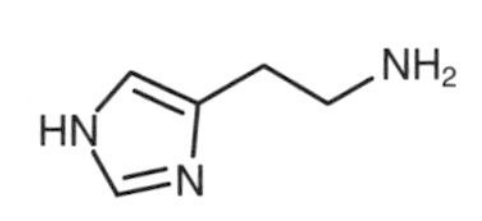

Histamine

Thyroxine

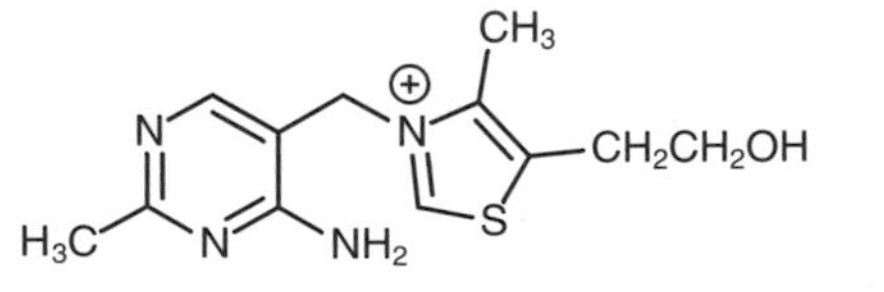

Thiamin

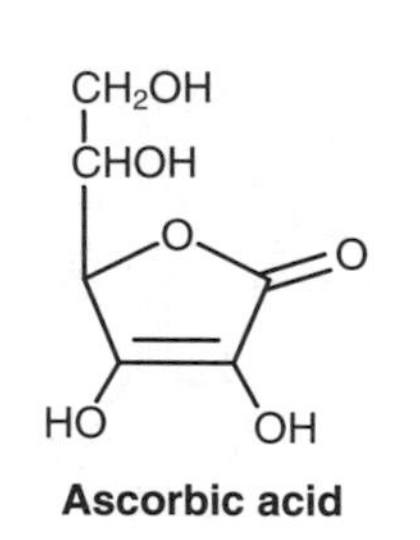

Ascorbic acid

Fig. 3. Endogenous compounds.

more information about the chemistry of the body, revealing the importance of neurotransmitters and hormones (e.g. **epinephrine**, **norepinephrine**, **histamine**, **thyroxine**, and **insulin**), and vitamins [e.g. **thiamin** (Vitamin B_1) and **ascorbic acid** (Vitamin C)]. Endogenous compounds are present in such minute quantities in the body that it was necessary to devise synthetic routes to these compounds in order to study them properly. Once synthetic routes were devised, it was then possible to synthesize analogs that could also be tested for pharmacological activity.

Regulation

At the end of the 19th century, drugs could be bought without prescription. Indeed, heroin could be purchased in grocery shops. However, the serious addiction that resulted from taking drugs such as cocaine, morphine and heroin soon demonstrated that **regulations** were necessary.

In the early years of the twentieth century, rules and regulations were introduced restricting the availability and use of addictive drugs. In America, the **Pure Food and Drugs Act** of 1906 required the correct labeling of food and drugs. However, there were still no regulations regarding how drugs should be tested for safety.

M4 THE DAWN OF THE ANTIBACTERIAL AGE (1930–1945)

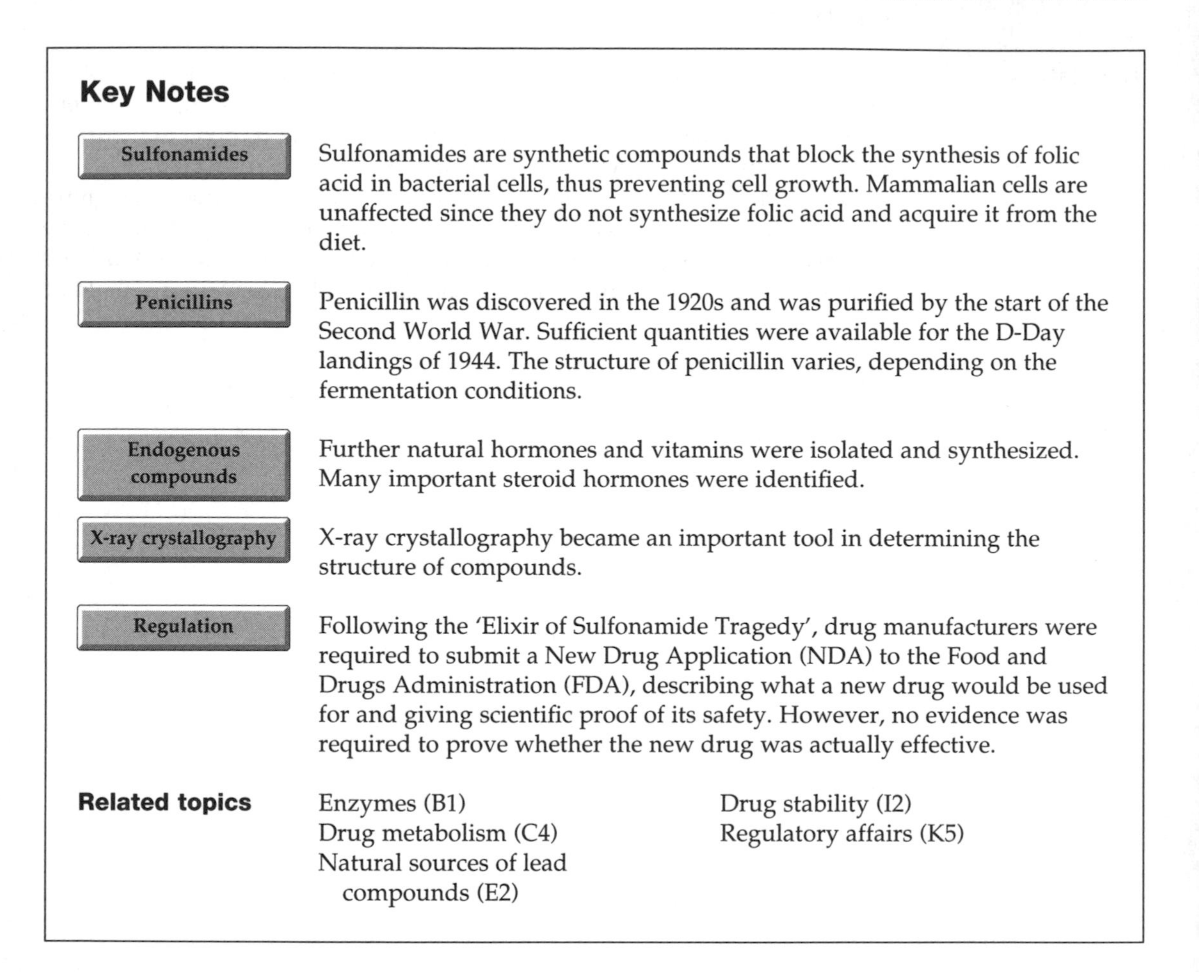

Key Notes

Sulfonamides	Sulfonamides are synthetic compounds that block the synthesis of folic acid in bacterial cells, thus preventing cell growth. Mammalian cells are unaffected since they do not synthesize folic acid and acquire it from the diet.
Penicillins	Penicillin was discovered in the 1920s and was purified by the start of the Second World War. Sufficient quantities were available for the D-Day landings of 1944. The structure of penicillin varies, depending on the fermentation conditions.
Endogenous compounds	Further natural hormones and vitamins were isolated and synthesized. Many important steroid hormones were identified.
X-ray crystallography	X-ray crystallography became an important tool in determining the structure of compounds.
Regulation	Following the 'Elixir of Sulfonamide Tragedy', drug manufacturers were required to submit a New Drug Application (NDA) to the Food and Drugs Administration (FDA), describing what a new drug would be used for and giving scientific proof of its safety. However, no evidence was required to prove whether the new drug was actually effective.

Related topics

Enzymes (B1)
Drug metabolism (C4)
Natural sources of lead compounds (E2)
Drug stability (I2)
Regulatory affairs (K5)

Sulfonamides

Despite the introduction of **salvarsan**, medicine had no effective agents for treating the many bacterial infections that plagued society. As a result, life was often a lottery. Even a simple wound could be a serious matter, often requiring amputation to prevent the spread of gangrene. Death from infection, especially among children and the aged, was common. Diseases such as pneumonia and tuberculosis were widespread and incurable, resulting in the establishment of special hospitals in the country to isolate patients from mainstream society. Surgery was risky and limited to operations that could be carried out speedily to minimize the risk of infection.

The first effective group of antibacterial drugs to be discovered was the sulfonamides. The **sulfonamides** are synthetic compounds, which were developed in the 1930s following the chance discovery that a synthetic dye called

prontosil had antibacterial activity. However, it was quickly discovered that prontosil itself had no antibacterial activity *in vitro*, but was converted in the body to a compound called **sulfanilamide**, which proved to be the active compound (*Fig. 1*).

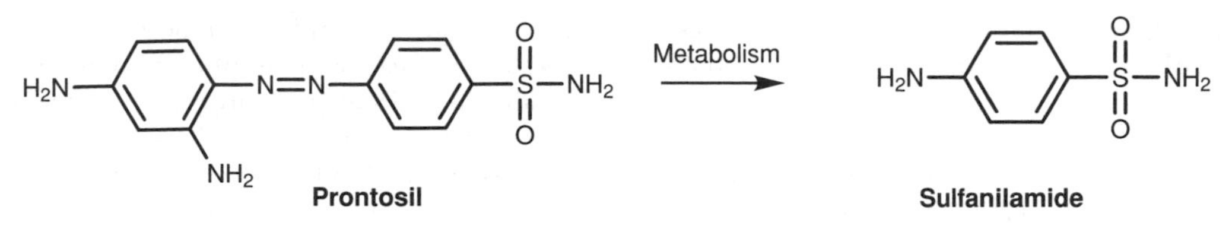

Fig. 1. Metabolism of prontosil to sulfanilamide.

A large range of sulfonamide analogs was synthesized, one of the most effective being **sulfapyridine** (M&B 693) (*Fig. 2*). This was the first effective drug to be used against pneumonia and was used to treat Winston Churchill when he fell seriously ill in North Africa during the Second World War.

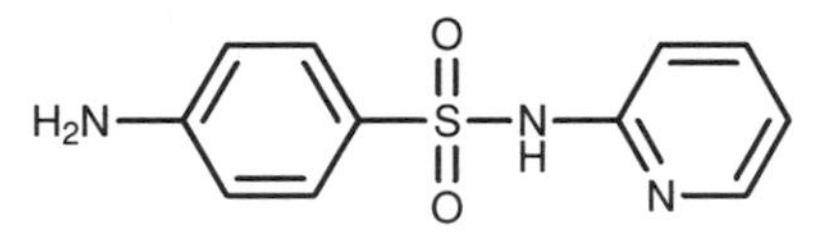

Fig. 2. Sulfapyridine.

Sulfonamides block the synthesis of an essential vitamin called **folic acid** in bacterial cells, thus inhibiting all the biosynthetic pathways requiring this compound. As a result, the bacterial cells stop growing and dividing, allowing the body's own defenses to wipe out the invader. The sulfonamides are called **bacteriostatic** agents since they inhibit cell growth but do not actively kill the cell. They are not toxic to human cells, as human cells acquire folic acid from the diet and do not synthesize it.

Penicillins

Despite the success of the sulfonamides, the real revolution in antibacterial therapy began with the discovery of **penicillin** (a fungal metabolite) in the 1920s. Penicillin was purified by the start of the Second World War, and a method of producing the compound in bulk was developed so that sufficient drug was available to treat all the casualties from the D-Day landings. It was soon discovered that the structure of penicillin varied depending on the conditions used to grow it. The early penicillins were given letters to distinguish them (e.g. **Penicillin G** and **Penicillin V**) (*Fig. 3*). The penicillins work by preventing

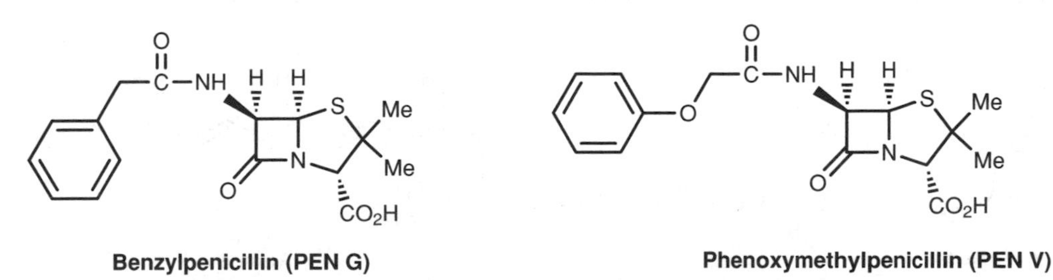

Fig. 3. Penicillins G and V.

the synthesis of bacterial cell walls. Water enters the cell by osmosis and the cell swells until it eventually bursts (lysis). Since penicillins actively kill bacterial cells, they are called **bactericidal** agents. They are nontoxic to human cells since the latter lack a cell wall.

Endogenous compounds

The 1930s saw great strides in the isolation, purification and synthesis of many important steroid hormones such as **estradiol, testosterone, progesterone** and **corticosterone**, as well as further vitamins [e.g. **riboflavin, nicotinamide, pyridoxine, biotin, folic acid**, and **retinal** (Vitamin A)] (*Fig. 4*). The challenges faced in devising syntheses for these compounds resulted in the development of new chemical reactions and an increased understanding of reactivity and stereochemistry in organic synthesis.

X-ray crystallography

During this period, **X-ray crystallography** was developed as a powerful tool for the identification of chemical structures. For example, the structures of **ergosterol** and **penicillin** were established during this period.

Regulation

Until the Second World War, drugs did not need to be tested for safety or effectiveness before they reached the market. The only legal aspect was that they

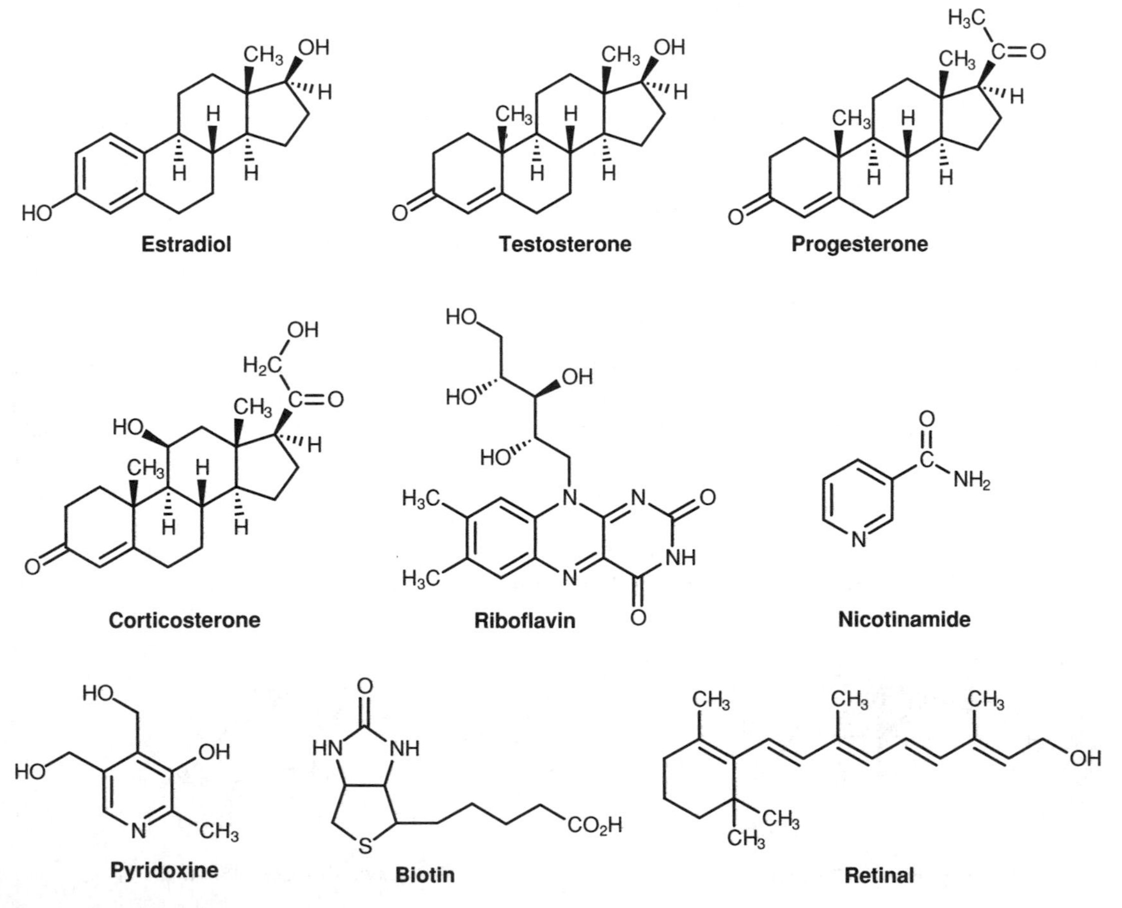

Fig. 4. Steroid hormones and vitamins.

should be correctly labeled. However, this situation changed following the **'Elixir of Sulfonamide Tragedy'** of 1937. An American manufacturer had reasoned that a solution of a sulfonamide might be useful for treating sore throats. Unfortunately, he used ethylene glycol (a highly toxic solvent used in antifreeze) to dissolve the drug. The elixir was put on the market without being tested and ended up killing 107 children before it was withdrawn. A public outcry led to the **Food, Drug and Cosmetic Act** of 1938, which required drugs to be tested for safety before they reached the market. From then on, drug manufacturers had to submit a **New Drug Application** (NDA) to a body known as the **Food and Drugs Administration** (FDA) before any new drug could be marketed. This application had to detail the drug's uses and provide scientific proof of its safety. However, there was no requirement to prove whether the drug was actually effective or not. Therefore, clinical trials were not required.

M5 THE ANTIBIOTIC AGE (1945–1970S)

Key Notes

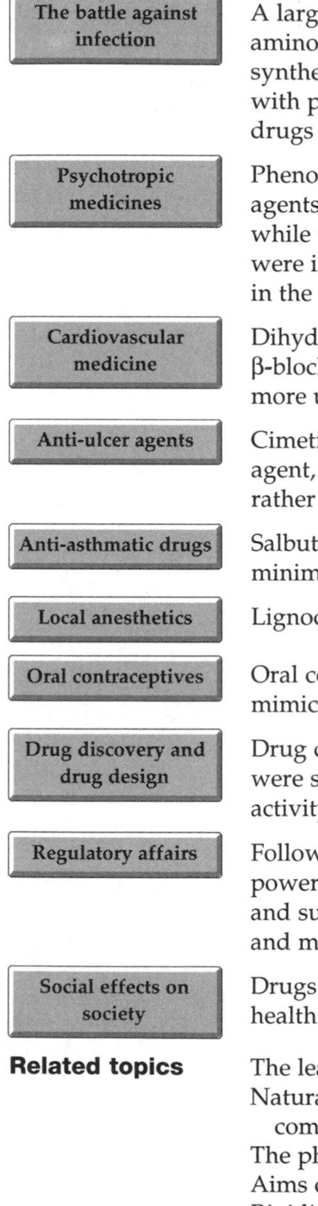

The battle against infection — A large range of antibiotics was discovered including tetracyclines, aminoglycosides, cephalosporins and macrolides. Important semi-synthetic penicillins were prepared and clavulanic acid was administered with penicillins to protect them from enzymatic breakdown. Synthetic drugs such as the quinolones and fluoroquinolones were discovered.

Psychotropic medicines — Phenothiazines and butyrophenones were introduced as antipsychotic agents. The benzodiazepines proved effective treatments for anxiety, while the monoamine oxidase inhibitors and tricyclic antidepressants were introduced as antidepressants. Lithium carbonate proved effective in the treatment of manic depression.

Cardiovascular medicine — Dihydropyridines were introduced for the treatment of angina. The β-blockers were originally designed to treat angina, but were found more useful as antihypertensives.

Anti-ulcer agents — Cimetidine was introduced in the 1970s as the first effective anti-ulcer agent, and for the first time, serious ulcers could be treated by drugs rather than by surgery.

Anti-asthmatic drugs — Salbutamol (Ventolin) was the first effective anti-asthmatic drug with minimal cardiovascular side effects.

Local anesthetics — Lignocaine replaced procaine as the local anesthetic of choice in dentistry.

Oral contraceptives — Oral contraceptives were introduced. These were steroid structures that mimicked the steroid hormone progesterone and prevented ovulation.

Drug discovery and drug design — Drug design was based on the structure of a lead compound. Analogs were synthesized with the aim of finding an analog with improved activity and/or selectivity. Several strategies were proven to be effective.

Regulatory affairs — Following the thalidomide disaster, regulatory bodies were given the power to approve whether a drug could be put forward for clinical trials and subsequently to market. More stringent safety tests were required and manufacturers had to prove that the drug was effective.

Social effects on society — Drugs have had important social consequences, other than increased health and life expectancy.

Related topics

The lead compound (E1)
Natural sources of lead compounds (E2)
The pharmacophore (G4)
Aims of drug design (H1)
Rigidification (H2)
Simplification of complex molecules (H3)
Conformational restraint (H4)
Drug stability (I2)
Regulatory affairs (K5)
The dawn of the antibacterial age (1930–1945) (M4)

The battle against infection

The discovery that a fungus could produce penicillin initiated a massive worldwide search in the 1940s for new fungal strains to see whether they contained novel antibiotics. This huge effort led to the discovery of the major antibiotics used in medicine today, including the **tetracyclines** (e.g. **chlortetracyclin**), **aminoglycosides** (e.g. **streptomycin**), **macrolides** (e.g. **erythromycin**) and the **cephalosporins** (*Fig. 1*).

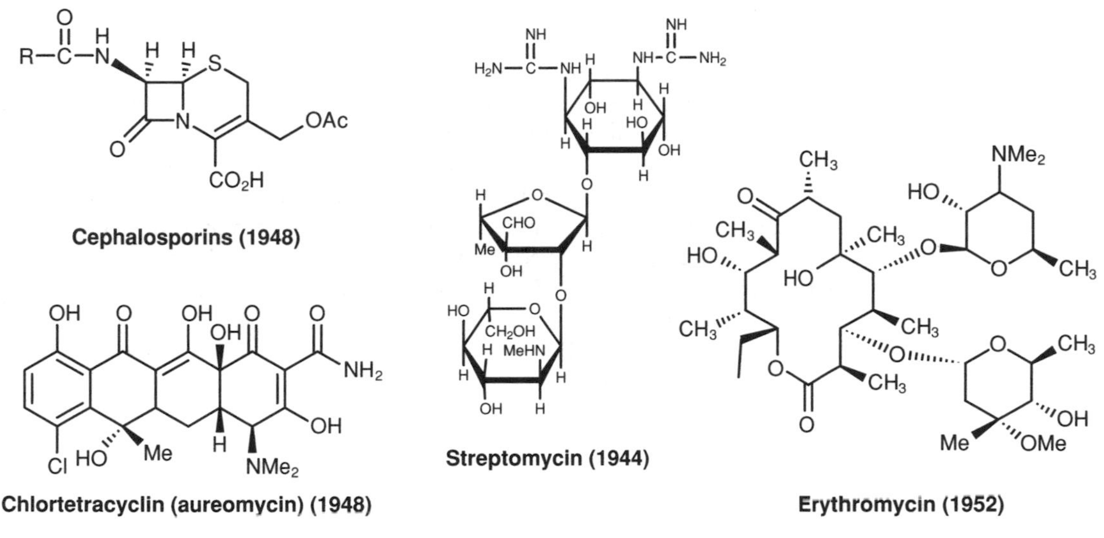

Fig. 1. Antibiotics.

Advances were also made in developing synthetic and semi-synthetic drugs, particularly in the area of **semi-synthetic penicillins**. The original penicillins (e.g. Penicillin G) had several drawbacks in that they had to be injected and had a limited range of activity. Therefore, efforts were made to synthesize analogs that would have better properties. These efforts were enhanced by Beecham's isolation in 1960 of a biosynthetic intermediate called **6-aminopenicillanic acid** (6-APA). This allowed a wide range of semi-synthetic penicillins to be prepared by acylating the amino group of 6-APA with various acid chlorides (*Fig. 2*). As a result, important semi-synthetic penicillins such as **ampicillin** and **amoxycillin** were discovered (*Fig. 3*).

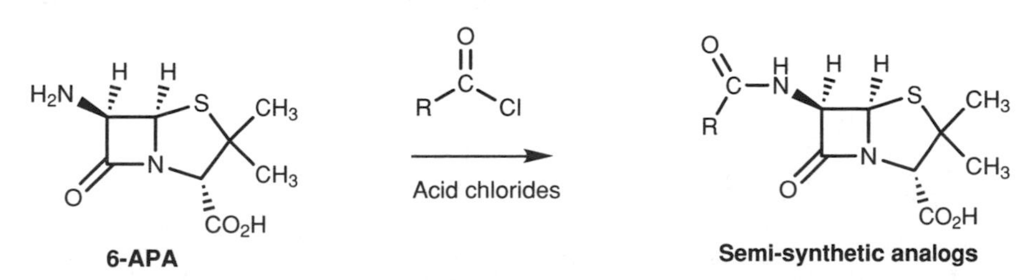

Fig. 2. Preparation of semi-synthetic penicillins.

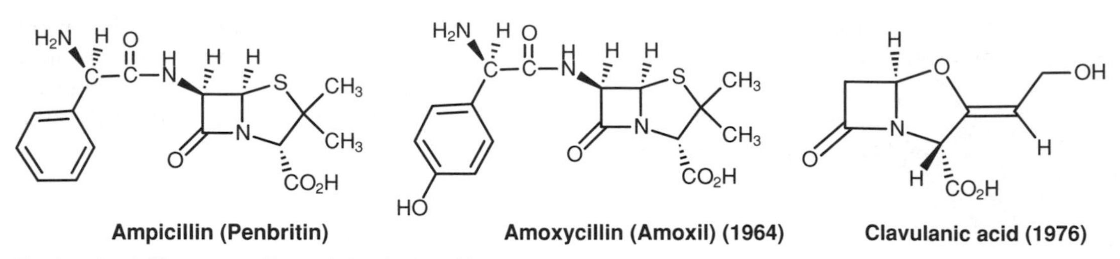

Fig. 3. Ampicillin, amoxycillin and clavulanic acid.

A further advance was made by Beecham's in 1976 with the discovery of **clavulanic acid** (*Fig. 3*), a fungal metabolite capable of inhibiting β-lactamases. **β-Lactamases** are bacterial enzymes that catalyze the hydrolysis of the β-lactam ring and thus deactivate penicillin (*Fig. 4*). As a result, bacteria having this enzyme are resistant to penicillins. Since clavulanic acid inhibits β-lactamases, it is administered along with a penicillin so that the latter is protected from attack and can be used against bacterial infections that were previously resistant to it. The combination of amoxycillin and clavulanic acid is known as **Augmentin**.

Purely synthetic agents were also discovered, such as the **quinolones** (e.g. **nalidixic acid**) (*Fig. 5*). These were the forerunners to the **fluoroquinolone** antibacterial agents used today.

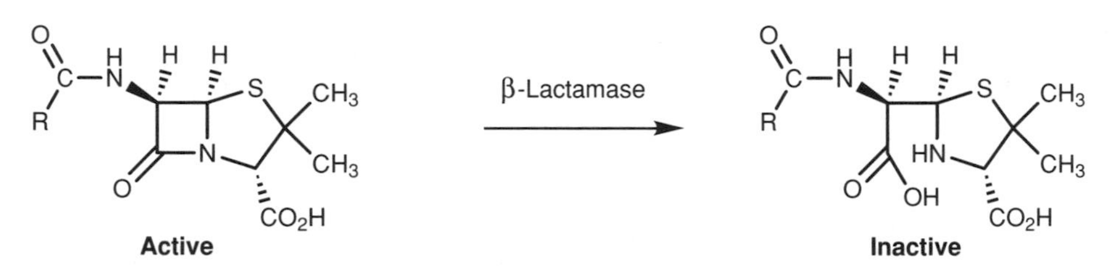

Fig. 4. Hydrolysis of penicillin by β-lactamase.

The discovery of a large arsenal of antibacterial agents has been one of medicine's great success stories and has removed the dread of infection, which was so common in previous generations. The antibiotic age has also been instrumental in allowing the many surgical advances that have taken place since the war. Complicated surgical procedures such as heart transplants would have been unthinkable before the availability of antibacterial agents.

Psychotropic medicines

The 1950s and 1960s saw great advances in medicines that act on the central nervous system, allowing for the first time treatment of schizophrenia, anxiety and depression. The first important antipsychotic agent to be discovered was

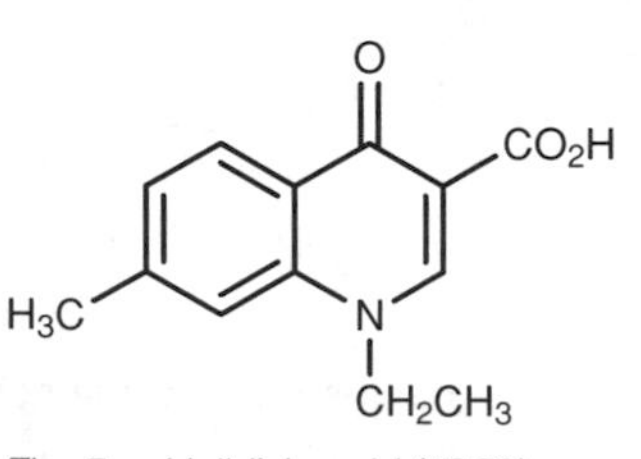

Fig. 5. Nalidixic acid (1962).

chlorpromazine, which revolutionized psychiatry and spawned a series of analogs known as the **phenothiazines**. In 1958, a more potent antipsychotic agent with fewer side effects was discovered. This was called **haloperidol** (Haldol) and belongs to a group of compounds called the **butyrophenones** (*Fig. 6*).

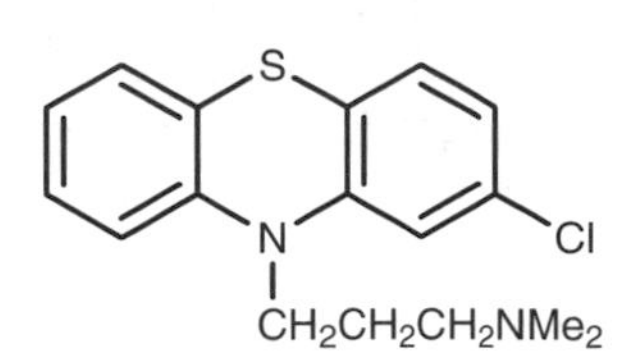

Chlorpromazine (1952)

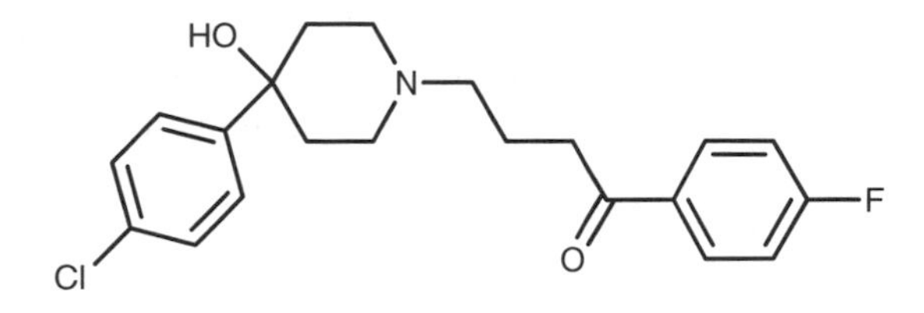

Haloperidol (1958)

Fig. 6. Antipsychotic agents.

Anti-anxiety (anxiolytic) agents were also discovered, first with **meprobamate** and then with the **benzodiazepines** (e.g. **Librium** and **Valium**) (*Fig. 7*).

Two important groups of antidepressant compounds were discovered during this period (*Fig. 8*). **Tricyclic antidepressants** (e.g. **imipramine**) were discovered when the central ring of the tricyclic phenothiazine structures was expanded. The other major group of antidepressant compounds to be discovered was the **monoamine oxidase inhibitors** (MOAI) (e.g. **iproniazid** and **phenelzine**), so called because they inhibit an enzyme called monoamine oxidase. This enzyme is involved in the metabolism of important neurotransmitters such as norepinephrine and serotonin. If the enzyme is inhibited, these important stimulatory neurotransmitters are only slowly metabolized and can activate their receptors for longer periods of time, thus countering the effects of depression. The use of

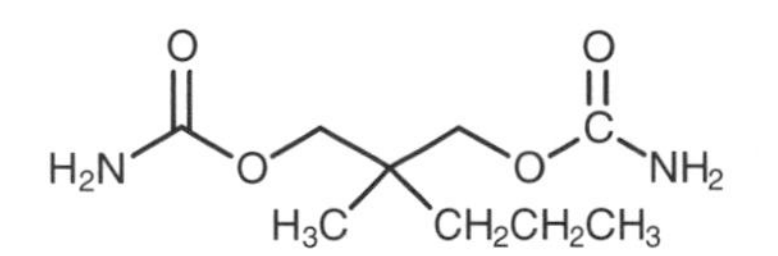

Meprobamate (1950)

Chlordiazepoxide (Librium) (1960)

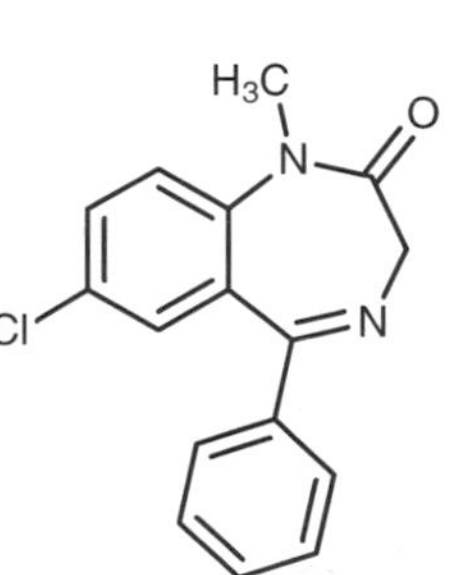

Diazepam (Valium) (1963)

Fig. 7. Anxiolytic agents.

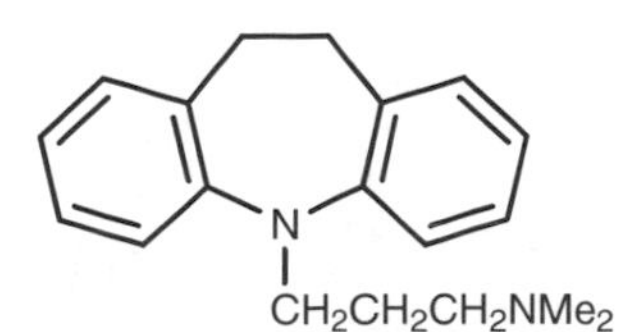

Imipramine (1956)

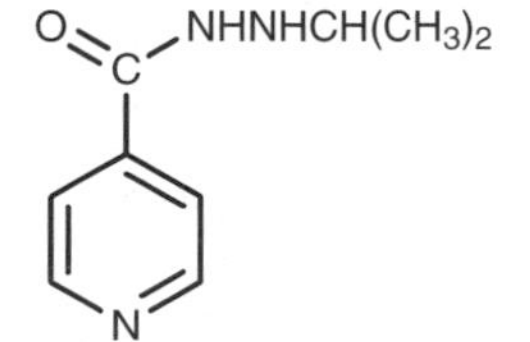

Iproniazid

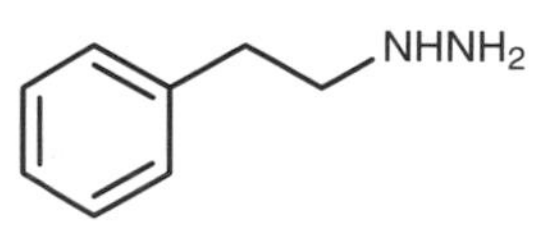

Phenelzine (Nardil)

Fig. 8. Antidepressants.

lithium ions (administered as **lithium carbonate**) was also found to be useful in treating manic depressive illness.

Cardiovascular medicine

In the 1960s and 1970s several important groups of cardiovascular drugs reached the market (*Fig. 9*). The **dihydropyridines** (e.g. **nifedipine** (Adalat)) were introduced for the treatment of angina. They act by blocking the movement of calcium through ion channels. Calcium ions are required for the release of neurotransmitters such as norepinephrine, so inhibiting calcium ion flow reduces the levels of norepinephrine in the heart and leads to relaxation of cardiac muscle.

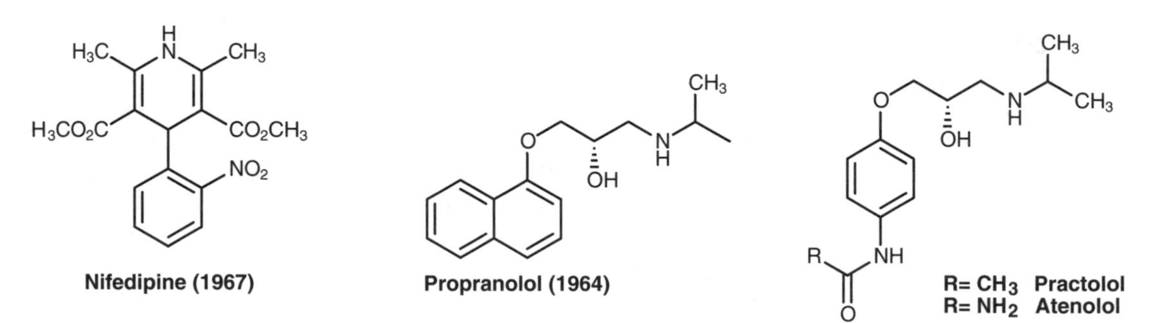

Fig. 9. Cardiovascular agents.

The first of the **β-blockers** (i.e. drugs that act as antagonists at β-adrenergic receptors of the heart) were developed in the 1960s. These compounds bind to the receptors for norepinephrine and prevent them being activated, leading to the relaxation of heart muscle. **Propranolol** was introduced in 1964 followed by **practolol** in 1970. The latter had to be withdrawn due to adverse side effects following long-term use, but other β-blockers such as **atenolol** took its place. The β-blockers were introduced originally to treat angina, but were found to be effective antihypertensive drugs (i.e. they lowered blood pressure).

Anti-ulcer agents

Until the 1970s, the treatment of ulcers was not particularly effective, consisting mainly of attempts to neutralize stomach acids with large doses of sodium bicarbonate. Serious ulcers could be life threatening and in severe cases, the only treatment was the surgical removal of part of the stomach. The first truly effective anti-ulcer agent to be introduced to the market was **cimetidine** (Tagamet) (*Fig. 10*). This agent was designed to prevent histamine interacting with histamine receptors present in the stomach wall, thus preventing the release of gastric acid into the stomach. Lowered levels of gastric acid resulted

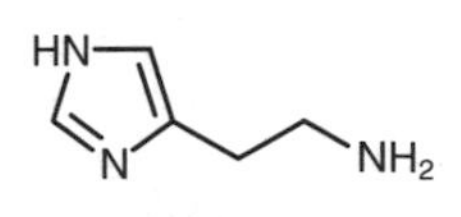

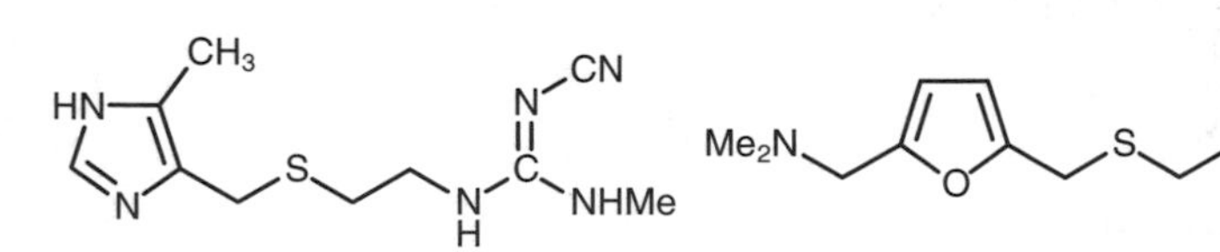

Ranitidine (Zantac) (1981)

Fig. 10. Histamine and anti-ulcer agents.

in less irritation of the stomach wall and quicker healing of existing ulcers. Such was the importance of this drug, that it became the largest selling prescription drug in the world until **ranitidine** (Zantac) overtook it in 1988.

Anti-asthmatic agents

Following 1945, research was carried out to find effective anti-asthmatic agents (*Fig. 11*). This research concentrated on synthesizing analogs of epinephrine, which could be used to switch on the adrenergic receptors in the airways and thus relieve the spasms caused by asthma. **Epinephrine** itself has been used to treat severe asthma, but epinephrine also has effects on the heart. The first effective drug to be discovered was **isoprenaline**, but this too has side effects on the heart, and so efforts were made to find an agent that would be selective for the lungs and not the heart. This led to the discovery of **salbutamol** (Ventolin).

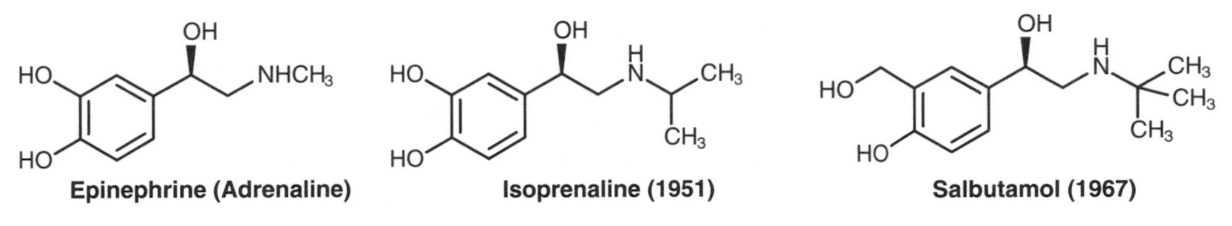

Fig. 11. Anti-asthmatic agents.

Local anesthetics

The discovery of **lignocaine** (xylocaine) (*Fig. 12*) saw it superseding procaine as the local anesthetic of choice in dentistry.

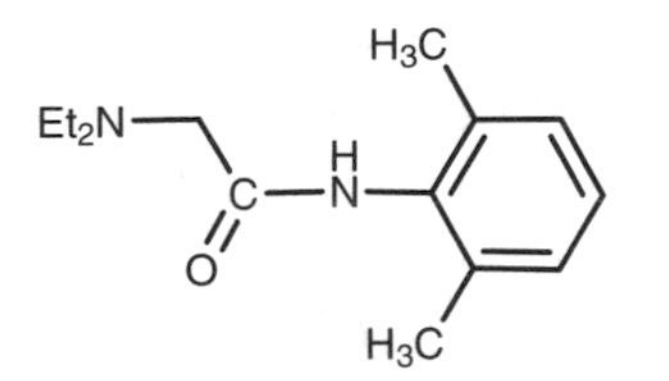

Fig. 12. Lignocaine (Xylocaine) (1948).

Oral contraceptives

Oral contraceptives were discovered in the 1960s. These compounds are steroid structures that mimic the natural hormone progesterone and prevent the release of ova during pregnancy. Progesterone itself is unsuitable for oral administration since it is rapidly metabolized in the liver. However, introducing an acetylenic group led to metabolically stable oral contraceptives such as **norethynodrel** (Enavid) and **ethynodiol diacetate** (*Fig. 13*).

Drug discovery and drug design

The 30-year period after the Second World War was one of rapid expansion in medicinal chemistry, with pharmaceutical companies synthesizing thousands of compounds in a bid to find novel drugs. As a result, many effective drugs were discovered and it has been estimated that over 90% of all the drugs used in 1964 were unknown in 1938. However, for every successful drug to reach the market, there were many thousands that did not. Drug design was still in its infancy, and medicinal chemistry was more of an art than a science. Moreover, the mechanisms by which drugs acted at the molecular level were poorly understood,

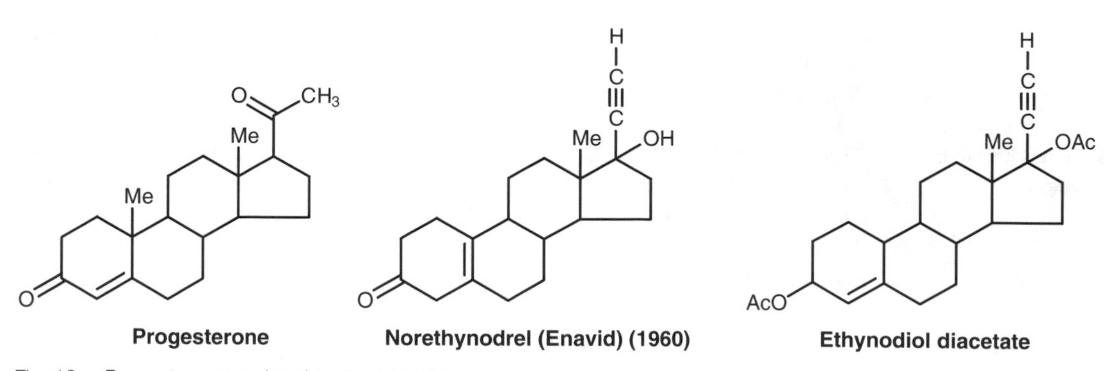

Fig. 13. Progesterone and oral contraceptives.

principally because little was known about the molecular mechanisms of the body itself. Medicinal chemistry relied on the discovery of a **lead compound** – an agent that had some sort of useful activity. A large variety of different analogs were then synthesized until an analog was discovered that had some advantage over the original compound. In this way, it was often found that certain portions of the lead compound were essential for activity, while others were not. This led to the concept of the **pharmacophore** – the crucial atoms required for activity. This in turn gave some indication about the structure of binding sites in target proteins, but, in the main, these targets were treated as 'black boxes'. Nevertheless, the studies carried out established many useful strategies in drug design. For example, complex molecules from natural sources could be significantly simplified in structure, yet still retain useful activity. At the other extreme, the activity and selectivity of very simple molecules could often be improved by making them more rigid.

To conclude, the emphasis of drug design during this period focused on the structure of the lead compound – **structure-orientated drug design.**

Regulatory affairs

In the late 1950s, a new drug called **thalidomide** was introduced to the European market as a safe sedative/hypnotic (*Fig. 14*). Thalidomide was widely prescribed and many pregnant women took the drug to treat morning sickness. However, by 1962, it became clear that these women were giving birth to children with severe deformities, particularly stunted limbs. The disaster occurred because thalidomide was a **teratogen** and affected growing fetuses in the womb. Although drugs were tested for safety, it was not appreciated that drugs might have such an effect and no teratogenic tests had been carried out. Since that time, regulatory bodies have tightened up the process by which drugs are approved and have required increasing numbers of toxicology and safety tests to be carried out. Furthermore, regulations were introduced requiring the

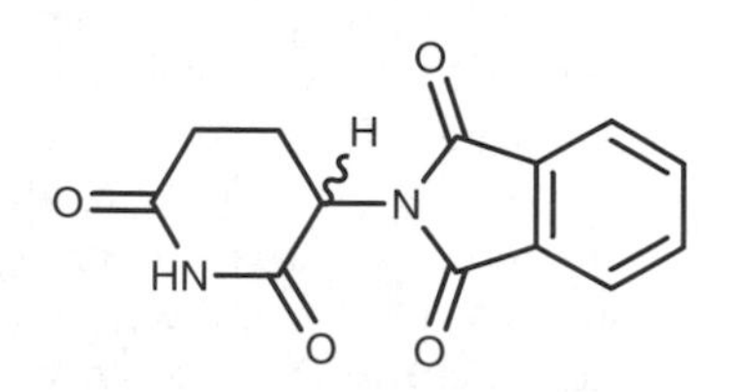

Fig. 14. Thalidomide.

manufacturer to show scientifically that the drug has the clinical effect claimed. The importance of carrying out extensive **clinical trials** was emphasized for the first time, and regulatory authorities were given the responsibility to approve new drugs rather then to review them. As a consequence, the cost and time required to bring a drug to market increased dramatically.

Social effects on society

The various medicines discovered since the Second World War have had a dramatic **social effect** beyond health alone. The antibiotic revolution has resulted in people living healthier and longer lives, with life expectancy almost doubling in the last 100 years. This has led to a rising world population, which has had knock on effects in terms of food resources, employment and environmental issues. The introduction of the contraceptive pill has also had a huge impact on society with respect to sexual freedom and family planning. This period also marked the beginning of the large-scale use of drugs for 'social recreation' rather than for medicine. The demand for drugs such as cocaine, heroin and the amphetamines spawned a black market trade, which led to wealthy crime syndicates and serious social problems on city streets.

M6 THE AGE OF REASON (1970S TO PRESENT)

Key Notes

Drug design

The focus of drug design has switched from structure orientated research to target orientated research. Early examples of rational drug design include the development of cimetidine and captopril. Attaining drug selectivity is now a high priority in drug design.

The biological revolution

Advances in biology have led to a better understanding of how the body works at the molecular level. The structure of macromolecular drug targets can be studied, allowing the design of drugs that will interact more effectively. Novel targets are being revealed by the human genome project.

The chemical revolution

Novel reagents and synthetic procedures permit the synthesis of the most complex of molecules. Asymmetric synthesis has been developed, allowing the selective synthesis of enantiomers. Robotics and solid phase synthetic techniques have resulted in the development of combinatorial chemistry. Miniaturization is a trend both for the synthesis and the analysis of new compounds.

The computer revolution

Computers are now an integral part of drug design.

Herbal remedies

Herbal remedies may offer some advantages over conventional medicines. A combination of different pharmacologically active compounds could have a beneficial synergistic effect that is greater than the sum of the individual components. However, it is wrong to assume that herbal remedies are safe because they are 'natural'.

Multinational pharmaceutical companies

Pharmaceutical companies have merged into 'super' multinationals, reflecting the globalization of medicinal chemistry and the rationalization of drug regulations between different countries.

Challenges for the future

Effective drugs are required for diseases such as cancer, viral infections, autoimmune disease and the diseases associated with old age. The problems of drug-resistant bacteria should not be ignored.

Related topics

Synthetic considerations (F1)
Stereochemistry (F2)
Combinatorial synthesis (F3)
Aims of drug design (H1)
Computer aided drug design (H2)
The antibiotic age (1945–1970s) (M5)

Drug design

This period has seen the 'coming of age' of medicinal chemistry. Until the 1970s, **drug design** was often a hit-and-miss affair where the emphasis was on synthesizing as many analogs of a lead compound as possible. Since then, the emphasis has increasingly been on rational drug design based on an understanding of a drug's mechanism of action and the structure of its target. A few examples of rational drug design had started to appear in the 1970s, including the development of the anti-ulcer agent **cimetidine** (Topic M5). Histamine was the lead compound for this project and various strategies were used to find an analog that would prevent it fitting its receptor. Once an antagonist was discovered, a theory was proposed on how it might interact with the histamine receptor at the molecular level. Further analogs were then synthesized to test the theory and the theory was continually modified as required. The successful design of cimetidine proved that a rational approach could work and was more efficient than synthesizing analogs and 'hoping for the best'.

Another important early example of rational drug design was the design and synthesis of the antihypertensive agents, **captopril** and **enalapril** (*Fig. 1*). This was one of the early projects to use models of a target binding site as an aid in drug design.

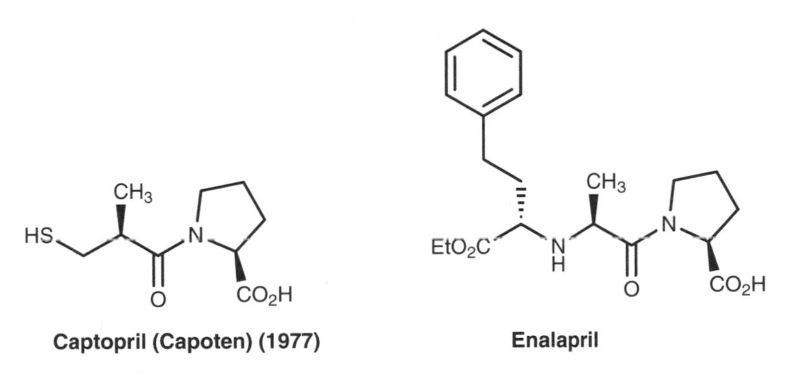

Fig. 1. Antihypertensive agents.

The aims of medicinal chemistry have also altered slightly in recent years and the main priority is often attaining the best selectivity of action rather than maximum activity. The more selective a compound is for a particular target, the more effective it is likely to be, and the fewer side effects it will have. Nowadays, rational drug design is the norm rather than the exception and this change in emphasis has been due to advances in various scientific disciplines.

The biological revolution

In the last few decades, there have been massive advances in scientific areas such as biochemistry, physiology, pharmacology and genetics, leading to a better understanding of the workings of the body at the molecular level. **Molecular biology** has led to a much better understanding of the structure and function of important drug targets, such as enzymes and receptors. This in turn has led to a better understanding of how drugs can be designed for these targets.

The complex molecular mechanisms by which cells process the messages received by receptors (**signal transduction**) is now known in far greater detail and more is being discovered each year (Topic B2). With each new discovery, it is possible to identify new targets that can be used in drug therapy.

Many of these advances owe a great deal to the advances made in genetic engineering. **Genetic engineering** has been used to produce human proteins and enzymes in fast growing microbial cells, allowing these molecules to be obtained in far greater yields than if they were extracted from human tissue. This makes it easier to study these macromolecules and to design drugs that will interact with them.

Of course, one of the most exciting projects in recent years has been the **human genome project**, which is mapping human DNA. This has immense implications for medicinal chemistry. Many new proteins and receptors are being discovered that offer potential new targets in medicinal chemistry. The challenge in future years will be in determining what the functions of these new proteins are, then designing drugs to interact with them.

The chemical revolution

There have been significant advances in **chemistry** over recent years. New synthetic agents and procedures have been devised which make it feasible to synthesize extremely complex molecules. Reagents have been developed that allow **asymmetric syntheses** to be carried out, whereby the single enantiomer (or mirror image) of a chiral compound can be selectively prepared. This is an important area in medicinal chemistry since life is inherently chiral and the drug targets within the body are chiral. As such, they can distinguish between the enantiomers of a chiral drug, so the use of a racemic drug is inherently wasteful, since only one enantiomer is ideally designed to interact with its target. Moreover, the existence of the 'wrong' enantiomer could create problems if it interacted with a different receptor, resulting in side effects.

In the field of analysis, techniques such as **X-ray crystallography** and **2-D nuclear magnetic resonance spectroscopy** have been developed that allow the structural determination of the most complex molecules.

Advances in chromatography and other **purification techniques** now make the separation of complex mixtures far easier than it was many years ago. Thus, the isolation of the active principle from a complex plant extract is a more efficient procedure.

Finally, the development of instrumentation that can automatically carry out solid phase synthesis has created a completely new field of chemistry called **combinatorial chemistry**. This has taken the pharmaceutical industry by storm and allows the synthesis of large numbers of novel structures in a much faster time period than was ever possible in the past. Allied to this, the drive towards microscale experimentation may ultimately lead to nanoscale technology, where new structures are synthesized and tested in 'wires' rather than flasks.

The computer revolution

Computers have become increasingly more powerful with the passing years. Software programs have also been developed that are user friendly and can be used by medicinal chemists on ordinary laboratory computers. As a result, medicinal chemists can now study the interaction of their drugs with target proteins in 3D, and devise new drugs that will interact more effectively.

Herbal remedies

Although there are ever increasing advances in devising selective agents, there has been an increasing interest in **herbal medicines**. This should not be seen as a retrograde step. Often, a herbal mixture contains a variety of different pharmacologically active compounds that work in a synergistic fashion. Indeed, many of the components may only have a very small beneficial effect when used on their own, but when mixed together in the extract, they complement each other to produce a clinically useful effect. A further advantage of using herbal extracts is that the active principles present are diluted and there is less chance of taking dangerous overdoses. On the other hand, it has to be realized that herbal medicines are not 'safer' than conventional medicines just because they are natural. They still contain pharmacologically active compounds, which could have side effects or interact adversely with other chemicals, whether these chemicals are present in food or medicines.

Multinational pharmaceutical companies

As the world shrinks, **pharmaceutical companies** have become increasingly multinational. In the past, research projects and clinical trials were usually carried out in one country and drugs were developed that were unique to that country. Nowadays, that work is often spread over a variety of countries, promoting a drive to common regulations. It is therefore important to ensure that one set of clinical trials will suffice for registration of a product worldwide.

Mergers between multinationals have been common in recent years, emphasizing the increasingly global character of the pharmaceutical industry. For example, in 1979, Beechams, Smith, Kline and French, Glaxo and Wellcome were four separate companies. They have now merged into one giant multinational.

Challenges for the future

Although medicinal chemistry has many triumphs to its credit, there are many challenges to be faced in the future. The fight against cancer goes on and there is also a need for effective drugs against a large number of viral infections, including influenza, the common cold and the AIDS virus. With an increase in life expectancy, the diseases of old age have become more significant, and research into novel drugs that can be used to treat Alzheimer's disease, arthritis, and senility are of increasing importance. Autoimmune diseases, such as multiple sclerosis and Huntingdon's chorea still require effective treatments.

Diseases that have been conquered should not be ignored either. In recent years, scientists have come to appreciate just how fragile the victory over bacterial infections has been. Bacteria are masters at evolving to cope with different environmental pressures, and their ability to acquire resistance to antibiotics is no exception. The threat of 'superbugs' evolving, which might be resistant to all known antibacterial drugs is a frightening, but realistic prospect, so it is crucial that medicinal chemistry continually searches for new strategies to defeat these adversaries.

FURTHER READING

General reading

Albert, A. (1987) *Xenobiosis*, Chapman and Hall.

Bowman, W.C. and Rand, M.J. (1980) *Textbook of Pharmacology*. Blackwell Science, Oxford, UK.

Foreman, J.C. and Johansen, T. (1996) *Textbook of Receptor Pharmacology*. CRC press, Boca Raton, FL, USA.

Ganellin, C.R. and Roberts, S.M. (1993) *Medicinal Chemistry – The Role of Organic Chemistry in Drug Research*. Academic Press.

King, F.D. (1994) *Medicinal Chemistry – Principles and Practice*. The Royal Society of Chemistry.

Mann, J. (1992) *Murder, Magic and Medicine*. Oxford University Press, Oxford, UK.

Nogrady, T. (1988) *Medicinal Chemistry: A Biochemical Approach*. Oxford University Press, New York.

Patrick, G.L. (2001) *An Introduction to Medicinal Chemistry – 2nd edition*. Oxford University Press, Oxford, UK.

Rang, H.P., Dale, M.M. and Ritter, J.M. (1999) *Pharmacology (4th edition)*. Churchill Livingstone.

Roberts, S.M. and Price, B.J. (eds) (1985) *Medicinal chemistry – The Role of Organic Chemistry in Drug Research*. Academic Press.

Sammes, P.G. (ed) (1990) *Comprehensive Medicinal Chemistry*. Pergamon Press, Oxford, UK.

Silverman, R. (1992) *The Organic Chemistry of Drug Design and Drug Action*. Academic Press.

Smith, C.M. and Reynard, A.M. (1992) *Textbook of Pharmacology*. W.B. Saunders.

Sneader, W. (1985) *Drug Discovery: The Evolution of Modern Medicine*. Wiley.

Stenlake, J.B. (1979) *Foundations of Molecular Pharmacology*. Volumes 1 and 3. Athlone Press.

Suckling, K.E. and Suckling, C.J. (1980) *Biological Chemistry*. Cambridge University Press.

Thomas, G. (2000) *Medicinal Chemistry – An Introduction*. Wiley.

Wermuth, C.G. (1996) *The Practice of Medicinal Chemistry*. Academic Press.

Wolff, M.E. (1995) *Burger's medicinal chemistry and drug discovery (5th edition)*. Wiley.

INDEX

Acetylcholine, 10, 15–16, 71, 113–114
Acetylcholinesterase, 3, 10, 144
  inhibitors, 3, 136–137
Acid anhydrides, 171
Acid chlorides, 171
Active conformation, 92, 121–122, 135, 144, 151–152
Active principles, 248–249, 268
Active sites *see* enzymes, active sites
Acylation, 158–159, 185–187
Adalat *see* nifedipine
Adenine, 29–30
Adenosine diphosphate, 21, 222
Adenosine triphosphate, 19–21, 222–223, 227, 234, 239–240, 242
Adenylate cyclase, 19–20
Adrenal medulla, 15
Adrenaline *see* epinephrine
Adrenergic receptors, 16, 152, 262–263
  agonists, 166
  antagonists, 81 *see also* β-blockers
  binding site, 108
Adrenoreceptors *see* adrenergic receptors
Aerosols, 56
Affinity, 64–67, 105–106, 155
Affinity constant, 65–6
African willow tree, 79
Agonists *see* receptor agonists
AIDS, 57, 269
AIDS drugs, clinical trials, 213–14
Alanine, 89
Albumin, 47
Alcohol, 47, 55, 212–213
Alcohols
  alkylation of, 86, 118–119, 158–159
  binding interactions, 111–112
  bromination of, 199–200
  conjugation reactions, 50
  conversion to alkyl chlorides, 194
  drug metabolism of, 172
  esterification, 86–87, 168
  hydrogen bonding, 165
  isotopic labeling, 208–209
  masking by prodrugs, 115, 167–168
  protection of, 210
  resolution of, 97–98
Aldehydes, 171
  binding interactions, 113–114
  isotopic labeling, 209
Aliphatic electronic substituent constants, 130
Alkaloids, 78, 248
Alkenes
  asymmetric epoxidation, 95–96
  asymmetric hydrogenation, 95–96
  binding interactions, 117
  conformational restraint, 153
  isotopic labeling, 209
Alkyl chlorides
  conversion to nitriles, 184
  synthesis from alcohols, 194
Alkyl fluorides, 118
Alkyl groups
  as steric blockers, 166
  binding interactions, 109, 118–119, 162
  drug metabolism of, 48–49
Alkyl halides, 109, 117–118, 171
  drug metabolism of, 50
Alkylating agents, 26, 32, 109
Alkylation, 118
  of alcohols, 86, 118–119, 158–159
  of amines, 86–87, 118–119, 158–159
  of morphine analogs, 88
  of phenols, 86, 159–159
Alkynes
  conformational restraint, 153
  isotopic labeling, 209
  reduction of, 184–185
Allosteric binding sites, 11, 16, 66
Alzheimer's disease, 136, 213, 269
Amides
  binding interactions, 115–116
  conformational restraint, 153
  drug metabolism of, 48–49, 172
  hydrogen bonding, 165
  hydrolysis of, 118
  protection from drug metabolism, 172–173
Amines
  alkylation of, 86–87, 118–119, 158–159
  binding interactions, 108, 112–113, 163–165
  conjugation reactions, 50
  demethylation of, 118–119
  drug metabolism of, 48–49
  in alkaloids, 78
  in dopamine pharmacophore, 122
  in opiate pharmacophore, 120–121
  pharmacokinetics, 165
  resolution of, 97
  use as resolving agents, 97
Amino acids
  chirality of, 93
  in chromatographic resolution, 96
  isotopically labeled, 208
γ-aminobutyric acid, 16
Aminoglycosides, 259
6-aminopenicillanic acid, 88, 259
Aminopeptidase, 98
Aminosalicylate, 170
Amoxycillin, 89, 259–260
Amphetamines, 25, 72, 265
Ampicillin, 89, 98, 259–60
Analgesics, 2 *see also* aspirin, fentanyl, morphine, opium
  in vivo test for, 70
Ancient remedies, 245
Angina, treatment of, 262
Angiotensin converting enzyme, 156
Angiotensin II, 225
Antagonists *see* receptor antagonists
Anti-anxiety drugs *see* anxiolytic agents
Anti-asthmatic agents *see* epinephrine, isoprenaline, salbutamol
  administration of, 56
  clinical trials for, 214
Antibacterial drugs, 34 *see also* aminoglycosides, amoxycillin, ampicillin, Augmentin, cephalosporins, chloramphenicol, chlortetracyclin, erythromycin, fluoroquinolones, macrolides, nalidixic acid, pencillins, penicillin methyl ester, polymyxin B, proflavine, prontosil, quinolones, streptomycin, sulfanilamide, sulfapyridine, sulfasalazine, sulfonamides, tetracyclines, vancomycin
  in vitro and *in vivo* testing, 67
Antibiotics, 259 *see also* antibacterial agents
Antibodies, 39
Anticancer drugs *see* CC-1065, combretastatin, calicheamycin, gliotoxin, taxol, uracil mustard, vincristine
  clinical trials of, 213–214
  in vivo testing of, 71
Anticholinesterases *see* acetylcholinesterase inhibitors
Anticoagulants, drug–drug interactions, 213
Antidepressant agents *see* imipramine, iproniazid, lithium carbonate, phenelzine
Antifungal drugs, 36 *see also* polymyxin B
Antigens, 39
Antihistamines, 3 *see also* fexofenadine, terfenadine
Antihypertensive agents *see* atenolol, captopril, enalapril, enalaprilate, practolol, propranolol
Antimalarial drugs *see* artemisinin, quinine
Antipsychotic agents *see* chlorpromazine, haloperidol
Antisense therapy, 33–34
Antitumor drugs, 32
Anti-ulcer drugs *see* cimetidine (Tagamet), ranitidine (Zantac)
Anxiety, treatment of, 79, 260
Anxiolytic agents *see* asperlicin, Librium, meprobamate, Valium
Aorta, 46
Arbusov reaction, 191
Arcyriaflavin A, 227–229
Arginine, 237
Aromatic rings
  acylation of, 158–159
  binding interactions, 109, 117, 162–163
  binding with cysteine residues, 237
  drug metabolism of, 48–49, 172
  in dopamine pharmacophore, 122
  in opiate pharmacophore, 120–121
  induced dipole interactions, 113
  isotopic labeling, 209
  reduction of, 87
  use in conformational restraint, 153
Arsenic, 251
Artemisinin, 78
Arteries, 46
Arthritis, 269
  treatment of, 28, 39
Ascorbic acid, 252–253
Asparagine, 237
Aspartic acid, 8, 108, 113–114
Aspergillus alliaceus, 79
Asperlicin, 79–80
Aspirin, 2, 72, 86
  drug–drug interactions, 213
  synthesis of, 85–86
Association of British Pharmaceutical Industry, 219
Asthma, treatment of, 16
Asymmetric centers, 92, 94–96, 99, 208
  epimerisation of, 174, 182
  in staurosporine, 227
  removal of, 149–50
Asymmetric synthesis, 94–95, 268
  of L-Dopa, 96
  of propranolol, 95
Asymmetry, 92 *see also* chirality
Atenolol, 262
Athrobacter, 99
Atracurium, 175
Atropine, 248
Augmentin, 260
Auto-immune diseases, 269
  treatment of, 39

Bacterial cell walls
  biosynthetic inhibition of, 36–37
Bactericidal, 67, 256
Bacteriorhodospin, 145–146
Bacteriostatic, 67
Baeyer–Villiger oxidation, 189–190
Barbiturates, 47 *see also* phenobarbitone, pentobarbital, pentobarbitone
Basicity in drug design, 163–165
Beechams, 260, 269
Benzaldehyde, 199
Benzodiazepines, 16, 261
Benzoic acid, ionization of, 129
Benzoic anhydride, 186
Benzoyl chloride, 186
Benzyl penicillin *see* penicillin G
Bile, 51
Bile duct, role in drug excretion, 51
Binding sites
  adrenergic receptor, 108
  allosteric, 11, 16
  ATP in kinases, 227
  binding interactions, 107–109, 111–119
  binding studies, 146, 155–159
  cholinergic receptor, 113–114
  drug design, 142
  EGF-receptor kinase, 233–244
  identification of, 122, 146
  of tyrosine kinases, 223
  pharmacophore identification, 122–123
Bio-equivalence studies, 213
Biotin, 256
Bisindolyl-maleimides, 227–229, 235
β-blockers *see* atenolol, practolol and propranolol
Blood brain barrier, 47, 57
Breast cancer, 222
Breast milk
  drug excretion, 52
Bromination, 199–200
Bromoanisole, 128
Brussel sprouts, 49
Butyllithium, 186, 199
Butyrophenones, 261

Caffeine, 72, 212, 248
Calicheamicin $\gamma 1^{I}$, 32–33
Camphorsulfonic acid, 96–98
Cancer, 269
  treatment of, 39, 221
Cancers, 22, 56, 222
  treatment of, 22, 28 *see also* anticancer drugs
Cannabis, 56
Capillaries, 46–47
Capoten *see* captopril
Captopril, 79, 118, 156, 267
Carbidopa, 174–175
Carbohydrates
  cell surface, 38–39
  in asymmetric synthesis, 95
Carboxylic acids
  attachment to Wang resin, 101
  binding interactions, 116–117
  esterification, 86, 118–119, 168
  in resolution of amines, 97
  isotopic labeling, 208
  masking by prodrugs, 115, 167–168
  resolution of, 97
Carcinogenicity tests, 203–204
Cardiac muscle, relaxation of, 262
Cardiovascular drugs, 262
Carrier proteins, 24–26, 35, 43–44, 46–47
CC-1065, 80
Cell assays, 225
Cell communication, 38
Cell division, 27–28
Cell membranes, 35, 38–39, 42–43
Cell recognition, 38
Cellulose, 27
Cephalosporins, 259
CGP 53353, 230–231
CGP 59326, 239
CGP52411, 230–233, 234–235
CGP54690, 232
CGP57198, 232
CGP58109, 230–231, 235
CGP58522, 232
CGP59326, 242
Chain contraction/extension *see* drug design
Chem3D, 144
ChemDraw, 144
Chemical development, 5, 178
  catalysts, 178, 184–185
  concentration, 184
  cost, 178, 182, 189, 196
  environment, 178, 189, 195–196
  experimental procedures, 192
  hazards, 180
  impurities, 198–201
  method of addition, 186, 199
  phase, 179–180
  process development, 180, 193–196
  promoters, 191
  purifications, 195, 198
  purity specifications, 178–179, 197–198
  reactants, 185–186, 190, 198
  reaction conditions, 199
  reaction optimisation, 182–187
  reaction pressure, 182
  reaction temperature, 182, 191, 199
  reaction time, 183
  reaction yield, 178, 182
  reagents, 178, 186, 189–190, 198
  removal of product, 186
  safety, 178, 182, 189, 194–195
  scale up, 188–192
  side products, 191
  solvents, 178, 183–184, 190–191, 195–196, 199
Chemistry
  birth of, 247–248
  recent advances, 268
Chemotherapy, principle of, 251
Chinese Hamster Ovarian cells, 64
Chiral auxiliaries, 96
Chirality, 92–99, 149, 268
Chloramphenicol, 34
Chlorine as metabolic blocker, 172
Chloroform, 249
Chloroperbenzoic acid, 189–190
Chlorpromazine, 261
Chlortetracyclin, 259
Cholesterol, 79
Cholinergic receptor, 16
  binding site, 113–114
  tests for agonists, 71
chromatography
  resolution of racemates, 96–97
Churchill, Winston, 255
Ciba Pharmaceuticals, 221, 241
Cigarettes, 56, 212
Cimetidine, 49, 76, 109, 262, 267
Cinchona bark, 78, 245, 248
Clavulanic acid, 260
Clinical trials, 5, 59, 61, 179, 203, 205–206, 212–217, 265, 269
Clonidine, 56
Coca leaves, 245, 248
Cocaine, 25, 47, 248–250, 252–253, 265
  administration of, 56
  in vivo test for local anesthesia, 70
Codeine, 248
Cofactor
  zinc, 118
Colchicine, 248
Collagen, 27
Combinatorial chemistry, 268
Combinatorial libraries, 103–104
Combinatorial synthesis, 100–104
  binding studies on mixtures by nmr, 69
  linkers, 101
  mix and split method, 103–104
  parallel synthesis, 101–102
  resins, 101
  source of lead compounds, 82–83
  synthesis of mixtures, 102–104
  Wang resin, 101
Combretastatin, 78, 153
Combretastatin A-4, 169
  phosphate prodrug, 169
Common cold, 269
Comparative molecular field analysis (CoMFA), 135–139
Compound banks/libraries, 83
Computer aided drug design, 143–147, 268, *see also* molecular modeling
  3D QSAR, 134
  model binding sites, 233
  QSAR, 126, 131
Conformational restraint *see* drug design
Conglomerates, 96
Conjugation reactions, 49–50, 172
Contraceptive 'pill', 265
Contraceptives, 39
Convergent synthesis, 194
Cope rearrangement, 182
Corticosterone, 256
Covalent bonds, 109
Craig plots, 132–133
Curacin A, 80
Cyclic AMP, 19–20
Cyclic AMP dependant protein kinase, 234
Cyclic dipeptides, 89
Cyclohexane, 91
Cyclopropanation, 189
Cysteine, 8–9, 109, 237
Cytochrome P450 enzymes, 48–49
Cytokines, 21
Cytosine, 29–30

D-Day landings, 255
10-deacetylbaccatin III, 88
Demethylation, 48–49, 118–119
Dentistry, local anesthetics, 252
Deoxygenation, 189
Deoxyribonucleic acid, 29–30
  alkylation of, 32
  base pairing, 30
  chain cutters, 32–33
  cross-linking, 32
  intercalating drugs, 31–32
Deoxyribose, 30
Depression, treatment of, 16, 25, 260
  see also antidepressant drugs, tricyclic antidepressants
Diabetes, treatment of, 3
Diacetylmorphine, 249
Diacylglycerol, 20
Diaminoethane, 185–186
Diamorphine, 249
Dianilino-phthalimides, 228–231, 234–237, 240
Diastereomers, 97–98
Diazonium salts, 190
Diels Alder reaction, 182, 228
Diethyl ether, 249
diethyl tartrate, 96
Dihydropyridines, 262
DIOP, 96
Dipole–dipole interactions, 109, 114
Dissociation binding constant, 65
Dissociation constant, 66–67, 129
Distomer, 93
Diuretics, 213
Docking *see* molecular modeling
Dopa, 26, 174–175
  asymmetric synthesis of, 96
Dopa decarboxylase, 175
Dopamine, 15–16, 25, 108, 122, 124
  pharmacophore, 122–123
Dopaminergic receptors *see* receptors, dopaminergic
Dose levels, 212–213
Dose ratio, 67
Dose response curves, 71–72
Drug absorption, 42–44 *see also* prodrugs
  labeling studies, 207, 213

Drug addiction, 249–250, 253
Drug administration, 54–57
Drug classifications, 2
Drug design, 5, 141–147
  aliphatic substitution, 162
  aromatic substitution, 161
  basicity, 163–165
  chain contraction/extension, 160–161, 230–231, 235, 243
  conformational restraint, 151–154, 264
  de nova, 142, 147
  enhancing binding interactions, 160–165
  extension, 155–159, 233, 243
  functional group transposition, 161
  rigidification *see* conformational restraint
  ring contraction/expansion, 161–162, 231–232, 235
  simplification, 148–150, 227–228, 243, 252, 264
  steric blockers, 153–154, 166
  variation of alkyl groups, 162
Drug discovery, 4–5
Drug distribution, 45–47
  labeling studies, 207, 213
Drug dosing, 58–59, 214
Drug excretion, 51–53
  labeling studies, 207, 213
Drug half life, 58–59, 175
Drug load, 205
Drug metabolism, 5, 48–50, 61 *see also* drug stability, prodrugs
  hydrolysis of penicillin methyl ester, 71
  labeling studies, 207–210, 213
  metabolic blockers, 172
  of CGP52411, 230–231
  of esters, 115
  of norepinephrine and serotonin, 261
  of progesterone, 263
  of prontosil, 255
  role in excretion, 53
  steric shields, 172
Drug preparations, stability of, 206
Drug regulation, 5
Drug solubility, 43, 167–170
Drug stability, 171–175
Drug targeting
  anti-asthmatic drugs, 54–56
  antibodies, 39
  sulfasalazine, 170
Drug targets, 4, 267–268
  basis for drug design, 142
Drug testing, 4
Drug tolerance, 59–60
Drug-drug interactions, 61, 212–213
Drug–food interactions, 213
Dust explosion test, 195
Dysentery, treatment of, 245

EC_{50}, 66–67
ED_{50}, 72–73
Efficacy, 64, 66–67, 71–72, 105–106
  clinical trials, 214
EGF-receptor, 233
EGF-receptor kinase, 222
  active site, 233–237, 239–244
  inhibitors, 221–225, 227–244
Ehrlich, Paul, 251
Electrostatic interactions, 108
Elixir of Sulfonamide Tragedy, 257
Emetine, 248
Enalapril, 267
Enalaprilate, 156
Enantiomeric excess (ee), 94, 96, 99, 178
Enantiomers, 92–99, 149, 178, 268
Enavid *see* norethynodrel
Endogenous compounds, 252–253, 256
Enzyme assays, 224–225
Enzyme inhibitors, 8, 46
  competitive, 10–11
  design of, 155–156
  in vitro testing, 63–64, 68
  irreversible, 12, 109
  non competitive, 11, 63–64
  reversible, 11–12
  selectivity, 12
Enzymes, 7, 267 *see also* acetylcholinesterase, adenylate cyclase, aminopeptidase, angiotensin converting enzyme, cyclic AMP dependant protein kinase, cytochrome P450 enzymes, dopa decarboxylase, epidermal growth factor receptor kinase, esterases, β-galactosidase, lactate dehydrogenase, lipases, monoamine oxidase, peptidases, phospholipase C, proteinases, protein kinase C, thymidylate kinase, tyrosine kinase linked receptors, zinc metalloproteinases
  active site, 8, 9
  chirality, 93
  competitive, 63
  in asymmetric synthesis, 95
  isozymes, 12
  mechanism of catalysis, 9, 10
  resolution of racemates, 98–99
  role in hydrolysis of penicillin G side chain, 88
  substrates, 7
Epidermal growth factor, 22, 222
Epidermal growth factor receptor *see* receptors, epidermal growth factor
Epinephrine, 15–16, 81, 166, 252–253, 263
  binding interactions, 108–109
Epoxidation
  asymmetric, 95–96
Epoxides, drug metabolism, 50
erbB1 gene, 221
Ergosterol, 256
Erythromycin, 34, 259
E_S, 130–132
Esterases, 49, 115, 168, 175
Esterification, 86–87, 118–119, 168, 211
  effect of pressure, 182
  of 6-APA, 88
Esters
  as prodrugs, 115
  binding interactions, 115
  drug metabolism of, 48–49, 172
  enzymatic hydrolysis, 115
  extended, 168
  hydrogen bonding, 165
  hydrolysis of, 98, 118–119
  hydrolysis with yeast, 210
  of amino acids, 169
  phosphate, 169
  protection of drug metabolism, 172–174
  resolution of, 98
  succinate, 169
Estradiol, 256
Estrogen, 22–23
  administration of, 56
Estrogen receptor, 22–23, 68
  testing for agonists and antagonists, 68
Ethanediamine, 186–187
Ethers
  demethylation of, 118–119
  hydrogen bonding, 165
Ethyl bromoacetate, 210
ethylene glycol, 257
Ethynodiol diacetate, 263–264
Etorphine, 94–95
Eudismic ratio, 93
Eutomer, 93
Eye drops, administration of, 56

F, 130–131, 137
Fentanyl, 55–56, 58
Fermentation
  generation of penicillin analogs, 89
  of 6-APA, 88
Fertilization, 39
Fetus, placental barrier, 47
Fever, treatment of, 245
Fexofenadine, 178–179
First pass effect, 55, 57
Flash point, 190
Flu, 269
Fluorine as a metabolic blocker, 172, 230–231
Fluoroquinolones, 101–102, 260
Fluvostatin, 198–199
Folic acid, 255–256
Folk medicine, 79
Food and Drugs Administration (FDA), 216–217, 257
Food, Drug and Cosmetic Act, 257
Formulation, 179, 205–206
Free-Wilson approach in QSAR, 133–134
Freud, Sigmund, 249
Friedal Crafts reaction, 178, 184–185
Fungi as sources of lead compounds, 79–80, 89

β-galactosidase, 68
gangrene, 254
Gas chromatography, 183
Gastric acid, 76, 262
General anesthetics, 35, 51, 56, 249
General Medical Council, 219
Generic drugs, 217, 219
Genetic disease, treatment of, 39
Genetic engineering, 63–64, 68, 71, 145, 225, 267
Genomic projects, 83, 123
Glaxo, 269
Gliotoxin, 89
Glomeruli, 52
Glucuronides, 50
Glutamic acid, 237
Glutamine, 234–237
Glutathione, 50
Glyceryl trinitrate, 55
Glycine, 93, 237
Glycoconjugates, 38
Good Clinical Practice (GCP), 219
Good Laboratory Practice (GLP), 218
Good Manufacturing Practice (GMP), 218–219
G-proteins, 18–20
Grapefruit juice, 49
Grignard reaction, 94–95, 200–201
Growth factors, 21
Guanine, 29–30, 32, 109
Guanosine diphosphate, 18
Gut infections, treatment of, 44, 170

Haldol *see* haloperidol
Haloperidol (Haldol), 261
Hammer test, 195
Hammett substitution constant, 129–133
Hansch equation, 131–133
Heart, 46
Henderson–Hasselbalch equation, 163
Hepatitis, 57
Herbal medicines, 245, 268
Heroin, 56, 85–86, 249–250, 253, 265
Hexene, 91
High performance liquid chromatography, 183, 198, 207
High throughput screening, 67–68
Histamine, 3, 16, 76, 81, 108, 252–253, 262, 267
Histamine receptors 76, 262, 267
  antagonists, 81 *see also* antihistamines
Histidine, 8–10
Hofmann elimination, 175
Homologous series, 158
Hormones, 14–15 *see* also angiotensin II, corticosterone, cytokines, epidermal growth factor, epinephrine, estradiol, growth hormones, insulin, progesterone, testosterone, thyroxine
Horner–Emmons reaction, 186, 200
Human genome project, 268
hunger, relief of, 245
Huntingdon's chorea, 269
Hydrogen bonding, 108–109

Hydrogen bonding (*contd.*)
of alcohols, 111–112, 165
of amides, 115–116, 165
of amines, 112–113
of ammonium ions, 164
of carboxylic acids, 116–117
of esters, 115, 165
of ethers, 165
of ketones, 165
of ketones and aldehydes, 114
of phenols, 111–112, 165
of thioethers, 165
of ureas, 165
hydrogenation, 95–96, 184
Hydrogenolysis, 183
Hydrolysis
effect of pressure, 182
of amides, 48–49, 118
of eters, 48–49, 98, 118–119, 210
Hydrophobicity
QSAR, 126–129, 270
Hypertensive drugs *see* clonidine
Hypnotics *see* phenobarbotine, pentobarbitone, thalidomide
Hypodermic syringe, 249

IC_{50}, 64, 66
Ignition temperature, 190
Imipramine, 261
in vitro testing, 61, 63, 70, 224–225
for lead compounds, 75–76
for structure activity relationships, 105–106
of antibacterial drugs, 67
pharmacology, 205
structure activity relationships, 111
toxicology, 204
in vivo testing, 61, 70–73, 225
for structure activity relationships, 106
of antibacterial drugs, 67
pharmacology, 205
toxicology, 204
Incas, 245
Induced dipole interactions, 113
induced fit, 9, 14–16, 146
infections, treatment of, 34, 36, 39, 44, 57, 170, 251, 254, 260, 269 *see also* antibacterial drugs, antifungal drugs
Inflammation, treatment of, 39
Infra red spectroscopy, 183
Inhibitory constant, 66
Inositol triphosphate, 20
Institutional Review Board (IRB), 219
Insulin, 21, 55, 57, 253
Intercalating drugs *see* deoxyribonucleic acid
Intramuscular injections, 57
Intrathecal injections, 57
Intravenous injections, 57
Ion channels, 17, 35, 262
Ionic bonding, 108
of amines, 112, 163–165
of carboxylic acids, 116–117
of quaternary ammonium salts, 113–114
Ipecacuanha, 245
Iproniazid, 261
Isomers, 91–92
Isoprenaline, 166, 263
Isotope effect, 209
Isotopic labeling, 207–210
Isozymes, 12

K_a, 65–66
K_d, 65–67
Ketalization, 186
Ketones
binding interactions, 113–114
hydrogen bonding, 165
isotopic labeling, 209
ketalization, 186
reduction of, 86, 114
K_i, 66
Kidneys, drug excretion, 52–53

β-lactamases, 260
lactate dehydrogenase, 93
lactic acid, 91–93
LD_{50}, 72–73, 195
lead compounds, 1, 5, 11, 85, 148, 155, 264, 267
addition of extra binding groups, 158–159
analogs of, 85–89
basis for drug design, 141–142
binding interactions, 160
definition of, 75
design of, 147
from combinatorial synthesis, 103
from pharmacophore searching, 123–124
pharmacophore, 120
pyrazolopyrimidines, 241–242
sources of, 75, 78–84
staurosporine, 227
structure activity relationships, 105–106
testing for, 75–76
Leucine, 234–237
Leukocytes, 39
Lewis acids, 185
Librium, 261
Lignocaine, 263
Linear synthesis, 194
Lineweaver Burk plot, 63
Lipases in resolution, 98
Lipid carriers, 36–37
Lithium carbonate, 262
Liver, 48
Local anesthetics, 55, 70 *see also* cocaine, lignocaine, procaine
Log*P*, 127–129, 131, 133
Lovastatin, 79–80
Lungs
drug excretion, 51

M&B 693, 255
Macrolides, 259
Magic bullet principle, 251
Malaria, treatment of, 245
Malic acid, 96–97
Manic depression, treatment of, 262
Marijuana, 56
Marine chemistry as source for lead compounds, 80–81
Marketing, 219
Mass spectroscopy, 198, 207
Medicinal chemistry degrees, 2
Medicine men, 245–246
Mephobarbital, 97
Meprobamate, 261
Mercapturic acid conjugates, 50
Mercuric acetate, 189
Metabolic blockers, 118, 172, 230–231
Methamphetamine, 56
Methionine, 234–237
Methyl ethanoate, hydrolysis of, 130
Methyl groups
as steric shields, 172
drug metabolism, 172
Methyl pseudomonate, 210
Methylcyclohexane, 92
Microtubules, 27–28
Minipumps, 57
Molar refractivity, 131
Molecular biology, 267
Molecular modeling, 233–234, 237, 239
3D structures, 144
alignments, 137–138
alignments for 3D QSAR, 144
binding site studies, 138, 146–147, 156–158, 166
design of lead compounds, 75
docking, 146, 234–235, 239, 242
energy minimization, 144
overlays, 124, 137–138, 144
pharmacophores, 83–84, 120–123
protein structure, 145–146
steric energies, 144
structure comparisons, 144

monoamine oxidase, 132, 261
monoamine oxidase inhibitors (MOAI), 261
morning sickness, treatment of, 264
morphine, 2, 72, 86, 124, 144, 248–249, 253
analogs, 88, 156, 161
demethylation of, 88
simplification of, 149
source of, 78
MR, 131, 137
m-RNA, 225
Mucous membranes, 55–56
Multiple sclerosis, 269
Mutagenicity tests, 203–204

Nalidixic acid, 260
Nanoscale technology, 268
Nardil *see* phenelzine
Nasal decongestants, administration of, 55–56
Nephrons, 52–53
Neuromuscular blocking agents *see* atracurium, suxamethonium
Neurotransmitters, 14–15 *see also* acetylcholine, dopamine, norepinephrine, serotonin
Neutrophils, 28
New Drug Application, 257
Nicotinamide, 256
Nicotine, 47, 52, 56, 248
patch, 56
Nifedipine (Adalat), 262
Nitriles, synthesis of, 184
Noradrenaline *see* norepinephrine
Norepinephrine, 15, 25, 108, 152, 252–253, 261–262
Norethynodrel (Enavid), 263–264
Nose drops, 56
Nuclear magnetic resonance spectroscopy, 198
2D, 268
analysis of reactions, 183
design of lead compounds, 75
in vitro binding studies, 68–69
isotopic labeling, 207
Nucleic acids, 46
alkylation of, 109
chirality, 93
Nucleophilic substitution, 99, 118, 182, 184

Oleic acid, 211
Opiate analgesics pharmacophore, 120–121, 123–124
Opium, 245, 248
Optical purity, 99
Optical yield, 94
Oral contraceptives, 263
Orphan drugs, 218
Ovarian cancer, 222
Overlays *see* molecular modelling
Oxidation, 48–49, 189
Ozonolysis, 210–211

pA_2, 67
Palladium chloride, 189
Panic attacks, treatment of, 79
Parkinson's disease, treatment of, 26, 174
Partition coefficient, 127–129, 131, 133
Patents, 5, 83, 177–178, 217
pD_2, 66
Penicillin G, 88–89, 255, 259
Penicillin V, 89, 255
Penicillins, 3, 47, 71, 255–256
methyl esters, 71, 168
semi-synthetic, 88, 259
Pentobarbital, 59–60
Pentobarbitone, 252
Peptidases, 49, 173
Peptides
combinatorial synthesis, 102–103
protection from drug metabolism, 173–174
Pfeiffer's rule, 93

Pharmacodynamics, 1, 14
Pharmacokinetics, 1, 41, 141
  clinical trials, 212–214
  drug solubility, 167–170
  drug stability, 171–175
  of amines, 165
  structure activity relationships, 106
  testing, 225
Pharmacology studies, 205
Pharmacophore, 83, 120–124, 141, 148–149, 167, 264
  comparisons of, 144
  in 3D QSAR, 135–136
  of 4-(phenylamino)pyrrolopyrimidines, 240
  of kinase inhibitors, 237–239
  of opiate analgesics, 149
Pharmacophore searching, 83
Pharmacophore triangles, 123–124
Phenelzine (Nardil), 261
Phenethylmorphine, 156, 161
Phenobarbitone, 49, 252
Phenols
  alkylation of, 86, 158–159
  binding interactions, 108, 111–112
  conjugation reactions, 50
  drug metabolism of, 172
  esterification of 86–87, 168
  hydrogen bonding, 165
  in dopamine pharmacophore, 122
  in opiate pharmacophore, 120–121
  isotopic labeling, 208
  masking by prodrugs, 115, 167–168
Phenothiazines, 261
Phenoxyacetic acid, 89
Phenoxymethyl penicillin *see* penicillin V
Phenylacetic acid, 89
Phenylalanine, 8, 89, 109, 122
4-(phenylamino)pyrrolopyrimidines, 239–240, 242
Phenylglycinamide, 98
Phenylglycine, 98
(Phenyloxyethyl)cyclopropylamines, 132
Phosphatidylcholine, 42–43
Phosphatidylinositol diphosphate, 20
Phospholipase C, 20
Phospholipid bilayer, 35, 42–43
Phospholipids, 42
Phosphonates, synthesis of, 191
Phosphonium salts, preparation of, 182
Phosphorylation, 21–22
  of angiotensin II, 225
  of serine, 225
  of threonine, 225
  of tryosine, 222–223, 225
Physical dependence, 60
Physostigmine, 248
Pinocytosis, 44
pK_a, 163–165, 167, 170
pK_b, 163
Placebo, 213–214, 219
Placebo effect, 245–246
Placental barrier, 47
Plane polarized light, 93
Plasma proteins, 46–47, 52, 58
Pneumonia, 254–255
Polymyxin B, 36
Poppies, 78, 245
Potency, 66, 71–72
Practolol, 262
Preclinical tests, 61
Preformulation, 205
Procaine, 252
Process development *see* chemical development
Prodrugs, 61, 115, 167–170
Proflavine, 31–32
Progesterone, 256, 263
Prontosil, 61, 82, 255
Propranolol, 262
  analogs of, 87
  synthesis of, 87, 95
Protection of alcohol groups, 210
protein biosynthesis, 30–31
  inhibition of, 34
Protein kinase C inhibitors, 227–228
Protein sequencing, 145
Protein structure, 145–146, 233
Proteinases, 173
Proteins
  protection from drug metabolism, 173
Protozoal infections, treatment of, 251
Prozac, 93
Pseudomonas infections, treatment of, 36
Pseudomonic acid, 210
Psoriasis, treatment of, 230
Psychotropic medicines, 260
Pure Food and Drugs Act, 253
Purity specifications *see* chemical development
Pyrazolopyrimidines, 241–243
Pyridinium chlorochromate, 189
Pyridoxine, 256

Quantitative structure activity relationships, 126–134, 136–137
  3D, 134–139, 144
  Free Wilson approach, 133–134
Quaternary ammonium salts, 113–114
Quaternization
  effect of pressure, 182
Queen Victoria, 249
Quinine, 78, 97, 248–249
Quinolones, 260

R, 130–131, 137
Racemates, 93–94, 96–99, 268
Racemization, 94, 98, 208
Radiochemical purity, 208
Radiolabeling synthesis, 207–210
Radioligand labeling, 64–68
Ranitidine (Zantac), 263
Receptor agonists, 16
  as lead compounds for antagonists, 76
  design of, 156
  in vitro testing, 64–67
  lead compounds for antagonists, 81
  structure activity relationships, 105–106
Receptor antagonists, 16, 109
  design of, 156–157
  from receptor agonists, 76, 81, 157
  in vitro testing of, 64–68
  structure activity relationships, 105–106
Receptors, 14, 267–268
  adrenergic, *see* adrenergic receptors
  binding site, 14–15
  chirality, 93
  cholinergic *see* cholinergic receptor
  dopaminergic, 16
  epidermal growth factor, 22, 221–223
  families, 17
  GABA, 16
  G-protein coupled, 18–20, 145–146
  histamine *see* histamine receptors
  intracellular, 22–23
  ligand gated ion channel, 17
  serotonin *see* serotonin receptors
  steroid, 46
  types and subtypes, 16
  tyrosine kinase linked, 20–21, 221–225, 227
Recombinant DNA technology, 225
Reduction, 86, 211
  of alkynes, 184–185
  of aromatic rings, 87
  of ketones, 114
Regulatory affairs, 216–219
  Abbreviated New Drug Application, 217
  fast tracking, 217–218
  Investigational Exemption to a New Drug Application (IND), 217
  labeling, 218
  New Drug Application (NDA), 217
  orphan drugs, 218
Regulatory authorities, 179, 212, 215–217
Reproduction abnormalities tests for, 203–204
Resolution, 96–99
Retinal, 256
Retinoids, 22
Riboflavin, 256
Ribonucleic acid, 29–30
  antisense therapy, 32, 34
  messenger, 30–34
  ribosomal, 30–31, 34
  transfer, 30–31
Ribose, 30, 93
Ribosomes, 30–31
Rigidification, *see* drug design
Rule of five, 41, 43
ρ_I, 130

Salbutamol (Ventolin), 263
Salicylic acid, 86
Salvarsan, 251, 254
Scatchard equation, 65
Scatchard plot, 65–66
Schild analyses, 67
Schild plot, 67
Schizophrenia, treatment of, 260
Secondary messengers, 19–20
Secondary metabolites, 78, 148, 155
Sedatives *see* etorphine, opium, pentobarbital, phenobarbitone, thalidomide
Selectins, 39
Selective toxicity, 251
Semi-synthetic drugs, 249
Semi-synthetic synthesis, 88–89
Senility, 269
Serine, 8, 122
  as a nucleophile, 9–10, 109
  in adrenergic receptor binding site, 108
  in gliotoxin biosynthesis, 89
  phosphorylation of, 225
Serine-threonine kinases, 227
Serotonin, 15, 154, 261
Serotonin receptors, 16
  antagonists, 154
Shamans, 245
Sharpless epoxidation, 95–96
Signal proteins, 22 *see also* G-proteins
Signal transduction, 267
Simplification *see* drug design
Smith Kline and French, 269
Sodium bicarbonate, 262
Sodium borohydride, 114
Sodium dimethyl phosphonate, 191
Solid phase synthesis, 83, 101, 268
Specific activity, 207–208
Spinal cord, drug administration, 57
Squamous cell carcinoma, 222
Stamina, increase of, 245
Standard operating procedures, 218
Staurosporine, 227
Steady state concentration, 58–59
Stereochemistry, 91–99
  role in drug stability, 174
Steric blocking, 153–154, 166
Steric shields, 117, 168, 172
Sterimol, 131
Steroids, 3, 15, 22, 50
Stimulants *see* amphetamine, caffeine
Streptomyces zelensis, 80
Streptomycin, 34, 259
Stroke, treatment of, 39
Structure activity relationships, 5, 85, 102, 105–109, 111–120, 142
  of adenine triphosphate, 234
  of dianilino-phthalimides, 230, 235
  of pyrazolopyrimidines, 243
Strychnine, 248
Subcutaneous implants, 57
Subcutaneous injections, 57
Substituent hydrophobicity constant, 128–129, 131–133, 137
Sulfanilamide, 61, 255
Sulfapyridine, 170, 255

Sulfates, 50
Sulfonamides, 3, 82, 170, 254–255 *see also* sulfanilamide, sulfapyridine, sulfasalazine
'Superbugs', 269
Suppositories, 56
Suxamethonium, 113–114
Sweat
  drug excretion, 52
Synaptic gap, 15
Synergism, 174
Syphilis, 251

Tacrine, QSAR study, 136–139
Taft's steric factor (E_S), 130, 132
Tagamet *see* cimetidine
Tartaric acid, 96–97
Taxol, 28, 78, 88
Teprotide, 79
Teratogens, 264
Terfenadine, 178–179
Testosterone, 256
Tetracyclines, 34, 259
Thalidomide, 264
Therapeutic index/ratio, 72–73
Thiamin, 252–253
Thin layer chromatography, 183
Thioethers, hydrogen bonding, 165
Thiols, binding interactions, 118
Thionyl chloride, 194
Threonine, 225, 234–237
Thrombosis, treatment of, 39
thymidylate kinase inhibitors, 147
Thymine, 29–30
Thyroid hormones, 22
Thyroxine, 252–253
Tolerance, 249
Topical drugs, 56
Topliss schemes, 134
Toxicology studies, 61, 179, 203–204
Toxins, 79
Tranquilizers *see* benzodiazepines
Transcription proteins, 22
Transgenic animals, 71
Transition states, diastereomeric, 98
Tricyclic antidepressants, 25, 261
Trifluoroacetic acid, 101
Trimethyl phosphite, 191
Triphenylphosphine, 194, 200
Tripos Software Ltd, 135
Tryptophan, 113–114
Tuberculosis, 254
Tubulin, 27
Tumors *see* cancers
  tests for antitumor activity, 225
Typhoid, treatment of, 34
Tyrosine
  in acetylcholinesterase active site, 10
  in cholinergic receptor binding site, 113–114
  phosphorylation of, 21–22, 222–223, 225
tyrosine kinases *see* receptors, tyrosine kinase linked and receptors, epidermal growth factor

UDP-glucose, 50
Uracil, 30
Uracil mustard, 26, 32
Ureas, hydrogen bonding, 165
Urethanes, drug stability, 174

Valine, 237
Valium, 261 *see* diazepam
van der Waals interactions, 109
  of alkenes, 117
  of alkyl groups, 118–119
  or aromatic rings, 117
Vancomycin, 36–37
Vapor density, 190
Vapor pressure, 190
Veins, 46
Venoms, 79
Ventolin *see* salbutamol
Verloop steric parameters, 131
Vinblastine, 28
Vincristine, 28
Vinyloxycarbonyl chloride, 88, 119
Viral infections, 269
vitamins *see* ascorbic acid (vitamin C), biotin, folic acid, nicotinamide, pyridoxine, retinal (Vitamin A, riboflavin, thiamin (vitamin B1)

Wang resin, 101–102
Warfarin, 47
Wellcome, 269
Witch doctors, 245–246
Withdrawal symptoms, 60
Wittig Horner reaction, 210
Wittig reaction, 200
Wittig reactions
  synthesis of, 194

X-ray crystallography, 122, 138, 142, 145–147, 156, 228, 233, 256, 268
Xylocaine *see* lignocaine

Yeast, 210
Yew tree, 78, 88

Zantac *see* ranitidine
Zinc metalloproteinase inhibitors, 118
Zinc powder, 189
Zinc–copper amalgam, 189